LONDON MATHEMATICAL SOCIETY STUDENT TEXTS

Managing Editor: Professor E.B. Davies, Department of Mathematics,
King's College, Strand, London WC2R 2LS, England

1 Introduction to combinators and lambda-calculus, J.R.HINDLEY & J.P.SELDIN
2 Building models by games, WILFRID HODGES
3 Local fields, J.W.S. CASSELS
4 An introduction to twistor theory, S.A.HUGGETT & K.P.TOD

London Mathematical Society Student Texts. 3

Local Fields

J.W.S. CASSELS
Department of Pure Mathematics and Mathematical Statistics
University of Cambridge

CAMBRIDGE UNIVERSITY PRESS
Cambridge
London New York New Rochelle
Melbourne Sydney

Published by the Press Syndicate of the University of Cambridge
The Pitt Building, Trumpington Street, Cambridge CB2 1RP
32 East 57th Street, New York, NY 10022, USA
10, Stamford Road, Oakleigh, Melbourne 3166, Australia

First published 1986

Printed in Great Britain at the University Press, Cambridge

Library of Congress cataloging in publication

Cassels, J.W.S. (John William Scott)
Local Fields
(London Mathematical Society student texts; 3)
Includes bibliographies.
1. Local Fields (Algebra)
I. Title II. Series
QA247.C34 1985 512'.3 85-47934

British Library cataloguing in publication data

Cassels, J.W.S.
Local Fields
(London Mathematical Society student texts; 3)
1. Fields, Algebraic
I. Title II. Series
512'.3 QA247

ISBN 0 521 30484 9 hardcover
ISBN 0 521 31525 5 paperback

PREFACE

My heart is inditing a good matter. Psalm 45.

After a general discussion of real-valued valuations of fields, attention will focus on the p-adic fields $\mathbb{Q}_p$ and their finite extensions. These provide the framework for much important and exciting research at the present day. They also give valuable insights at a humbler level, and not infrequently, provide remarkable easy and natural solutions to problems which apparently have no relation to p-adic fields and which otherwise can be resolved, if at all, only by deep and arduous methods.

The book supplies a self-contained introduction at the level of an MSc or beginning graduate student, though much will be of interest and accessible to the mathematical undergraduate or amateur. The aim is not to bring the reader to the frontiers of knowledge but, rather, to illustrate the versatility, power and naturalness of the approach. We therefore break the orderly exposition from time to time to make applications, some of which it is hoped that the reader will find striking. At the ends of the chapters are numerous exercises, ranging from the five-finger kind to substantial results of independent interest. The author will have failed if he does not persuade the reader that the p-adic numbers are every bit as natural and worthy of study as the reals and complexes.

In some of the applications and exercises we have assumed that the reader has access to a programmable calculator or home computer. The advent of the computer has immeasurably enhanced the armoury of the research mathematician by enabling him to experiment numerically in a way which was previously unthinkable. What is not so widely recognized, is the effect on the tactics of mathematical exposition and proof. Nowadays it is preferable, where possible, to replace a tedious and detailed consideration of cases, subcases and sub-sub-cases --- "watchmaker's mathematics", which nobody reads anyway --- by the verification of a large number of similar simple cases which can be checked easily with a simple computer program, even though this would be infinitely boring by hand.

There is some disagreement in the literature on the precise connotation of the term "local field". Serre (1962) defines it to be a field complete with respect to a discrete valuation but in his contribution to Cassels and Fröhlich (1967) adds the condition that the residue class field must be finite. Weil (1967) takes this more restrictive definition but includes the reals and the complexes.

If one excepts the reals and complexes (which are, in any case, completely anomalous) the earliest examples of local fields are the p-adics, which were introduced by Hensel (1902) in a paper whose title translates as "The development of algebraic numbers in power series". He pursued the analogy with the power series treatment of algebraic curves in Hensel and Landsberg (1902), which generalizes Weierstrass' use of power series expansions for analytic functions of a complex variable. These latter did not form a complete valued field, however, since there was always a condition that the series should converge in some neighbourhood. Hensel used the p-adic fields and their finite extensions (the $\mathfrak{p}$-adic fields) to provide a treatment of algebraic numbers alternative to the original approach of Kummer. This was expounded in two books, Hensel (1908, 1913).

The general notion of valuation was introduced by Kürschák (1913), and the set of all valuations of the rationals was determined by Ostrowski (1918) in a paper which made other important advances.

In the early 1920's, at the very commencement of his career, Hasse, inspired by a truly delphic postcard from Hensel (reproduced in vol I of Hasse (1975)), formulated the "local-global principle", nowadays usually called the "Hasse principle". In a series of 5 papers in Crelle's Journal for 1923 and 1924, all reproduced in Hasse (1975), he showed that it holds for quadratic forms over the rationals or, more generally, over any number field. His formulation succinctly subsumes a mass of earlier results, often with quite complicated enunciations. Nevertheless the merits of this approach took a long time to percolate into the collective mathematical consciousness. As late as 1930 L.E. Dickson could write a monograph on quadratic forms (Studies in the theory of numbers) and Mordell could review it (Math. Gazette 15 (1930-31), 361-362) without either of them betraying the least awareness of the p-adic viewpoint.

Skolem (1933) introduced a new and powerful method of attacking certain diophantine equations. He appears to have been ignorant of existing work on local fields and it seems to have been left to Mahler in his review of the paper in the Zentralblatt to point out that, in effect, Skolem was expanding functions of a p-adic variable in power series and studying their zeros. Mahler (1935) went on to use similar methods to prove a very general theorem about the values taken by a recurrence relation, only very special cases being already known. Later Skolem's p-adic method was further elaborated, notably by Skolem himself, Ljunggren and Chabauty.

By the 1930's, however, local fields came increasingly into the mainstream of mathematics. Mahler proved analogues for $\mathfrak{p}$-adic numbers of results about the transcendance of certain real or complex numbers. Witt showed that the theory of quadratic forms over number fields takes on a methodologically much simpler shape if one starts from the consideration of forms over local fields. A local theory of non-commutative associative algebras was developed which also had important consequences for algebraic number theory: there is an excellent contemporary account in Deuring (1935), or see Weil (1967). The advantages of the $\mathfrak{p}$-adic approach to algebraic number theory were increasingly recognized. In the succeeding decades the key theorems of class field theory assumed a brief and memorable formulation and, later, it was realized that the proofs, too, became simpler and more natural if one takes the local situation as the starting point.

As we approach the present day, p-adic methods have become a natural and indispensible tool in many areas not merely number theory but also, for example of representation theory and algebraic topology. A survey would be impossible, but some aspects will be encountered in the pages which follow.

I should like to express my gratitude to Mrs. J. Bunn for the rapid and accurate way in which she has transformed raw manuscript into elegant typescript.

CONTENTS

Paragraphs which can be omitted at a first reading are marked with an asterisk. But if you do skip them, then you miss some of the lollipops. The logical dependence of the chapters is given by the Leitfaden.

L E I T F A D E N

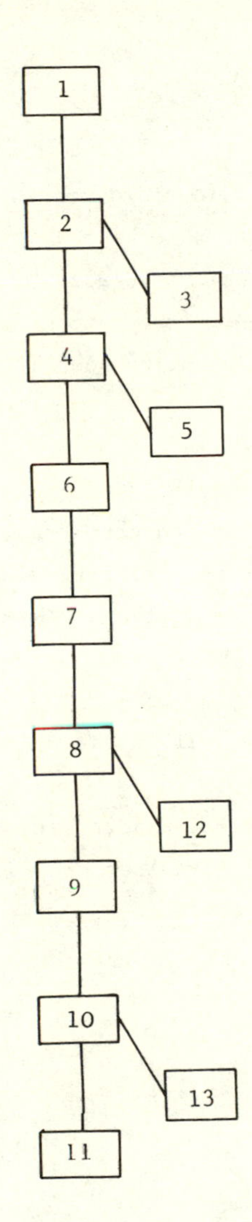

This shows diagrammatically the logical dependence between the chapters.

NOTATIONAL CONVENTIONS

Most of our notational conventions are too standard to require comment.

Following Bourbaki, we denote by $\mathbb{C}$, $\mathbb{R}$, $\mathbb{Q}$, $\mathbb{Z}$ respectively the complex numbers, the reals, the rationals and the rational integers.

We normally write congruences in the shape

$$a \equiv b \quad (m)$$

where, say, $a,b,m \in \mathbb{Z}$. This indicates that $b - a$ is divisible by m. If there is no doubt what modulus is intended, the "(m)" may be omitted. We have occasionally written "(mod m)" instead of "(m)" where this seemed clearer.

By $a|b$ and $a \nmid b$ we mean "a divides b" and "a does not divide b" respectively. If p is prime, $p^n \| b$ means that $p^n|b$ but $p^{n+1} \nmid b$.

A list of the books, papers, etc., referred to is given at the end of the book. In the text they are cited by giving the name(s) of the author(s) and the date.

CHAPTER ONE: INTRODUCTION

1 VALUATIONS

These are generalizations of the ordinary absolute value on the field $\mathbb{C}$ of complex numbers. A valuation is a real valued function on a field k satisfying certain axioms. We leave formal definitions to the next Chapter, but here give more examples of valuations to illustrate some salient features.

(a) $k = \mathbb{C}$. For $a = u + iv$ with $u, v \in \mathbb{R}$, the ordinary absolute value is $|a| = \sqrt{(u^2 + v^2)}$. Then

$$\text{(i)} \quad |a| \geq 0, \text{ with } = 0 \text{ only for } a = 0. \tag{1.1(i)}$$

$$\text{(ii)} \quad |ab| = |a||b|. \tag{1.1(ii)}$$

$$\text{(iii)} \quad |a + b| \leq |a| + |b|. \tag{1.1(iii)}$$

The inequality (iii) is usually called the triangle inequality.

(b) $k = k_o(T)$, where k_o is any field and T is transcendental over k_o. We first define $|\,|$ on the ring of polynomials $k_o[T]$. Let $c > 1$ be fixed arbitrarily. If

$$f = f(T) = f_o + f_1 T + \dots + f_n T^n \quad (f_j \in k_o, f_n \neq 0),$$

we put

$$|f| = c^n, \quad |0| = 0.$$

Any element h of $k_o(T)$ is of the form $f(T)/g(T)$ with $f(T), g(T) \in k_o[T]$ and $g(T) \neq 0$. We put

$$|h| = |f|/|g|.$$

Then for $f, g \in k_o(T)$ we have

(i) $|f| \geq 0$, with $= 0$ only for $f = 0$

(ii) $|fg| = |f||g|$

(iii)* $|f + g| \leq \max\{|f|, |g|\}$,

as is easily verified (do it first for $f, g \in k_o[T]$).

We call (iii)* the ultrametric inequality. It is clearly stronger than the triangle inequality (iii).

(c) $k = \mathbb{Q}$. Let p be a (positive) prime and let $\gamma \in \mathbb{R}$, $0 < \gamma < 1$. Any nonzero $r \in \mathbb{Q}$ can be written

$$r = p^{\rho}u/v,$$

where $\rho, u, v \in \mathbb{Z}$ and $p \nmid u$, $p \nmid v$. By unique factorization in $\mathbb{Z}$, the number ρ depends only on r. We put

$$|r|_p = \gamma^{\rho}, \quad |0|_p = 0.$$

Then

$$\text{(i)} \quad |r|_p \geq 0 \text{ with equality only for } r = 0 \qquad (1.2\text{(i)})$$

$$\text{(ii)} \quad |rs|_p = |r|_p \, |s|_p \qquad (1.2\text{(ii)})$$

$$\text{(iii)*} \quad |r + s|_p \leq \max\{|r|_p, |s|_p\}. \qquad (1.2\text{(iii)})$$

Here (i), (ii) are trivial. To check (iii)* we may suppose that $r \neq 0$, $s \neq 0$ and without loss of generality that $|r|_p \geq |s|_p$. Then

$$r = p^{\rho}u/v, \quad s = p^{\sigma}x/y$$

for $\rho, \sigma, u, v, x, y \in \mathbb{Z}$ with

$$p \nmid uvxy,$$

and

$$\sigma = \rho + \tau \quad \text{with} \quad \tau \geqslant 0.$$

Now

$$r + s = p^{\rho}U/V,$$

where

$$V = vy \in \mathbb{Z}, \quad U = uy + p^{\tau}vx \in \mathbb{Z}.$$

Clearly $p \nmid V$. It is, however, quite possible that $p|U$ (at least when $\tau = 0$), say $U = p^{\lambda}W$, $\lambda \geqslant 0$, $p \nmid W$. Then

$$|r + s|_p = \gamma^{\rho+\lambda} \leqslant \gamma^{\rho} = \max\{|r|_p, |s|_p\}.$$

Note. It is usual to take $\gamma = p^{-1}$, when we have the p-adic valuation of $\mathbb{Q}$.

(d) $k = k_o(T)$, as in (b). Let $p(T)$ be an irreducible element of $k_o[T]$ and, as before, let $0 < \gamma < 1$. Every $h(T) \in k_o(T)$ can be written

$$h(T) = \{p(T)\}^{\rho} f(T)/g(T),$$

where $f(T), g(T) \in k_o[T]$ are not divisible by $p(T)$ and where $\rho \in \mathbb{Z}$ depends only on h (and, of course, p). We put

$$|h|_p = \gamma^{\rho}, \quad |0|_p = 0$$

and the reader will readily check that (i), (ii) and (iii)* hold.

The examples (b), (d) are closely related. Indeed on replacing T by T^{-1} in (b) we obtain (d) for the special polynomial $p(T) = T$.

We note also that except for (a) we always have the

ultrametric inequality (iii)*, not just the triangle inequality (iii). We shall see in Chapter 3 that (a) is indeed essentially the only valuation for which the ultrametric inequality fails.

2 REMARKS

Working with general valuations may require psychological adjustment. Consider first some consequences of the ultrametric inequality

$$|a + b| \leq \max(|a|, |b|). \tag{2.1}$$

It implies by an obvious induction that

$$|a_1 + \ldots + a_n| \leq \max_j |a_j| \tag{2.2}$$

and so (replacing a_j by $a_{j+1} - a_j$) that

$$|a_n - a_1| \leq \max_j |a_{j+1} - a_j|. \tag{2.3}$$

Let now b be a point of the "disc"

$$D = \{x : |x - a| < 1\}$$

of "centre" a. Then by (2.3)

$$|x - b| \leq \max\{|x - a|, |b - a|\} < 1,$$

for every $x \in D$. Conversely $|x - b| < 1$ implies that $x \in D$. Hence

$$D = \{x : |x - b| < 1\}.$$

Every point of the disc has an equal right to be regarded as a centre!

Again, consider the sequence (Conway and Sloane)

$$a_1 = 4, \quad a_2 = 34, \quad a_3 = 334, \ldots, a_n = 3\ldots34, \ldots$$

of integers. Then with respect to the 5-adic valuation we have

$$|a_m - a_n|_5 = 5^{-n} \qquad (m > n).$$

In particular, the sequence $\{a_n\}$ has the properties of what in ordinary real or complex analysis we call a <u>fundamental sequence</u> (Cauchy sequence). In this case we have

$$3a_1 = 12, \quad 3a_2 = 102, \quad 3a_3 = 1002, \ldots, 3a_n = 10\ldots02, \ldots,$$

and so

$$|3a_n - 2|_5 = 5^{-n}$$

Hence $\{a_n\}$ tends to the limit $2/3$ in a 5-adic sense. (We shall be defining these notions formally in the next Chapter).

On the other hand, fundamental sequences occur very naturally which do not have a rational limit. For example, one can find (in many ways) a sequence $\{a_n\}$ of integers such that

$$a_n^2 + 1 \equiv 0 \quad (5^n) \tag{2.4}$$

$$a_{n+1} \equiv a_n \quad (5^n) \tag{2.5}$$

for all $n \geq 1$. We take $a_1 = 2$. If a_n has already been chosen, we have to find an integer b such that

$$a_{n+1} = a_n + b5^n$$

satisfies (2.4), that is

$$(a_n + b5^n)^2 + 1 \equiv 0 \quad (5^{n+1}).$$

This is easily seen to be equivalent to

$$c_n + 2a_n b \equiv 0, \quad (5) \tag{2.6}$$

where $a_n^2 + 1 = 5^n c_n$. Since a_n is clearly not divisible by 5, we can satisfy (2.6). Hence we have an a_{n+1}, and the inductive process continues.

By (2.2) we have

$$|a_m - a_n|_5 \leq 5^{-n}, \quad (m \geq n)$$

so again $\{a_n\}$ is a fundamental sequence. Suppose that it has a limit $e \in \mathbb{Q}$. By (2.4) we have

$$|a_n^2 + 1|_5 \leq 5^{-n}$$

from which it easily follows that

$$|e^2 + 1|_5 = 0.$$

Hence

$$e^2 + 1 = 0 \tag{2.7}$$

by (1.2(i)). But there is no $e \in \mathbb{Q}$ satisfying (2.7).

In the next chapter we shall show that any field may be "completed" with respect to a valuation on it in the same way as the real numbers are constructed from the rationals by completing with respect to the ordinary absolute value. The completion of $\mathbb{Q}$ with respect to a p-adic valuation $|\ |_p$ is the field $\mathbb{Q}_p$ of p-adic numbers. The argument above will then show that $\mathbb{Q}_5$ contains a solution e of (2.7).

3 AN APPLICATION

Here we show that the bare definition of the p-adic valuation provides a natural proof of an interesting result. Nothing in this section is used later, so it can be omitted if desired. In Chapter 12 we shall, however, indicate that it is not an isolated result but, rather, has been the starting point of much recent work.

We recall that the Bernoulli numbers B_k are given by the formal power-series expansion

$$\frac{X}{e^X - 1} = B_o + \frac{B_1 X}{1!} + \ldots + \frac{B_k X^k}{k!} + \ldots . \tag{3.1}$$

Hence

$$B_0 = 1, \qquad B_1 = -1/2 .$$

Further,

$$B_k = 0 \quad (k \text{ odd}, > 1), \tag{3.1 bis}$$

since

$$\frac{X}{e^X - 1} + \frac{X}{2} = \frac{X(e^{\frac{1}{2}X} + e^{-\frac{1}{2}X})}{2(e^{\frac{1}{2}X} - e^{-\frac{1}{2}X})}$$

is unchanged by the substitution $X \to -X$.

Clearly the B_k are rational. The first few values are:

$B_2 = 1/6$	$B_{12} = -691/2730$
$B_4 = -1/30$	$B_{14} = 7/6$
$B_6 = 1/42$	$B_{16} = -3617/510$
$B_8 = -1/30$	$B_{18} = 43867/798$
$B_{10} = 5/66$	$B_{20} = -174611/330$

We shall present Witt's proof of

<u>THEOREM 3.1</u> (von Staudt-Clausen). <u>Let</u> k <u>be even</u>. <u>Then</u>

$$B_k + \sum_{\substack{q \text{ prime} \\ (q-1)|k}} q^{-1} \in \mathbb{Z} . \tag{3.2}$$

For example, the only primes q such that $q - 1$ divides 2 are $q = 2, 3$. In accordance with the Theorem $B_2 + \frac{1}{2} + \frac{1}{3} = 1$. Again, when $k = 20$ the relevant primes are $q = 2,3,5,11$ and one checks that

$$B_{20} + \frac{1}{2} + \frac{1}{3} + \frac{1}{5} + \frac{1}{11} = -528.$$

Put

$$S_k(n) = 1^k + 2^k + \dots + (n-1)^k. \tag{3.3}$$

On comparing coefficients on both sides of

$$1 + e^X + \dots + e^{(n-1)X} = \frac{e^{nX} - 1}{X} \frac{X}{e^X - 1}$$

using (3.1), we rapidly obtain the once well-known formula

$$S_k(n) = \sum_{r=0}^{k} \binom{k}{r} \frac{B_r}{k+1-r} n^{k+1-r}, \tag{3.4}$$

expressing $S_k(n)$ as a polynomial in n. Here $\binom{k}{r}$ is the binomial coefficient.

It follows that

$$B_k = \lim_{n \to 0} n^{-1} S_k(n). \tag{3.5}$$

With the ordinary definition of limit $n \to 0$ for positive integers n, this is a nonsense. If, however, we choose a prime p and work with the p-adic valuation $|\ |_p$, then it makes perfectly good sense; for example n can run through the sequence

$$p, p^2, p^3, \dots, p^m, \dots . \tag{3.6}$$

We therefore compare $p^{-m-1} S_k(p^{m+1})$ and $p^{-m} S_k(p^m)$. Every integer j in

$$0 \leq j < p^{m+1}$$

is uniquely of the form

$$j = up^m + v \quad (0 \leq u < p, \quad 0 \leq v < p^m).$$

Hence

$$S_k(p^{m+1}) = \sum_j j^k$$

$$= \sum_u \sum_v (up^m + v)^k$$

$$\equiv p \sum_v v^k + kp^m \sum_u u \sum_v v^{k-1} \quad (p^{2m})$$

on expanding by the binomial theorem. Here $\sum_v v^k = S_k(p^m)$. Further,

$$2 \sum_u u = p(p - 1) \equiv 0 \quad (p).$$

Hence

$$S_k(p^{m+1}) \equiv pS_k(p^m), \quad (p^{m+1}) \tag{3.7}$$

where for $p = 2$ we have used the hypothesis that k is even.

On dividing by p^{m+1}, we can write (3.7) as

$$|p^{-m-1} S_k(p^{m+1}) - p^{-m} S_k(p^m)|_p \leq 1.$$

By the consequence (2.3) of the ultrametric inequality (1.2(iii)), it follows that

$$|p^{-\ell} S_k(p^\ell) - p^{-m} S_k(p^m)|_p \leq 1 \tag{3.8}$$

for any positive integers ℓ, m. Put $m = 1$ and let ℓ tend to infinity in the conventional sense, so $p^\ell \to 0$ in the p-adic sense. Then

$$|B_k - p^{-1} S_k(p)|_p \leq 1 \tag{3.9}$$

by (3.5).

Now

$$S_k(p) = \sum_0^{p-1} j^k$$

$$\equiv \begin{cases} -1 & \text{if} \quad (p-1)|k \quad (p) \\ 0 & \text{otherwise.} \end{cases} \tag{3.10}$$

Hence and by (3.9),

$$\left.\begin{array}{ll} |B_k + p^{-1}|_p \leq 1 & \text{if} \quad (p-1)|k \\ |B_k|_p \leq 1 & \text{otherwise} \end{array}\right\} . \tag{3.11}$$

Put

$$W_k = B_k + \sum_{\substack{q \text{ prime} \\ (q-1)|k}} q^{-1} . \tag{3.12}$$

If p is any prime, we have

$$W_k = (B_k + p^{-1}) + \sum_{q \neq p} q^{-1} \tag{3.13(i)}$$

or

$$W_k = B_k + \sum_q q^{-1} \tag{3.13(ii)}$$

according as p is, or is not, a q. In both cases (3.11) implies

$$|W_k|_p \leq 1, \quad \text{(all primes } p) \tag{3.14}$$

on using the consequence (2.2) of the ultrametric inequality. But (3.14) implies that the rational number W_k has no primes in its denominator, i.e. $W_k \in \mathbb{Z}$, as asserted.

Notes

§3 Theorem 3.1 was announced briefly by Clausen (1840). This prompted the paper of von Staudt (1840), who said he had known the result for some time. Witt's proof seems to have entered the folklore without being published by him. See also Chapter 12, §5.

Exercises

1. Let p be a prime and s a positive integer. Show that

$$|s|_p \geq s^{-1},$$

for the p-adic valuation, with equality precisely when s is a power of p.

2. Let $k \geq 4$ be even and $p \neq 2,3$. Show that

$$|p^{-m} S_k(p^m) - B_k|_p \leq p^{-2m+\varepsilon}$$

for all $m \geq 1$, where $\varepsilon = 1$ if $(p-1)|(k-2)$ and $\varepsilon = 0$ otherwise.

[Note. A numerical example is $p = 5$, $k = 4$, $m = 1$, so $\varepsilon = 0$. Then $S_4(5) = 354$, and

$$5^{-1} S_4(5) - B_4 = 425/6.$$

Hint. By (3.4)

$$|p^{-m} S_k(p^m) - B_k|_p \leq \max_{r=0}^{k-1} \left|\frac{B_r}{k+1-r}\right|_p p^{m(r-k)}$$

Now estimate the $|B_r|_p$ by (3.1 bis) and Theorem 3.1, and use Exercise 1.]

3. For positive integer m let N_m be the integral part of $(1+\sqrt{3})^{2m+1}$. Show that $|N_m|_2 = 2^{-m-1}$.

CHAPTER TWO: GENERAL PROPERTIES

1 DEFINITIONS AND BASICS

DEFINITION 1.1. Let k be a field. A real-valued function $|b|$ defined for $b \in k$ is a valuation if there is some $C \in \mathbb{R}$ such that the following hold:

(i) $|b| \geq 0$, with equality only for $b = 0$.

(ii) $|bc| = |b||c|$ for all $b, c \in k$.

(iii) $|b| \leq 1 \Rightarrow |1 + b| \leq C$.

A particular example is the trivial valuation $|\ |_0$ given by

$$|b|_0 = \begin{cases} 0 & \text{if } b = 0 \\ 1 & \text{otherwise.} \end{cases}$$

We shall sometimes tacitly exclude the trivial valuation from general enunciations.

We have adopted the form (iii) of the third axiom (following Artin) for two reasons. First, it gives some formal simplifications (e.g. in the enunciation of the first lemma below). Second, in some contexts it is substantially simpler to show that it holds than it is to do this for, say, the triangle inequality.

From the definition we have at once:

COROLLARY 1. $|1| = 1$.

Proof. $1^2 = 1$, so $|1|^2 = |1|$ but $|1| \neq 0$.

COROLLARY 2. *If* $|a^n| = 1$, *then* $|a| = 1$.

COROLLARY 3. $|-1| = 1$, *so* $|-a| = |a|$.

COROLLARY 4. *If* k *is a finite field, then* $|\ |$ *is trivial.*

Proof. Follows from Corollary 2.

LEMMA 1.1. *If* $|\ |$ *is a valuation on* k *and* $\lambda > 0$ *is real, then* $|a|_1 = |a|^\lambda$ *is a valuation.*

Proof. Trivial. The corresponding constant is $C_1 = C^\lambda$.

DEFINITION 1.2. If $|\ |$, $|\ |_1$ are related as in the lemma then they are *equivalent*.

It is easy to see that this is an equivalence relation.

LEMMA 1.2. *A necessary and sufficient condition that a valuation* $|\ |$ *satisfies the triangle inequality is that one can take* $C = 2$ *in the definition of a valuation.*

Proof. (i) If the triangle inequality holds and $|a| \leq 1$, then $|1 + a| \leq |1| + |a| \leq 2$.

(ii) Suppose that $C = 2$. If $a_1, a_2 \in k$ with say $|a_1| \geq |a_2|$, $a_2 = aa_1$, $|a| \leq 1$ then

$$|a_1 + a_2| = |a_1(1 + a)| = |a_1||(1 + a)| \leq 2|a_1|;$$

that is

$$|a_1 + a_2| \leq 2 \max\{|a_1|, |a_2|\}.$$

By induction,

$$|a_1 + \ldots + a_{2^n}| \leq 2^n \max |a_j| \quad (1 \leq j \leq 2^n).$$

Now let $a_1, \ldots, a_N$ be given in k. Fix n by $2^{n-1} < N \leqslant 2^n$ and define $a_{N+1} = \ldots = a_{2^n} = 0$. Then

$$|a_1 + \ldots + a_N| \leqslant 2^n \max|a_j| \leqslant 2N \max|a_j|.$$

In particular, taking $a_j = 1$ $(1 \leqslant j \leqslant N)$ we get

$$|N| \leqslant 2N \qquad (N, \text{ positive integer}).$$

Now let $b, c \in k$ and let n be a positive integer. Then

$$\begin{aligned}
|b + c|^n &= |(b + c)^n| \\
&= \left|\sum_{r=0}^{n} \binom{n}{r} b^r c^{n-r}\right| \\
&\leqslant 2(n + 1) \operatorname*{Max}_r \left|\binom{n}{r} b^r c^{n-r}\right| \\
&\leqslant 2(n + 1) \operatorname*{Max}_r \left|\binom{n}{r}\right| |b|^r |c|^{n-r} \\
&\leqslant 4(n + 1) \operatorname*{Max}_r \binom{n}{r} |b|^r |c|^{n-r} \\
&\leqslant 4(n + 1) \sum_r \text{ditto} \\
&= 4(n + 1)(|b| + |c|)^n.
\end{aligned}$$

On taking the n-th root and letting $n \to \infty$, we get

$$|b + c| \leqslant |b| + |c|,$$

as required.

<u>COROLLARY</u>. <u>Every valuation is equivalent to one satisfying the triangle inequality</u>.

For most purposes we shall be concerned only with properties of equivalence classes of valuations and so we shall be able to use the triangle inequality and its familiar properties. In particular, we note for later use:

LEMMA 1.3. *Suppose that* $| \ |$ *satisfies the triangle inequality. Then*

$$\big| |b| - |c| \big|_{\infty} \leq |b - c|,$$

where $| \ |_{\infty}$ *is the absolute value on the reals.*

Proof. Since $c = b + (c - b)$ we have

$$|c| \leq |b| + |c - b|.$$

Similarly

$$|b| \leq |c| + |b - c|.$$

But $|b - c| = |c - b|$.

We now consider the ultrametric inequality.

DEFINITION 1.3. A valuation $| \ |$ on a field k is *non-archimedean* (*non-arch.*) if one can take $C = 1$ in Definition 1.1.

LEMMA 1.3. bis. *The valuation* $| \ |$ *on* k *is non-arch. if and only if it satisfies the ultrametric inequality*

$$|b + c| \leq \max\{|b|, |c|\}.$$

Further, a valuation which is equivalent to a non-arch. valuation is itself non-arch.

Proof. Clear.

The following trivial lemma is often vital.

LEMMA 1.4. *Suppose that* $| \ |$ *is non-arch. and that* $|c| < |b|$. *Then*

$$|b + c| = |b|.$$

Proof. Clearly $|b + c| \leq |b|$. On the other hand, $b = (b + c) + (-c)$, so

$|b| \leq \max\{|b + c|, |c|\}$.

The following criterion is useful:

LEMMA 1.5. *Let* $|\ |$ *be a valuation on the field* k. *Then* $|\ |$ *is non-archimedean if and only if* $|e| \leq 1$ *for all* e *in the ring generated by* 1 *in* k.

Note. The wording takes care of the possibility that k has non-zero characteristic.

Proof. The condition is clearly necessary. To prove sufficiency, we argue as in the proof of Lemma 1.2. By Lemma 1.2, Corollary, we may suppose that $|\ |$ satisfies the triangle inequality. For $b, c \in k$ and positive integer n we now have

$$\begin{aligned} |b + c|^n &= \left|\sum_r \binom{n}{r} b^r c^{n-r}\right| \\ &\leq \sum \left|\binom{n}{r}\right| |b|^r |c|^{n-r} \\ &\leq \sum |b|^r |c|^{n-r} \\ &\leq (n + 1) \{\max(|b|, |c|)\}^n. \end{aligned}$$

Now take the n-th root and let $n \to \infty$.

COROLLARY 1. *Let* $k \subset K$ *be fields and let* $|\ |$ *be a valuation on* K. *Then* $|\ |$ *is non-archimedean on* K *precisely when its restriction to* k *is non-arch.*

COROLLARY 2. *Let* k *be of prime characteristic. Then every valuation* $|\ |$ *on* k *is non-arch.*

Proof. For k contains a finite field.

2 VALUATIONS ON THE RATIONALS

THEOREM 2.1 (Ostrowski). *Every non-trivial valuation on* $\mathbb{Q}$ *is equivalent either to a p-adic valuation or to the ordinary absolute value.*

Proof. By Lemma 1.2, Corollary, we may suppose that $|\ |$ satisfies the triangle inequality.

Let $a > 1$ and $c > 0$ be integers. We can write c in the scale of a:

$$c = c_m a^m + c_{m-1} a^{m-1} + \dots + c_o \tag{2.1}$$

where

$$m = m(c,a), \quad c_m, c_{m-1}, \dots c_o \in \{0,1,\dots,a-1\}, \ c_m \neq 0.$$

Clearly

$$m \leq \log c/\log a. \tag{2.2}$$

By the triangle inequality applied to (2.1), we have

$$|c| \leq (m+1)\, M \max\{|a|^m, 1\}, \tag{2.3}$$

where

$$M = \max\{|1|, |2|, \dots, |a-1|\}$$

is independent of c. Now let $b > 1$ be an integer, and put $c = b^n$ for some integer $n > 0$. By (2.2), (2.3) we have

$$|b|^n \leq \{n(\log b/\log a) + 1\}\, M \max\{|a|^{n \log b/\log a}, 1\}.$$

On extracting the n-th root and letting $n \to \infty$, we have

$$|b| \leq \max\{|a|^{\log b/\log a}, 1\}. \tag{2.4}$$

We now distinguish two cases.

(i) There is an integer $b > 1$ with $|b| > 1$. Then by (2.4) we have $|a| > 1$ for all $a > 1$. On interchanging a,b in (2.4), we get

$$|b|^{1/\log b} = |a|^{1/\log a}.$$

Since this is true for all pairs a,b, it follows that

$$|b| = b^{\lambda}$$

for all integers $b > 1$ and for some λ independent of b. It readily follows that

$$|x| = |x|_{\infty}^{\lambda}$$

for all $x \in Q$, where $|\ |_{\infty}$ is the ordinary absolute value.

(ii) $|b| \leq 1$ for all $b > 1$. Then $|\ |$ is non-arch. by Lemma 1.5. If $|b| = 1$ for all $b > 1$, then we have the trivial valuation. Otherwise, there is some $b > 1$ with $|b| < 1$. Choose b minimal. If $b = cd$ with $c > 1$, $d > 1$, we have $1 > |b| = |c||d|$; and so either $|c| < 1$ or $|d| < 1$ contrary to the minimality of b. Hence $b = p$ (say) is prime.

Let $c \in \mathbb{Z}$, $p \nmid c$. Then

$$c = up + v, \quad 0 < v < p.$$

Now $|v| = 1$ by the above minimality condition, but $|up| = |u||p| < 1$. Hence $|c| = 1$ by Lemma 1.4. From all this, it readily follows that $|\ |$ is equivalent to the p-adic valuation.

3 INDEPENDENCE OF VALUATIONS

We want to show that valuations of the same field k are either equivalent or are independent in a very strong sense. The main result is in Theorem 3.1, but we need some preliminaries.

<u>LEMMA 3.1</u>. <u>Let</u> $|\ |_1$, $|\ |_2$ <u>be two valuations of</u> k. <u>Suppose that</u> $|\ |_1$ <u>is non-trivial and that</u>

$$|a|_1 < 1 \Rightarrow |a|_2 < 1. \tag{3.1}$$

<u>Then</u> $|\ |_1$ <u>and</u> $|\ |_2$ <u>are equivalent</u>.

<u>Proof</u>. By (3.1) with a^{-1} for a, we have

$$|a|_1 > 1 \Rightarrow |a|_2 > 1.$$

Now suppose, if possible, that there is a $b \in k$ with

$$|b|_1 = 1, \quad |b|_2 \neq 1,$$

say

$$|b|_2 > 1.$$

Since $|\ |_1$ is non-trivial by hypothesis, there is a nonzero $c \in k$ with

$$|c|_1 < 1.$$

Then

$$|cb^n|_2 = |c|_1 |b|_1^n < 1$$

for all $n \geq 0$, but

$$|cb^n|_2 = |c|_2 |b|_2^n > 1$$

for large enough n, contrary to (3.1) for $a = cb^n$. Hence

$$|a|_1 = 1 \Rightarrow |a|_2 = 1.$$

To sum up so far,

$$|a|_1 \gtreqless 1 \tag{3.2}$$

according as the corresponding inequality holds for $|a|_2$.

Now let b, c be non-zero elements of k and apply this remark to $a = b^m c^n$ with $m, n \in \mathbb{Z}$. On taking logarithms we have

$$m \log|b|_1 + n \log|c|_1 \gtreqless 0$$

according as

$$m \log|b|_2 + n \log|c|_2 \gtreqless 0.$$

Assuming, further, that $|c|_1 \neq 1$, it readily follows that

$$\log|b|_1 = \lambda \log|b|_2, \tag{3.3}$$

where $\lambda = \log|c|_1/\log|c|_2 > 0$ depends only on c. Hence

$$|b|_1 = |b|_2^\lambda \quad (\text{all } b \in k),$$

as required.

We now observe that a valuation $|\ |$ on k induces a topology, a basis for the open sets being the

$$U(b,\delta) = \{c: |c - b| < \delta\}.$$

Equivalent valuations obviously induce the same topology. If $|\ |$ satisfies the triangle inequality, the topology is that induced by the metric

$$d(b,c) = |b - c|.$$

Clearly the induced topology is the <u>discrete topology</u> (all sets are open) if and only if $|\ |$ is the trivial valuation.

<u>LEMMA 3.2</u>. <u>Let</u> $|\ |_1$ <u>and</u> $|\ |_2$ <u>induce the same topology on</u> k. <u>Then they are equivalent</u>.

<u>Proof</u>. After the preceding remark, we may suppose that $|\ |_1$ and $|\ |_2$ are non-trivial. We have the following chain of equivalent statements.

$$|b|_1 < 1 \iff |b^n|_1 \to 0 \quad (n \to \infty)$$

$$\iff b^n \text{ tends to } 0 \text{ with respect to the topology}$$

$$\iff |b^n|_2 \to 0 \quad (n \to \infty)$$

$$\iff |b|_2 < 1.$$

We can now invoke Lemma 3.1.

We continue to show that inequivalent valuations behave very independently. The next result will be superseded almost immediately.

LEMMA 3.3. *Let* $|\ |_1, \dots, |\ |_J$ *be non-trivial valuations on a field* k. *Suppose that no two are equivalent. Then there is an* $a \in k$ *with*

$$|a|_1 > 1, \quad |a|_j < 1 \quad (1 < j \leqslant J).$$

Proof. Induction on J.

J = 2. Since $|\ |_1$ is not trivial and $|\ |_1$, $|\ |_2$ are not equivalent, by a Lemma 3.1 there is a $b \in k$ with

$$|b|_1 < 1 \quad |b|_2 \geqslant 1.$$

Similarly there is a $c \in k$ with

$$|c|_2 < 1 \quad |c|_1 \geqslant 1.$$

Then

$$a = cb^{-1}$$

will do.

J > 2. By the induction hypothesis there is a $b \in k$ with

$$|b|_1 > 1, \quad |b|_j < 1 \quad (2 \leqslant j \leqslant J - 1).$$

Also $(J = 2)$ there is a $c \in k$ with

$$|c|_1 > 1, \quad |c|_J < 1.$$

We distinguish three cases

(i) $|b|_J < 1$. Take $a = b$.

(ii) $|b|_J = 1$. Then $a = b^n c$ will do for large enough positive integer n.

(iii) $|b|_J > 1$. Take $a = \frac{b^n}{1 + b^n} c$. We have

$$\frac{b^n}{1 + b^n} = \frac{1}{1 + b^{-n}} \to \begin{cases} 1 & \text{for } |\ |_1 \text{ and } |\ |_J. \\ 0 & \text{for } |\ |_j \quad (j \neq 1, J). \end{cases}$$

Hence a will do for large enough n.

THEOREM 3.1 Let $|\ |_j$ $(1 \leq j \leq J)$ be pairwise inequivalent non-trivial valuations. Let $b_1, \ldots, b_J \in k$ be given arbitrarily and let real $\varepsilon > 0$ also be arbitrary. Then there is an $a \in k$ such that simultaneously

$$|a - b_j|_j < \varepsilon \quad (1 \leq j \leq J).$$

Proof. By the preceding lemma there are $c_j \in k$ such that

$$|c_j|_j > 1 \qquad |c_j|_i < 1 \quad (i \neq j).$$

Then consider

$$\sum_j \frac{c_j^n}{1 + c_j^n} b_j$$

as $n \to \infty$.

Note. Theorem 3.1 is closely related to the "Chinese Remainder Theorem" of elementary number theory. Let $p_1, \ldots, p_J$ be distinct primes and let $m(j)$ be positive integers. One form of the Chinese Remainder Theorem asserts that given any $b_j \in \mathbb{Z}$ $(1 \leq j \leq J)$ there is an $a \in \mathbb{Z}$ such that simultaneously

$$a \equiv b_j \quad \left(p_j^{m(j)}\right) \quad (1 \leq j \leq J) \ . \tag{3.2}$$

We can write this as

$$|a - b_j|_j \leq p_j^{-m(j)}, \tag{3.3}$$

where $|\ |_j$ is the p_j-adic valuation. Theorem 3.1 merely asserts the existence of an $a \in \mathbb{Q}$ satisfying (3.3). In the context of a general field k there is no analogue of the additional information provided by the Chinese Remainder Theorem, namely that a can be chosen to be an integer. We shall see later, however, (Chapter 10, §4) that there is an analogue of the Chinese Remainder Theorem when k is an algebraic number field.

4 COMPLETENESS

Let k be a field with valuation $|\ |$. We say that a sequence

$$\{a_n\} = \{a_1, a_2, \dots, a_n, \dots\} \tag{4.1}$$

tends to b as a <u>limit</u> (with respect to $|\ |$) if for every $\varepsilon > 0$ there is an $n_o(\varepsilon)$ such that $|a_n - b| < \varepsilon$ for all $n > n_o$. Clearly the limit of a sequence is unique, if it exists. We say that the sequence $\{a_n\}$ is <u>fundamental</u> if for every $\varepsilon > 0$ there is an $n_1(\varepsilon)$ such that

$$|a_m - a_n| < \varepsilon \quad (\text{all } m, n > n_1). \tag{4.2}$$

If $\{a_n\}$ has a limit, then it is clearly fundamental. In Chapter 1 §2 we saw that the converse does not necessarily hold.

DEFINITION 4.1. The field k is <u>complete</u> with respect to the valuation $|\ |$ if every fundamental sequence has a limit.

Let k be a field with valuation $|\ |$ and let K be a field containing k. We say that a valuation $\|\ \|$ on K <u>extends</u> $|\ |$ if it takes the same values on k.

DEFINITION 4.2. Let k be a field with valuation $|\ |$. We say that a field K together with a valuation $\|\ \|$ extending $|\ |$ is a <u>completion</u> of k (with respect to k) if both

(i) K is complete

(ii) K is the closure of k with respect to (the topology induced by) $\|\ \|$

For example, the real field $\mathbb{R}$ is the completion of $\mathbb{Q}$ with respect to the ordinary absolute value.

THEOREM 4.1 *Let* k *be a field with valuation* $|\ |$. *A completion exists and any two completions are canonically isomorphic.*

Note By the last clause we mean that if K_j are completions with valuations $\|\ \|_j$ $(j = 1,2)$ and embeddings

$$k \hookrightarrow K_j$$

then there is a unique bijection between K_1 and K_2 which (i) is the identity on k (ii) respects the field structure and (iii) preserves the values of the $\|\ \|_j$.

Proof. By taking an equivalent valuation we may suppose that $|\ |$ satisfies the triangle inequality and so gives k the structure of a metric space. Let K be the completion of k with respect to the metric. Let D be the metric on K and put $\|\alpha\| = D(\alpha,0)$ for $\alpha \in K$. We shall show that K can be given a field structure and that $\|\ \|$ is a valuation on it.
[The exercises suggest an alternative proof of Theorem 4.1 which does not presuppose the existence of the completion of a metric space].

Let $\alpha, \beta \in K$, so that they are respectively the limits of sequences $\{a_n\}$, $\{b_n\}$ of elements of k. Then $a_n + b_n$ is a fundamental sequence and so has a limit γ (say) $\in K$. Similarly $a_n b_n$ has a limit δ (say). We define $\gamma = \alpha + \beta$; $\delta = \alpha\beta$. It is straightforward to check that the ring axioms are satisfied.

Now let $\alpha \in K$, $\alpha \neq 0$. Then $\|\alpha\| \neq 0$. Let $\{a_n\}$ be a sequence of elements of k with limit α. Then $|a_n| \to \|\alpha\|$ since distance on a metric space is a continuous function with respect to the topology it induces. Hence $a_n = 0$ for only finitely many n and we may suppose that $a_n \neq 0$ (all n). Put

$$b_n = a_n^{-1}.$$

Then

$$|b_m - b_n| = \frac{|a_m - a_n|}{|a_m|\,|a_n|} \to 0 \quad (m,n \to (\infty,\infty))$$

since $|a_m - a_n| \to 0$, $|a_m| \to \|\alpha\| \neq 0$, $|a_n| \to \|\alpha\| \neq 0$. Hence by completeness $\{b_n\}$ has a limit, which we define to be α^{-1}. It is then easy to check that K satisfies the field axioms.

Finally, by continuity, $\|\ \|$ on K satisfies the valuation axioms since $|\ |$ does in k.

It remains to show that K is unique up to isomorphism. Let L be any field complete with respect to a valuation $|||\ |||$ for which there is an embedding

$$\psi:\ k \hookrightarrow L$$

respecting the valuations $|\ |$, $|||\ |||$. Then ψ extends uniquely to an embedding of K in L since $\{\psi(a_n)\}$ is a fundamental sequence precisely when $\{a_n\}$ is one. Clearly $\psi(K)$ is the closure $\overline{\psi(k)}$ of $\psi(k)$ in L. If we now suppose that L is a completion of k, then $\overline{\psi(k)} = L$; and we have established an isomorphism between K and L.

For later use we note that we have proved the

COROLLARY. *Let* L *be a complete valued field and let* ψ *be an embedding of the valued field* k *in* L. *Then the closure* $\overline{\psi(k)}$ *is a completion of* k.

We can now put the independence results of §3 in a picturesque shape.

THEOREM 4.2 *Let* k *be a field and let* $|\ |_j$ $(1 \leq j \leq J)$ *be non-trivial pairwise inequivalent valuations on* k. *Let* k_j *be the respective completions, and let*

$$\Delta:\ k \hookrightarrow \prod_j k_j$$

be the diagonal map. Then $\Delta(k)$ *is everywhere dense (i.e. its closure is the whole of* $\prod_j k_j$).

Here $\prod_j k_j$ is, of course, the product of topological spaces.

Proof. Without loss of generality the $| \ |_j$ satisfy the triangle inequality. Let α_j be any elements of k_j $(1 \leq j \leq J)$. Then by the definition of completion there are $a_j \in k$ so that

$$|a_j - \alpha_j|_j < \varepsilon,$$

where $\varepsilon > 0$ is given. By Theorem 3.1 there is a $b \in k$ so that

$$|b - a_j|_j < \varepsilon \quad (1 \leq j \leq J).$$

Hence

$$|b - \alpha_j|_j < 2\varepsilon \quad (1 \leq j \leq J).$$

5 FORMAL SERIES AND A THEOREM OF EISENSTEIN

In this section we consider a valued field whose completion is especially easy to construct, and use it to prove a theorem of Eisenstein about the arithmetic nature of the coefficients in the power series expansions of rational functions.

Let $k(T)$ be the field of rational functions of the indeterminate T with coefficients in some field k and let $0 < \gamma < 1$. We denote by $| \ |$ the valuation introduced in Chapter 1, §1(d) in the special case $p(T) = T$: that is if

$$h(T) = T^{\rho} f(T)/g(T) \tag{5.1}$$

where $\rho \in \mathbb{Z}$, $f(T), g(T) \in k[T]$ and $f(0) \neq 0$, $g(0) \neq 0$, then we put

$$|h| = \gamma^{\rho}. \tag{5.2}$$

Now let $N \in \mathbb{Z}$ (possibly negative) and let $\{f_n\}$ be any sequence of elements of k for $n \geq N$. Then

$$f^{(m)} = \sum_{n=N}^{m} f_n T^n$$

is a fundamental sequence of elements of $k(T)$, since

$$|f^{(M)} - f^{(m)}| \leqslant \gamma^{m+1} \quad (M > m).$$

We denote the limit in the completion $k((T))$ of $k(T)$ by

$$f = f(T) = \sum_{N}^{\infty} f_n \, T^n. \tag{5.3}$$

If we had taken any $N_1 < N$ and extended the definition of f_n by putting $f_n = 0$ for $N_1 \leqslant n < N$, then we should have got the same element of $k((T))$. It is thus convenient, when we are not interested in some particular value of N, to write

$$f(T) = \sum_{n >> -\infty} f_n \, T^n \tag{5.4}$$

instead of (5.3), where now f_n is given for all $n \in \mathbb{Z}$ but the notation implies that $f_n = 0$ for all n less than some N (depending on $f(T)$).

It is clear how to add and multiply expressions of the type (5.4) and that they are a commutative ring with 1. We now show that any element of the type (5.4) of $k((T))$ other than 0 has an inverse of the same type. In fact we have

$$f(T) = T^{\rho} \, b \, (1 + \sum_{n \geqslant 1} g_n \, T^n)$$

for some $b \neq 0$ in k and some g_n in k. Let

$$\begin{aligned} h(T) &= 1 + \sum_{m \geqslant 1} (- \sum_{n \geqslant 1} g_n \, T^n)^m \\ &= 1 + \sum_{n \geqslant 1} h_n \, T^n \end{aligned}$$

where the h_n are given in the obvious way. Then

$$(1 + \sum_{n \geqslant 1} g_n \, T^n)(1 + \sum_{n \geqslant 1} h_n \, T^n) = 1:$$

which gives an inverse for $f(T)$. We have thus proved

LEMMA 5.1. *The completion* $k((T))$ *of* $k(T)$ *is just the set of expressions* (5.4) *together with* 0.

We denote by $k[[T]]$ the set of $f(T) \in k((T))$ for which $|f(T)| \leq 1$. Clearly $f(T) \in k[[T]]$ precisely when it can be written

$$f(T) = \sum_{n \geq 0} f_n T^n. \tag{5.5}$$

Further, $k[[T]]$ is a ring, the ring of <u>formal power series</u>.

We now put $k = \mathbb{Q}$.

<u>DEFINITION 5.1</u>. $f(T) \in \mathbb{Q}[[T]]$ is said to satisfy <u>Eisenstein's condition</u> if there are $u, v \in \mathbb{Z}$, $u \neq 0$, $v \neq 0$ such that (in the notation (5.5))

$$u v^n f_n \in \mathbb{Z} \quad \text{(all } n). \tag{5.6}$$

An equivalent formulation is that there is a $w \in \mathbb{Z}$, $w \neq 0$ such that $w^n f_n \in \mathbb{Z}$ for all $n > 0$.

<u>THEOREM 5.1</u> ("Eisenstein"). <u>Let</u> $f = f(T) \in \mathbb{Q}[[T]]$ <u>and suppose that there are</u> $g_j = g_j(T) \in \mathbb{Q}[T]$ <u>not all</u> 0 <u>such that</u>

$$\sum_{0 \leq j \leq J} g_j f^j = 0 . \tag{5.7}$$

<u>Then</u> f <u>satisfies Eisenstein's condition.</u>

<u>Proof</u>. For indeterminates X, Y write

$$H(X) = \sum_j g_j(T) X^j \in \mathbb{Q}[T,X] , \tag{5.8}$$

and

$$H(X + Y) = H(X) + H_1(X) Y + \ldots + H_J(X) Y^J, \tag{5.9}$$

where

$$H_j(X) \in \mathbb{Q}[T,X]. \tag{5.10}$$

By hypothesis,

$$H(f) = 0. \tag{5.11}$$

We may suppose without loss of generality that $H_1(f) \neq 0$, since otherwise we could operate with $H_1(X)$ instead of $H(X)$. Define m by

$$|H_1(f)| = \gamma^m. \tag{5.12}$$

Put

$$f(T) = u(T) + T^{m+1} v(T), \tag{5.13}$$

where

$$u(T) = f_o + \dots + f_m T^m + f_{m+1} T^{m+1} \in \mathbb{Q}[T]$$

and

$$v(T) = 0 + f_{m+2} T + f_{m+3} T^2 + \dots \in \mathbb{Q}[[T]].$$

It is clearly enough to show that $v(T)$ satisfies Eisenstein's condition. By (5.9), (5.11), (5.13) we have

$$0 = H(u + T^{m+1} v)$$

$$= H(u) + T^{m+1} H_1(u)v + \sum_{j \geq 2} T^{(m+1)j} H_j(u)v^j,$$

where

$$H(u),\ H_1(u),\ H_j(u) \in \mathbb{Q}[T].$$

Here all the summands except, possibly, the first are divisible by T^{2m+1} by (5.12); and so $H(u)$ is divisible by T^{2m+1} (in $\mathbb{Q}[T]$). On dividing by T^{2m+1} we obtain

$$0 = h + h_1 v + h_2 v^2 + \dots + h_J v^J, \tag{5.14}$$

where

$$h,\ h_1,\ \dots,\ h_J \in \mathbb{Q}[T]$$

and

$$h_j(0) = 0 \quad (j > 1) \tag{5.15}$$

but

$$h_1(0) \neq 0, \tag{5.16}$$

by (5.12).

On multiplying through by an appropriate integer, we may suppose that

$$h, h_1, \ldots, h_J \in \mathbb{Z}[T] .$$

Let

$$\ell = h_1(0) \quad (\neq 0).$$

We have constructed $v = v(T)$ so that its constant term is 0. We shall show by induction that $\ell^n v_n \in \mathbb{Z}$, where v_n is the coefficient of T^n. Indeed on equating coefficients of T^n in (5.14) we obtain that ℓv_n is the sum of terms of the type

$$e \prod_{m<n} v_m^{\mu(m)},$$

where $e \in \mathbb{Z}$ and $\sum m\mu(m) < n$ by (5.15).

Now $\ell^n v_n \in \mathbb{Z}$ follows readily by induction.

Notes

§1 There is a more general definition of valuation in which the values are taken in an arbitrary ordered group, see e.g. Schilling (1950). In this context the valuations with which we are concerned in this book are termed "rank one valuations".

§5 For Eisenstein's theorem and related results see Pólya and Szegő (1976), Section 8, Chapter 2. The theorem of Chapter 13 can be regarded as a partial converse.

Exercises

1. Fill in the details of the following alternative proof of Theorem 4.1. Here k is a field with valuation $|\ |$ satisfying the triangle inequality.

(i) The set F of fundamental sequences is a ring under the definitions

$$\{a_n \pm b_n\} = \{a_n\} \pm \{b_n\}$$

$$\{a_n b_n\} = \{a_n\}\{b_n\}.$$

(ii) The function

$$\|\{a_n\}\| = \lim_{n\to\infty} |a_n| \qquad (\{a_n\} \in F)$$

is well-defined, where the limit is a limit of real numbers with the usual topology.

(iii) Define $\{a_n\}$ to be a null-sequence if $\|\{a_n\}\| = 0$. Then the set N of null sequences is an ideal in F.

(iv) N is a maximal ideal and so the quotient ring F/N is a field.

(v) $\|\{a_n\}\|$ depends only on $\{a_n\} \in F$ modulo N and so $\|\ \|$ induces a well-defined function (also denoted by $\|\ \|$) on F/N.

(vi) The map which takes an element $b \in k$ into the fundamental sequence all of whose elements are b induces an embedding of k into F/N.

(vii) The function $\|\ \|$ on F/N satisfies the valuation axioms.

(viii) F/N is complete with respect to $\|\ \|$ and is the closure of k under the embedding given by (vi).

2. Let $g_j = g_j(T) \in \mathbb{Q}[[T]]$ for $0 \leq j \leq J$ be not all zero, and suppose that $f = f(T) \in \mathbb{Q}[[T]]$ satisfies

$$\sum_j g_j f^j = 0.$$

If the g_j satisfy Eisenstein's condition, show that f does so also.

3. Let

$$f(T) = \sum_{n\geq 1} n^{-1} T^n \in \mathbb{Q}[[T]].$$

Show that f does not satisfy any polynomial equation with coefficients in $\mathbb{Q}[T]$.

4. Let k be a field of characteristic $p > 0$ and let $f(T) \in k[[T]]$.

(i) Show that $\{f(T)\}^p \in k[[T^p]]$.

(ii) If q is a power of p, show that $\{f(T)\}^q \in k[[T^q]]$.

(iii) Suppose, further, that $f(0) = 1$ (i.e. $f(T) = 1 + \sum_{j\geq 1} f_j T^j$). For positive integer N put

$$g_N(T) = \prod_{0\leq n\leq N} \{f(T)\}^{p^n} .$$

Show that $g_N(T)$ is a convergent sequence in $k[[T]]$. Show, further, that

$$f(T) \{g(T)\}^{p-1} = 1,$$

where $g(T)$ is the limit of the $g_N(T)$.

CHAPTER THREE: ARCHIMEDEAN VALUATIONS

1 INTRODUCTION

If a valuation is not non-archimedean we say that it is *archimedean*. In this chapter we show that the only archimedean valuations are essentially the ordinary absolute value on the field $\mathbb{C}$ of complex numbers and the valuations which it induces in subfields. The rest of the book will then be confined to non-archimedean valuations. More precisely we prove

THEOREM 1.1 (Ostrowski). *Let k be a field complete with respect to an archimedean valuation $|\ |$. Then k is (isomorphic to) either $\mathbb{R}$ or $\mathbb{C}$ and $|\ |$ is equivalent to the ordinary absolute value.*

Note If $|\ |$ is an archimedean valuation on a field k but k is not complete then k is contained in its completion $\bar{k}$ (Chapter 2, §4) and Theorem 1.1 applies to $\bar{k}$.

The proof will be carried out in subsequent sections. We briefly indicate the strategy here. In the first place, k must be of characteristic 0 (Chapter 2, Lemma 1.4, Cor. 2) and so k contains the rationals $\mathbb{Q}$. The valuation induced by $|\ |$ on $\mathbb{Q}$ must be non-archimedean (Chapter 2, Lemma 1.5, Cor. 1) and so is equivalent to the ordinary absolute value $|\ |_\infty$ (Chapter 2, Theorem 2.1). Since k is complete it therefore contains the completion $\mathbb{R}$ of $\mathbb{Q}$ with respect to $|\ |_\infty$ (Chapter 2, Theorem 4.1, Cor.).

We must now distinguish cases. Suppose, first, that k contains a solution i of $i^2 = -1$. Then k contains the complex field $\mathbb{C}$. We have to show, however, that the valuation induced by $|\ |$ on $\mathbb{C}$ is equivalent to $|\ |_\infty$. This done, we derive a contradiction from the supposition that k is strictly larger than $\mathbb{C}$.

If k does not contain a solution i of $i^2 = -1$, then

we adjoin one. We show that the valuation $|\ |$ on k can be extended to $k(i)$ and are then reduced to the previous case.

Some of the lemmas are special cases of results valid for all valuations which will be proved later and the proofs are on somewhat similar lines. However later we shall be able to assume that the valuations concered are non-archimedean; which will simplify matters.

2 SOME LEMMAS

LEMMA 2.1. Any archimedean valuation $|\ |$ on $\mathbb{C}$ is equivalent to the absolute value $|\ |_\infty$.

Proof. Without loss of generality $|\ |$ satisfies the triangle inequality. By the remark at the end of §1, the valuations induced by $|\ |$ and $|\ |_\infty$ on $\mathbb{R}$ are equivalent, say

$$|a| = |a|_\infty^\lambda \qquad (\text{all } a \in \mathbb{R})$$

for some λ, $0 < \lambda < \infty$.

Let

$$\alpha = a + ib \qquad (a,b \in \mathbb{R})$$

Then

$$|a|_\infty \leq |\alpha|_\infty,$$

$$|b|_\infty \leq |\alpha|_\infty$$

and so

$$\begin{aligned} |\alpha| &\leq |a| + |ib|. \\ &= |a| + |b| \\ &\leq 2|\alpha|_\infty^\lambda \quad . \end{aligned} \tag{2.1}$$

If $|\ |$ and $|\ |_\infty$ were inequivalent, this would contradict

Theorem 3.1 of Chapter 2 on simultaneous approximation.

LEMMA 2.2. *Let* k *be complete with respect to a valuation* $|\ |$. *Suppose that* $T^2 + 1$ *is irreducible in* $k[T]$. *Then there is a* $\Delta > 0$ *such that*

$$|a^2 + b^2| \geq \Delta \max\{|a|^2, |b|^2\},$$

for all $a,b \in k$.

Note Compare Hensel's Lemma (Chapter 4, §3).

Proof. We may suppose that $|\ |$ satisfies the triangle inequality, and then show that

$$\Delta = |4|/(1 + |4|) \tag{2.2}$$

will do.

By homogenity, we have to show that if there is a $c_1 \in k$ with

$$|c_1^2 + 1| = \delta_1 (\text{say}) < \Delta, \tag{2.3}$$

then $T^2 + 1$ is reducible. We shall construct a $c^* \in k$ with $c^{*2} + 1 = 0$ by successive approximation.

By (2.3) and the triangle inequality (cf. Lemma 1.3 of Chapter 2), we have

$$|c_1^2| \geq 1 - \delta_1 \tag{2.4}$$

Put

$$c_2 = c_1 + h_1$$

for some $h_1 \in k$. Then

$$c_2^2 + 1 = c_1^2 + 1 + 2h_1c_1 + h_1^2.$$

We choose

$$h_1 = -(c_1^2 + 1)/2c_1$$

to eliminate the linear terms. Then

$$\begin{aligned}\delta_2 \text{ (say)} &= |c_2^2 + 1| \\ &= |h_1|^2 \\ &= |c_1^2 + 1|^2/|4||c_1|^2 \\ &\leqslant \theta\delta_1\end{aligned}$$

where by (2.2), (2.3), (2.4) we can take

$$\theta = \delta_1/|4|(1 - \delta_1) < 1.$$

On repeating the process, we obtain a sequence of elements $c_n \in k$ such that

$$\begin{aligned}\delta_n \text{ (say)} &= |c_n^2 + 1| \\ &\leqslant \theta\delta_{n-1} \\ &\leqslant \theta^{n-1}\delta_1.\end{aligned}$$

Further,

$$|c_{n+1} - c_n|^2 = |c_n^2 + 1|^2/|4||c_n|^2 = \delta_{n+1} \leqslant \theta^n\delta_1. \tag{2.5}$$

It is readily checked that (2.5) implies that $\{c_n\}$ is a fundamental sequence, so c_n tends to some $c^* \in k$, by completeness. Now

$$\begin{aligned}|c^{*2} + 1| &= \lim_n |c_n^2 + 1| \\ &= 0,\end{aligned}$$

and so $c^{*2} + 1 = 0$, as required.

<u>LEMMA 2.3</u>. <u>Let</u> k <u>be complete with respect to a valuation</u> $|\ |$. <u>Suppose that</u> $T^2 + 1$ <u>is irreducible in</u> $k[T]$. <u>Then there is an extension of</u> $|\ |$ <u>to</u> $k(i)$, <u>where</u> $i^2 = -1$.

<u>Note</u> For the extension of valuations to algebraic extensions of complete fields in general, see Chapter 7.

<u>Proof</u>. Without loss of generality, $|\ |$ satisfies the triangle inequality. We put

$$||a + ib|| = |a^2 + b^2|^{\frac{1}{2}}.$$

It is easy to check that this coincides with $|\ |$ on k and that parts (i), (ii) of the definition of a valuation (Chapter 2, Definition 1.1) are satisfied by $||\ \ ||$. It remains to verify (iii).

Suppose that

$$||a + ib|| \leq 1.$$

Then

$$|a|, |b| \leq \Delta^{-\frac{1}{2}}$$

by Lemma 2.2. Hence

$$\begin{aligned} ||1 + (a + ib)||^2 &= |(1 + a)^2 + b^2| \\ &\leq 1 + |2||a| + |a|^2 + |b|^2 \\ &\leq 1 + |2||\Delta|^{-\frac{1}{2}} + 2\Delta^{-1} \\ &= C^2 \quad \text{(say)}, \end{aligned}$$

which is what was required.

<u>Note</u> It is not asserted that $k(i)$ is complete with respect to the

extended valuation. That it is complete is a special case of a result proved later (Lemma 2.1, Corollary 2 of Chapter 7). For our present purposes it is enough to observe that $k(i)$ is contained in its completion with respect to the extended valuation.

3 COMPLETION OF PROOF

To complete the programme laid down at the end of §1 we need only

LEMMA 3.1. *Let k be complete with respect to the archimedean valuation $|\ |$ and suppose there is an $i \in k$ with $i^2 = -1$. Then $k = \mathbb{C}$, and $|\ |$ is equivalent to the absolute value $|\ |_\infty$.*

Proof. Without loss of generality, $|\ |$ satisfies the triangle inequality.

We have seen that k contains $\mathbb{R}$ and so it contains $\mathbb{R}(i) = \mathbb{C}$. By Lemma 2.1 the valuation induced by $|\ |$ on $\mathbb{C}$ is equivalent to $|\ |_\infty$.

Suppose, if possible, that $k \neq \mathbb{C}$, and let $\alpha \in k$, $\alpha \notin \mathbb{C}$. Then $|\alpha - a|$ is a continuous function of $a \in \mathbb{C}$ (by Lemma 1.3 of Chapter 2), and so attains its lower bound, say at $b \in \mathbb{C}$. Put $\beta = \alpha - b$. Then $|\beta| > 0$ since $\beta \neq 0$, and

$$0 < |\beta| = \inf_{a \in \mathbb{C}} |\beta - a|. \tag{3.1}$$

Now let

$$c \in \mathbb{C}, \quad 0 < |c| < |\beta| \tag{3.2}$$

and let n be a natural number. Then

$$\frac{\beta^n - c^n}{\beta - c} = \prod_{\substack{\varepsilon^n = 1 \\ \varepsilon \neq 1}} (\beta - \varepsilon c)$$

and

$$|\beta - \varepsilon c| \geq |\beta|$$

by (3.1). Hence

$$\frac{|\beta - c|}{|\beta|} \leq \frac{|\beta^n - c^n|}{|\beta|^n}$$
$$= |1 - (c/\beta)^n|$$
$$\leq 1 + |c/\beta|^n$$
$$\to 1 \quad (n \to \infty),$$

by (3.2). Thus $|\beta - c| \leq |\beta|$, and so

$$|\beta - c| = |\beta|$$

by (3.1).

In particular, we may take $\beta - c$ for β in (3.1) and repeat the process. Hence

$$|\beta - mc| = |\beta|$$

for all natural numbers m. But then

$$|m||c| \leq |\beta| + |\beta - mc|$$
$$\leq 2|\beta|$$

is bounded, contrary to the assumption that $|\ |$ is archimedean (Chapter 2, Lemma 1.4).

Notes

The results of this Chapter are due to Ostrowski (1918).

There is a generalization, for which we require a definition. Let k be a field with valuation $|\ |$. A field $K \supset k$ together with a function $\|\ \|$ on K is called a normed field over k if (I) K, $\|\ \|$ is a normal vector space over k (Definition 2.1 of Chapter 7) and (II) we have $\|\underline{ab}\| \leq \|\underline{a}\|\|\underline{b}\|$ for all $\underline{a},\underline{b} \in K$. Then the only normed fields over $\mathbb{R}$ are $\mathbb{R}$ and $\mathbb{C}$. If we permit skew fields, the only extra possibility is the hamiltonian quaternions. See Tornheim (1952), where there are references to the earlier literature and, for a different approach, Loomis (1953), Theorem 22F (p.68).

CHAPTER FOUR: NON-ARCHIMEDEAN VALUATIONS. SIMPLE PROPERTIES

1 DEFINITIONS AND BASICS

Let $|\ |$ be a non-arch. valuation on the field k. The set

$$\mathfrak{o} = \{a: |a| \leq 1\} \tag{1.1}$$

is clearly a ring. It is called the ring of (valuation-) integers. The set

$$\mathfrak{p} = \{a: |a| < 1\} \tag{1.2}$$

is clearly a maximal ideal in $\mathfrak{o}$. Hence the quotient ring $\mathfrak{o}/\mathfrak{p}$ is a field, the residue class field. The notation $\mathfrak{o},\mathfrak{p}$ will be standard. If $|a| = 1$ we say that a is a (valuation)-unit.

Let $\bar{k}$ be the completion of k with respect to $|\ |$ and let $\bar{\mathfrak{o}},\bar{\mathfrak{p}}$ be the corresponding ring of integers and maximal ideal. Clearly

$$\mathfrak{o} = \bar{\mathfrak{o}} \cap k; \qquad \mathfrak{p} = \bar{\mathfrak{p}} \cap k. \tag{1.3}$$

Hence there is a natural map

$$\mathfrak{o}/\mathfrak{p} \rightarrow \bar{\mathfrak{o}}/\bar{\mathfrak{p}} \tag{1.4}$$

of residue classfields induced by the inclusion of $\mathfrak{o}$ in $\bar{\mathfrak{o}}$.

LEMMA 1.1. *The map (1.4) is an isomorphism.*

Proof. We have only to show that it is an epimorphism.

If $\alpha \in \bar{\mathfrak{o}}$, then by the definition of $\bar{k}$ there is an $a \in k$ such that $|\alpha - a| < 1$. Then $a \in \mathfrak{o}$ and $\alpha - a \in \bar{\mathfrak{p}}$.

The set of non-zero values taken by $|\ \ |$, that is

$$\{|a| : \quad a \in k^*\}, \tag{1.5}$$

is a subgroup of the group of positive reals. Here we use the standard notation k^* for the multiplicative group of non-zero elements of a field k. We call (1.5) the valuation group. It is clearly the same for k and $\bar{k}$ (Chapter 2, Lemma 1.4). We say that the valuation is discrete if the valuation group is discrete in the real topology i.e. if there is some $\delta > 0$ such that

$$1 - \delta < |a| < 1 + \delta \Rightarrow |a| = 1. \tag{1.6}$$

LEMMA 1.2. The valuation is discrete precisely when the maximal ideal $\mathfrak{p}$ is principal.

Proof. (i) Suppose that $\mathfrak{p} = (\pi)$ is principal. Then

$$|a| < 1 \Rightarrow a \in \mathfrak{p} \Rightarrow a = \pi b \quad (b \in \mathfrak{o}) \Rightarrow |a| \leq |\pi|.$$

Similarly $\quad |a| > 1 \Rightarrow |a| \geq |\pi|^{-1}$.

(ii) Suppose that $|\ \ |$ is discrete. Then the set

$$\{|a| : |a| < 1\}$$

attains its upper bound, say at $a = \pi$. Then

$$|a| < 1 \Rightarrow a = \pi b, \quad |b| \leq 1 \quad (\text{i.e., } \ b \in \mathfrak{o}).$$

If $\mathfrak{p} = (\pi)$, we say that π is a prime element for the for the valuation.

Most of the valuations we shall be concerned with are discrete. If $|\ \ |$ is discrete and $b \in k^*$ there is an $n \in \mathbb{Z}$ such that

$$|b| = |\pi|^n.$$

We define n to be the <u>order</u> of b and write

$$n = \text{ord } b.$$

Clearly ord b does not depend on the choice of prime element π or, indeed, on the choice of valuation within an equivalence class. The axioms of a non-arch. valuation are equivalent to:

$$\left.\begin{array}{l} \text{ord}(b + c) \geq \min\{\text{ord } b, \text{ord } c\} \\ \\ \text{ord}(bc) = \text{ord } b + \text{ord } c \end{array}\right\} , \qquad (1.7)$$

and it is conventional to put

$$\text{ord } 0 = + \infty.$$

Although the use of the order function instead of the valuation symbolism has advantages in several contexts, we shall generally prefer the use of valuations, partly because of the psychological suggestion of the analogy with the ordinary absolute value.

We shall be extending the notions of elementary analysis to the context of valued fields, especially complete ones. In general we shall find the situation simpler than in the case of $\mathbb{R}$ and $\mathbb{C}$. We have already introduced the notion of limits in Chapter 2. We shall say that the infinite sum

$$\sum_{0}^{\infty} a_n \quad (a_n \in k) \qquad (1.8)$$

<u>converges</u> to the <u>sum</u> s if

$$s = \lim_{N\to\infty} s_N, \qquad (1.9)$$

where

$$s_N = \sum_{0}^{N} a_n. \qquad (1.10)$$

Clearly the non-archimedean property is inherited by infinite sums:

$$\left|\sum_0^\infty a_n\right| \leq \max_n |a_n|. \tag{1.11}$$

LEMMA 1.3. *Suppose that* k *is complete. Then* $\sum a_n$ *converges precisely when* $a_n \to 0$.

Proof. (i) Suppose that $\sum a_n$ converges. Then

$$\lim a_N = \lim(s_N - s_{N-1}) = \lim s_N - \lim s_{N-1} = s - s = 0.$$

(ii) Suppose that $a_n \to 0$ and let $M > N$. Then

$$|s_M - s_N| = |a_{N+1} + \ldots + a_M| \leq \max_{N<n\leq M} |a_n|$$

$$< \varepsilon \quad (N \geq N_o(\varepsilon)).$$

Hence $\{s_N\}$ is a fundamental sequence; so it converges by completeness.

We can now give a useful explicit description of the integers in a field complete with respect to a discrete valuation:

LEMMA 1.4. *Suppose* k *is complete with respect to the discrete valuation* $|\ |$ *and let* π *be a prime element* (Lemma 1.2). *Let* $A \subset \mathfrak{o}$ *be a set of representatives of* $\mathfrak{o}/\mathfrak{p}$. *Then every* $a \in \mathfrak{o}$ *is uniquely of the form*

$$a = \sum_0^\infty a_n \pi^n \quad (a_n \in A). \tag{1.12}$$

Conversely, the right hand side of (1.12) always converges to give an $a \in \mathfrak{o}$.

Proof. (i) The converse is trivial after Lemma 1.3, since $|a_n \pi^n| \leq |\pi|^n$, and so $a \in \mathfrak{o}$ by (1.11).

(ii) Now let $a \in \mathfrak{o}$ be given. There is precisely one $a_o \in A$ such that $|a - a_o| < 1$ and then $a = a_o + \pi b_1$ for some $b_1 \in \mathfrak{o}$. There is precisely one $a_1 \in A$ such that $|b_1 - a_1| < 1$ and then

$b_1 = a_1 + \pi b_2$ for some $b_2 \in \mathfrak{o}$. Continuing in this way we get for every N

$$a = a_0 + a_1\pi + \ldots + a_N \pi^N + b_{N+1} \pi^{N+1}$$

with $a_n \in A$ and $b_{N+1} \in \mathfrak{o}$. Now $b_{N+1} \pi^{N+1} \to 0$, and so (1.12) holds.

In the particular case when $k = \mathbb{Q}_p$ is the field of p-adic numbers, the ring $\mathfrak{o}$ of integers is denoted $\mathbb{Z}_p$ and called the ring of p-adic integers. We can take $\pi = p$ (say) and $A = \{0,1,\ldots,p-1\}$ by Lemma 1.1. The reader should familiarize him/her-self with operating with the representation (1.12) e.g. by expressing the sum or product of two elements of $\mathbb{Z}_p$ given in the shape (1.12) in the same shape.

Although it is not necessary for the truth of Lemma 1.4 it is usually convenient to assume that $0 \in A$ (representing, of course, the 0 of $\mathfrak{o}/\mathfrak{p}$). In particular we have the

COROLLARY *Suppose, in addition to the hypotheses of Lemma 1.4, that* $0 \in A$. *Then every* $a \in k^*$ *is uniquely of the shape*

$$a = \sum_N^\infty a_n \pi^n \quad (a_n \in A, \quad a_N \neq 0)$$

for some $N \in \mathbb{Z}$.

Proof. For $\pi^{-N} a \in \mathfrak{o}$ for some N.

We conclude with a lemma which plays a central role in some accounts of the subject but which we will seldom, if ever, need.

LEMMA 1.5. *Suppose that* k *is complete with respect to a discrete valuation* $|\ |$ *and that the residue class field* $\mathfrak{o}/\mathfrak{p}$ *is finite. Then* $\mathfrak{o}$ *is compact*.

Proof. Since $|\ |$ makes $\mathfrak{o}$ a metric space, compactness is equivalent to sequential compactness. We have therefore to show that every sequence $\{a^{(j)}\}$ of elements of $\mathfrak{o}$ has a convergent subsequence. The proof applies the well-known "diagonal process" to the representation

$$a^{(j)} = \sum_0^\infty a_{jn} \pi^n \quad (a_{jn} \in A)$$

furnished by Lemma 1.4. Since A is finite, there is some a_o^* which occurs as a_{jo} for infinitely many j. For the $a^{(j)}$ with $a_{jo} = a_o^*$, there is some a_1^* which occurs as a_{j1} for infinitely many j. For the $a^{(j)}$ with $a_{jo} = a_o^*$, $a_{j1} = a_1^*$ there is an a_2^* which occurs as a_{j2} infinitely often. And so on. There is then a subsequence tending to

$$a^* = \sum a_n^* \pi^n .$$

COROLLARY *Let $|\ |$ be a (non-arch.) valuation on a field k. A necessary and sufficient condition that k be locally compact with respect to it, is that all three of the following conditions be full-filled (i) k is complete (ii) $|\ |$ is discrete (iii) the residue class field is finite.*

Proof. The three conditions are trivially necessary. Sufficiency follows from the Lemma.

Note When $|\ |$ is archimedean, the field k has to be complete, and so the only locally compact fields are $\mathbb{R}$ and $\mathbb{C}$.

2 AN APPLICATION TO FINITE GROUPS OF RATIONAL MATRICES

In this section we use the ideas already introduced to prove a theorem which has no apparent relation with valued fields. We denote the multiplicative group of invertible $n \times n$ matrices with elements from a ring R by $GL_n(R)$. The unit matrix is I and the zero matrix (all of whose elements are 0) is 0.

LEMMA 2.1. *Let $p \neq 2$ and $A \in GL_n(\mathbb{Z}_p)$. If*

$$A \equiv I(p) \ ; \quad A \neq I, \tag{2.1}$$

Then A is of infinite order.

Note For the situation when $p = 2$, see Exercises.

<u>Proof</u>. It is enough to show that $A^q \neq I$ for every prime q and every A satisfying (2.1). We write

$$A = I + B$$

where B has elements $b_{ij} \in \mathbb{Z}_p$ $(1 \leq i \leq n,\ 1 \leq j \leq n)$. By (2.1) there are u,v such that

$$0 < \delta \text{ (say)} = |b_{uv}| = \max_{i,j} |b_{ij}| \leq p^{-1},$$

where $|\ | = |\ |_p$ is the p-adic valuation. We have

$$A^q = (I + B)^q = 1 + \binom{q}{1}B + \binom{q}{2}B^2 + \ldots + \binom{q}{q}B^q.$$

We distinquish two cases

(i) $q \neq p$. All the elements of the matrices

$$\binom{q}{j}B^j \quad (j \geq 2)$$

have value at most δ^2. On the other hand, $\binom{q}{1}B$ contains the element qb_{uv} which has value δ. Hence (since the p-adic valuation is non-arch.) $A^q - I \neq 0$.

(ii) $q = p$. The binomial coefficients $\binom{p}{j}$ $(2 \leq j \leq p - 1)$ are all divisible by p and so the elements of $\binom{p}{j}B^j$ $(2 \leq j \leq p - 1)$ all have value $\leq p^{-1}\delta^2$ (with the usual normalization $|p| = p^{-1}$ of the p-adic valuation). The elements of $\binom{p}{p}B^p$ have value $\leq \delta^p \leq \delta^3$ since $p \neq 2$. On the other hand, $\binom{p}{1}B$ contains the element pb_{uv} whose value is $p^{-1}\delta$. But $\delta \leq p^{-1}$ since δ is in the value group of $\mathbb{Q}_p$, and so

$$p^{-1}\delta > \max(p^{-1}\delta^2, \delta^3).$$

Hence $A^q - I \neq 0$, as before.

<u>LEMMA 2.2</u>. <u>Let</u> $p \neq 2$ <u>and let</u> G <u>be a finite subgroup of</u> $GL_n(\mathbb{Z}_p)$. <u>Then the order of</u> G <u>divides</u>

$$(p^n - p^{n-1})(p^n - p^{n-2}) \dots (p^n - 1). \tag{2.3}$$

Proof. Let $\mathbb{F}$ be the finite field of p elements. The residue class map $\mathbb{Z}_p \to \mathbb{F}$ induces a group homomorphism

$$GL_n(\mathbb{Z}_p) \to GL_n(\mathbb{F}). \tag{2.4}$$

Let $A \in G$ be in the kernel of (2.4). Then $A \equiv I\ (p)$ but A is of finite order. Hence $A = I$ by Lemma 2.1, and so (2.4) gives an isomorphism of G with a subgroup of $GL_n(\mathbb{F})$. Since (2.3) is the order of $GL_n(\mathbb{F})$, the Lemma is established.

For completeness, we prove the statement about the order of $GL_n(\mathbb{F})$. Let $B \in GL_n(\mathbb{F})$. The rows of B, say $\underline{b}_1,\dots,\underline{b}_n$ are any set of n linearly independent vectors. The first vector $\underline{b}_1$ can be anything other than $\underline{0}$ so there are $p^n - 1$ possibilities. Given $\underline{b}_1$ we can take $\underline{b}_2$ to be any vector independent of $\underline{b}_1$, so there are $p^n - p$ possibilities. Next $\underline{b}_3$ is any vector independent of $\underline{b}_1,\underline{b}_2$, so $p^n - p^2$ possibilities. And so on.

THEOREM 2.1 Let $G \subset GL_n(\mathbb{Q})$ have finite order g. Then g divides

$$g^*(n) = \prod_{q \text{ prime}} q^{\beta(q)},$$

where

$$\beta(2) = n + 2[n/2] + [n/2^2] + [n/2^3] + \dots$$

$$\beta(q) = [n/(q - 1)] + [n/q(q - 1)] + [n/q^2(q - 1)] + \dots$$

$$(q \neq 2.$$

Proof. Since G is finite, there is only a finite set S of primes which occur in the denominators of elements of the matrices of G. For $p \notin S$ we have $G \subset GL_n(\mathbb{Z}_p)$. By Lemma 2.2 if $p \neq 2$, $p \notin S$, then g divides (2.3).

We now apply Dirichlet's theorem on primes in arithmetic progressions which asserts that if a,b are positive integers without common factor, then $a + bm$ is a prime for infinitely many positive integers m (see any analytic number theory text).

Let $q \neq 2$ be a prime. By Dirichlet's theorem there are infinitely many primes p which are primitive roots modulo q^2: so there is certainly such a p which is not in S. By elementary number theory, p is then a primitive root modulo q^j for every $j > 0$ and it is then easy to see that $q^{\beta(q)}$ is the exact power of q dividing (2.3). For $q = 2$ one takes $p \notin S$, $p \equiv 3 \pmod 8$ and again $2^{\beta(2)}$ is the precise power of 2 dividing (2.3). This concludes the proof.

3 HENSEL'S LEMMA

This is a generic term for results which infer the existence of a solution of an equation from the existence of an approximate solution. The proofs are by successive approximation and are analogous to Newton's method for the reals. Here we give one of the simplest versions and make some applications. We will meet other versions later.

LEMMA 3.1. ("Hensel's Lemma"). *Let* k *be complete with respect to* $|\ |$ *and let*

$$f(X) \in \mathfrak{o}[X]. \tag{3.1}$$

Let $a_0 \in \mathfrak{o}$ *satisfy*

$$|f(a_0)| < |f'(a_0)|^2, \tag{3.2}$$

where $f'(X)$ *is the (formal) derivative. Then there is an* $a \in \mathfrak{o}$ *such that*

$$f(a) = 0. \tag{3.3}$$

Proof. Let $f_j(X)$ $(j = 1,2,\ldots)$ be defined by the identity

$$f(X + Y) = f(X) + f_1(X)Y + f_2(X)Y^2 + \ldots \tag{3.4}$$

for independent indeterminates X,Y. Then

$$f_1(X) = f'(X). \tag{3.5}$$

By (3.2) there is a $b_o \in \mathfrak{o}$ such that

$$f(a_o) + b_o f_1(a_o) = 0. \tag{3.6}$$

Then by (3.4) we have

$$|f(a_o + b_o)| \leqslant \max_{j \geqslant 2} |f_j(a_o)\, b_o^j|.$$

Here $|f_j(a_o)| \leqslant 1$ since $f_j(X) \in \mathfrak{o}[X]$ and $a_o \in \mathfrak{o}$. Hence

$$\begin{aligned} |f(a_o + b_o)| &\leqslant |b_o^2| \\ &= |f(a_o)|^2/|f'(a_o)|^2 \\ &< |f(a_o)| \end{aligned} \tag{3.7}$$

by (3.2).

Similarly, but more simply,

$$|f_1(a_o + b_o) - f_1(a_o)| \leqslant |b_o| < |f_1(a_o)|,$$

and so

$$|f_1(a_o + b_o)| = |f_1(a_o)|. \tag{3.8}$$

We now put $a_1 = a_o + b_o$ and repeat the process. In this way we get a sequence of $a_n = a_{n-1} + b_{n-1}$ such that

$$|f_1(a_n)| = |f_1(a_o)| \quad \text{(all n)}$$

and

$$\begin{aligned} |f(a_{n+1})| &\leqslant |f(a_n)|^2/|f_1(a_n)|^2 \\ &= |f(a_n)|^2/|f_1(a_o)|^2. \end{aligned}$$

Then $f(a_n) \to 0$. Further,

$$|a_{n+1} - a_n| = |b_n|$$

$$= |f(a_n)|/|f_1(a_n)|$$

$$= |f(a_n)|/|f_1(a_o)| \tag{3.9}$$

$$\to 0.$$

Thus $\{a_n\}$ is a fundamental sequence. By completeness it has a limit a and $f(a) = 0$.

COROLLARY 1. *We have*

$$|a - a_o| \leq |f(a_o)|/|f'(a_o)|. \tag{3.10}$$

Further, there is only one solution of $f(a) = 0$ *satisfying (3.10).*

Proof. We have

$$a - a_o = \sum b_n,$$

so (3.10) follows from (3.6). Suppose now that there is an $a^* \neq a$ such that

$$f(a^*) = 0, \quad |a^* - a_o| \leq |f(a_o)|/|f'(a_o)|.$$

Put $a^* = a + b^*$. Then

$$0 = f(a + b^*) - f(a) = b^* f_1(a) + b^{*2} f_2(a) + \ldots \tag{3.11}$$

Here

$$|b^*| \leq |f(a_o)|/|f_1(a_o)| < |f_1(a_o)| = |f_1(a)|$$

by (3.2) and (3.8). Since $|f_j(a)| \leq 1$ for $j \geq 2$, the first term in the right hand side of (3.11) has a strictly greater value than the others. This is impossible for a non-arch. valuation.

COROLLARY 2. *Let* $f(X) \in \mathfrak{o}[X]$ *have discriminant* D *and let* $a_o \in \mathfrak{o}$ *satisfy*

$$|f(a_o)| < |D|^2 \tag{3.12}$$

Then $f(X)$ *has a root* a *in* $\mathfrak{o}$:

Proof. We recall (Appendix A) that D is a polynomial in the coefficients of f with coefficients in $\mathbb{Z}$, so $D \in \mathfrak{o}$. Further, there are $u(X), v(X) \in \mathfrak{o}[X]$ such that

$$u(X)f(X) + v(X)f'(X) = D. \tag{3.13}$$

Now $|u(a_o)| \leq 1$, $|v(a_o)| \leq 1$ and $|f(a_o)| < |D|^2 \leq |D|$ by hypothesis. Hence (3.13) with $X \mapsto a_o$ implies

$$|f'(a_o)| \geq |D|. \tag{3.14}$$

Hence and by (3.12) the conditions of the Lemma are satisfied.

It is worthwhile to consider explicitly the case when

$$f(X) = f_o + f_1X + f_2X^2 \tag{3.15}$$

is a quadratic. Here

$$D = f_1^2 - 4f_of_2. \tag{3.16}$$

The identity (3.13) in this case is

$$-4f_2f(X) + (f_1 + 2f_2X)f'(X) = D. \tag{3.17}$$

We conclude this section with some concrete applications of Hensel's Lemma.

LEMMA 3.2. *Let* $p \neq 2$. *Let* $b \in \mathbb{Z}_p$, $|b| = 1$ *and suppose that there is an* $a_o \in \mathbb{Z}_p$ *such that* $|a_o^2 - b| < 1$. *Then* $b = a^2$ *for some* $a \in \mathbb{Z}_p$.

Proof. Follows from Lemma 3.1 with $f(X) = X^2 - b$ since

$|f'(a_o)| = |2a_o| = 1$. Alternatively, one can invoke Corollary 2 since $|D| = |-4b| = 1$.

COROLLARY. *The group* $\mathbb{Q}_p^*/(\mathbb{Q}_p^*)^2$ *has order 4 and exponent 2. Representatives of the cosets can be taken as* 1, p, c, pc *where* c *is any quadratic non-residue.*

LEMMA 3.3. (p = 2). *If* $b \in \mathbb{Z}_2$, $b \equiv 1$ (8), *then* $b = a^2$ *for some* $a \in \mathbb{Z}_2$.

Proof. In lemma 1 take $f(X) = X^2 - b$, so $|f(1)| \leq 2^{-3}$, $|f'(1)| = 2^{-1}$.

COROLLARY. $\mathbb{Q}_2^*/(\mathbb{Q}_2^*)^2$ *has order 8 and exponent 2. Representatives of a set of generators are* $-1, 5, 2$.

LEMMA 3.4. *Let* $p \neq 3$ *and let* $b \in \mathbb{Z}_p$, $|b| = 1$. *Suppose that* $b \equiv c^3$ (p) for some $c \in \mathbb{Z}_p$. Then $b = a^3$ for some $a \in \mathbb{Z}_b$.

Proof. Apply Lemma 3.1 to $X^3 - b$.

LEMMA 3.5. (p = 3). *A necessary and sufficient condition for a 3-adic unit* b *to be a cube is that* $b \equiv \pm 1$ (9).

Proof. For there is an $e \in \{0, \pm 1\}$ such that $b \equiv \pm(1 + 3e)^3$ (27). Now apply Lemma 3.1 to $f(X) = X^3 - b$ with $a_o = \pm(1 + 3e)$.

3 bis. APPLICATION TO DIOPHANTINE EQUATIONS

A *diophantine equation* is one in which the unknowns are required to lie in some specified field (e.g. $\mathbb{Q}$) or ring (e.g. $\mathbb{Z}$). Local fields have many applications to diophantine equations and indeed are indispensible for their deep study. Here we make some general observations about equations requiring solutions in $\mathbb{Q}$. We shall make further applications of what we are doing to diophantine equations from time to time and will return to them more systematically in Chapter 11.

It is obvious that the only solution in $\mathbb{Q}$ of

$$x^2 + y^2 + z^2 = 0$$

is the trivial solution $x = y = z = 0$, since that is true already for solutions in $\mathbb{R} = \mathbb{Q}_\infty$. One can use the $\mathbb{Q}_p$ in a similar way. For example, the only solution in $\mathbb{Q}$ of

$$3x^3 + 2y^2 - z^2 = 0$$

is the trivial solution, since that is true already for solutions in $\mathbb{Q}_3$, as follows at once from the next lemma.

It is generally a fairly straightforward matter to decide whether an explicitly given diophantine equation has a solution in a given p-adic field $\mathbb{Q}_p$. As an example we shall consider a quadratic form

$$F(\underline{X},\underline{Y}) = F(X_1,\ldots,X_n,Y_1,\ldots,Y_m) \tag{3 bis. 1}$$

$$= \sum_{j=1}^{n} a_j X_j^2 + \sum_{i=1}^{m} p\, b_i Y_i^2$$

where the a_j and b_i are p-adic units. This is more general than it looks, since any non-singular quadratic form can be brought to this shape by a non-singular linear transformation of the variables.

LEMMA 3 bis. 1. *Let* $p \neq 2$ *and let* F *be given by (3 bis. 1), where* $|a_j| = |b_i| = 1$ *for all* j,i. *A necessary and sufficient condition that there exist* $x_1,\ldots,x_n$, $y_1,\ldots,y_m \in \mathbb{Q}_p$, *not all zero, such that*

$$F(\underline{x},\underline{y}) = F(x_1,\ldots,x_n, y_1,\ldots,y_m) = 0 \tag{3 bis. 2}$$

is that (at least) one of the two following conditions holds:

either (i) *there are* $c_1,\ldots,c_n \in \mathbb{Z}$, *not all divisible by* p, *such that*

$$\sum a_j c_j^2 \equiv 0 \quad (p) \tag{3 bis. 3}$$

or (ii) *there are* $d_1,\ldots,d_m \in \mathbb{Z}$, *not all divisible by* p, *such that*

$$\sum_i b_i d_i^2 \equiv 0 \quad (p) \tag{3 bis. 4}$$

Proof. Suppose, first, that x_j, y_i exist satisfying (3 bis. 2). By multiplying them all simultaneously by a suitable power of p we may suppose without loss of generality that

$$\max\{|x_1|, \ldots, |x_n|, |y_1|, \ldots, |y_m|\} = 1 \qquad (3 \text{ bis. } 5)$$

There are now two cases. Suppose that

$$\max|x_j| = 1.$$

On choosing any $c_j \equiv x_j \ (p)$ it is clear that (3 bis. 2) implies (3 bis. 3). Otherwise we have

$$\max|x_j| \leq p^{-1}, \quad \max|y_i| = 1.$$

Then similarly (3 bis. 4) holds for any $d_i \equiv y_i \ (p)$.

We now prove the converse. Suppose that (3 bis. 3) holds where, without loss of generality $c_1 \not\equiv 0 \ (p)$. On applying Hensel's Lemma to

$$G(X) = a_1 X^2 + \sum_{j \neq 1} a_j c_j^2$$

we obtain an x_1 such that $\sum a_j x_j^2 = 0$ with $x_j = c_j \quad (j \neq 1)$. Similarly for (3 bis. 4).

COROLLARY. *For $p = 2$ the lemma continues to hold provided that the conditions (i) (ii) are replaced by*

(i_2) *There are* $c_1, \ldots, c_n \in \mathbb{Z}$ *not all even and* $d_1, \ldots, d_m \in \mathbb{Z}$ *such that*

$$\sum a_j c_j^2 + 2 \sum b_i d_i^2 \equiv 0 \quad (8).$$

(ii_2) *There are* $d_1, \ldots, d_m \in \mathbb{Z}$ *not all even and* $c_1, \ldots, c_n \in \mathbb{Z}$ *such that*

$$\sum b_i d_i^2 + 2 \sum a_j c_j^2 \equiv 0 \quad (8).$$

The proof of the Corollary is left to the reader.

It is convenient to introduce some general terminology. The field $\mathbb{Q}$ is an example of a <u>global field</u> and the $\mathbb{Q}_p$ (including $\mathbb{Q}_\infty = \mathbb{R}$) are the corresponding <u>local</u> fields (or <u>localizations</u>). We shall say that a diophantine equation has a solution <u>globally</u> if it has a solution in $\mathbb{Q}$ and that it has a solution <u>everywhere locally</u> if it has a solution in all the localizations $\mathbb{Q}_p$ (including $\mathbb{Q}_\infty$). In this lingo an obviously necessary condition for the existence of a global solution is that there should be a solution everywhere locally. (When the equation is homogeneous, we naturally exclude from consideration the <u>trivial solution</u> in which all the variables take the value 0).

In the present context the discussion of solubility everywhere locally is equivalent (apart from solubility in $\mathbb{R}$) to what are usually described as "congruence considerations"; and it may be felt that we have gained nothing except an aura of mystery. Even at the present level of discourse, however, the use of local fields has substantial advantages once the worker has familiarized himself with it. One of the reasons is that the $\mathbb{Q}_p$ are fields, while for congruence arguments one works with rings $\mathbb{Z}/m\mathbb{Z}$ with divisors of zero. The doubting reader is invited to accept the following challenge. Let

$$F(X) \in \mathbb{Q}[X]$$

with indeterminates $\underline{X} = (X_1,\ldots,X_n)$, and put

$$G(\underline{X}) = F(\underline{Y}),$$

where

$$Y_i = \sum_j t_{ij} X_j \qquad (1 \leq i \leq n)$$

with

$$t_{ij} \in \mathbb{Q}, \quad \det(t_{ij}) \neq 0.$$

Clearly $F(\underline{X}) = 0$ has solutions everywhere locally precisely when $G(\underline{X}) = 0$ has solutions everywhere locally. The reader should formulate this statement precisely in the language of congruences and prove his statement in that language (covering both the case when F is homogeneous and that when it is not).

It is natural to ask whether a diophantine equation which is soluble everywhere locally necessarily has a global solution. This is true for certain types of equation, as we shall see in Chapter 11. It is not, however, true in general, and we conclude this section by giving two examples.

LEMMA 3 bis. 2. *The equation*

$$(X^2 - 2)(X^2 - 17)(X^2 - 34) = 0$$

has a solution everywhere locally but not globally.

Proof. There is clearly no global solution. There are obviously solutions in $\mathbb{Q}_\infty$. Further, $2 \in (Q_{17}^*)^2$ and $17 \in (Q_2^*)^2$. If $p \neq 2, 17, \infty$, then $2, 17, 34$ are p-adic units and at least one of them is a quadratic residue mod p. This gives a root in $\mathbb{Q}_p$ by Lemma 3.2.

This example may be felt artificial because the left hand side is reducible in $\mathbb{Q}[X]$. For functions of a single variable we need reducibility but that is not the case in several variables:

LEMMA 3 bis. 3. *There are rational solutions of*

$$X^4 - 17 = 2Y^2 \qquad (3\text{ bis. } 6)$$

everywhere locally but not globally.

Proof. We check first solubility everwhere locally. There are clearly real solutions. For $\mathbb{Q}_2$ there is a solution with $Y = 0$ and for $\mathbb{Q}_{17}$ there is one with $X = 1$. For $p \neq 2, 17, \infty$, the theory of equations over finite fields (Appendix D) shows that there are $a,b \in \mathbb{Z}$ (depending on p) such that

$$a^4 - 17 \equiv 2b^2 \quad (p);$$

and this gives a solution in $\mathbb{Q}_p$ by Hensel's Lemma.

It remains to show that there is no global solution. Suppose if possible that there are $x,y \in \mathbb{Q}$ satisfying (3 bis. 6). It is readily deduced that then

$$x = a/c, \quad y = b/c^2,$$

where

$$a,b,c \in \mathbb{Z}, \quad \gcd(a,b,c) = 1. \tag{3 bis. 7}$$

Then

$$a^4 - 17c^4 = 2b^2,$$

which can be rewritten as

$$(5a^2 + 17c^2)^2 - 17(a^2 + 5c^2)^2 = (4b)^2,$$

or, again, as

$$(5a^2 + 17c^2 + 4b)(5a^2 + 17c^2 - 4b) = 17(a^2 + 5c^2)^2 \tag{3 bis. 8}$$

The two bracketed expressions on the left hand side are clearly positive. We claim that any common factor is a power of 2. For suppose, if possible, that an odd prime p divides both expressions. Then $p|b$ and $p|(5a^2 + 17c^2)$. Further, $p|(a^2 + 5c^2)$ by (3 bis. 8) and so p divides

$$(5a^2 + 17c^2) - 5(a^2 + 5c^2) = -8c^2;$$

that is, $p|c$. Similarly, $p|a$. This is a contradiction to (3 bis. 7) and so the claim is proved.

On considering the factorization of the bracketed expressions in (3 bis. 8) in the light of these remarks, we deduce that there are $u,v \in \mathbb{Z}$ such that one of the two following cases holds for an appropriate choice of sign:

	Case I	Case II
$5a^2 + 17c^2 \pm 4b =$	$17u^2$	$34u^2$
$5a^2 + 17c^2 \mp 4b =$	v^2	$2v^2$
$a^2 + 5c^2 =$	uv	$2uv$

In Case I we have

$$10a^2 + 34c^2 = 17u^2 + v^2 \tag{3 bis. 9}$$

$$a^2 + 5c^2 = uv \tag{3 bis. 10}$$

We show that this pair of homogeneous equations has no nontrivial solution in $\mathbb{Q}_{17}$. By homogeneity we may suppose that

$$\max\{|a|, |c|, |u|, |v|\} = 1, \tag{3 bis. 11}$$

where $| \ | = | \ |_{17}$. Then

$$|a| < 1, \quad |v| < 1$$

by (3 bis. 9) because 10 is a quadratic nonresidue mod 17. Then $|c| < 1$ by (4 bis. 10) and finally $|u| < 1$ by (3 bis. 9): and this contradicts (3 bis. 11).

In Case II we have

$$5a^2 + 17c^2 = 17u^2 + v^2,$$

$$a^2 + 5c^2 = 2uv.$$

A very similar argument shows that there is again no solution in $\mathbb{Q}_{17}$.

4 ELEMENTARY ANALYSIS

We continue to explore the concepts of elementary analysis in the context of a field k complete with respect to a non-arch. valuation $| \ |$.

The question of rearrangement of series and summation of double series, which is somewhat traumatic in $\mathbb{R}$ and $\mathbb{C}$, is quite straightforward in the present context.

LEMMA 4.1. *Let* $b_{ij} \in k$ *for* $i,j = 0,1,2,\ldots$. *Suppose that for every* $\varepsilon > 0$ *there is a* $J(\varepsilon)$ *such that* $|b_{ij}| < \varepsilon$ *whenever* $\max(i,j) \geq J(\varepsilon)$. *Then the series*

$$\sum_i(\sum_j b_{ij}), \quad \sum_j(\sum_i b_{ij}) \tag{4.1}$$

both converge, and their sums are equal.

Proof. Clearly $\sum_j b_{ij}$ converges for every i, and

$$\left|\sum_j b_{ij}\right| < \varepsilon \quad (i \geq J(\varepsilon))$$

by (1.11). Hence the first sum in (4.1) converges. It is easily seen that

$$\left|\sum_{i=o}^{J(\varepsilon)}\left(\sum_{j=o}^{J(\varepsilon)} b_{ij}\right) - \sum_i^{\infty}\left(\sum_j^{\infty} b_{ij}\right)\right| < \varepsilon . \tag{4.2}$$

On using also the inequality corresponding to (4.2) with i and j interchanged, we conclude that the difference between the two double sums in (4.1) is at most ε in value. As ε is arbitrary, they must be equal.

The notion of *radius of convergence* of a power series

$$f(X) = f_o + f_1X + \ldots + f_nX^n + \ldots \tag{4.3}$$

applies in our context, and again the situation is simpler than for $\mathbb{R}$ and $\mathbb{C}$. Put

$$R = \frac{1}{\limsup_n |f_n|^{1/n}}, \tag{4.4}$$

so

$$0 \leq R \leq +\infty \tag{4.5}$$

with the obvious conventions.

LEMMA 4.2. *Let* $\mathcal{D}$ *be the set of* $a \in k$ *for which the series* (4.3) *converges. Then*

(i) *if* $R = 0$, *then* $\mathcal{D}$ *consists of* 0 *alone.*

(ii) *if* $R = \infty$, *then* $\mathcal{D}$ *consists of all of* k.

(iii) *If* $0 < R < \infty$ *and*

$$|f_n| R^n \to 0 \tag{4.6}$$

then

$$\mathcal{D} = \{a \in k : |a| \leq R\} \tag{4.7}$$

(iv) *Otherwise*

$$\mathcal{D} = \{a \in k : |a| < R\} \tag{4.8}$$

Proof. By Lemma 1.3, $\mathcal{D}$ is precisely the set of $a \in k$ for which $f_n a^n \to 0$. The proof is now immediate.

Note If R is not in the value group of k, the sets (4.7) and (4.8) coincide. It is, however, useful to maintain the distinction e.g. when one is considering convergence in fields K containing k. Sometimes (4.7), (4.8) are referred to as the "closed" and "open" disc of radius R but then the quotation marks are essential. In fact each of (4.7), (4.8) is both open and closed in the ultrametric topology.

We next consider the expansion of a function defined by a power series about a point in its domain of convergence.

LEMMA 4.3. *Let* $f(X)$, $\mathcal{D}$ *be as in Lemma* 4.2 *and let* $c \in \mathcal{D}$. *For* $0 \leq m < \infty$ *put*

$$g_m = \sum_{n \geq m} \binom{n}{m} f_n c^{n-m} \tag{4.9}$$

Then the series

$$g(X) = \sum_m g_m X^m \tag{4.10}$$

<u>has domain of convergence</u> $\mathcal{D}$ <u>and</u>

$$f(b + c) = g(b) \tag{4.11}$$

<u>for all</u> $b \in \mathcal{D}$.

<u>Proof</u>. We note, first, that the series in (4.9) clearly converges. Let $b \in \mathcal{D}$. Then

$$\begin{aligned} f(b + c) &= \sum_n f_n(b + c)^n \\ &= \sum_n \sum_{m \leq n} \binom{n}{m} f_n\, c^{n-m}\, b^m. \end{aligned}$$

It is easy to see that Lemma 4.1 applies, and (4.11) follows on interchanging the order of summation. Hence the domain of convergence of $g(X)$ contains that of $f(X)$. That it cannot be larger follows on reversing the rôles of f and g.

<u>COROLLARY</u>. <u>A function</u> $f(X)$ <u>defined by a power series is continuous in its domain</u> $\mathcal{D}$ <u>of convergence</u>.

<u>Proof</u>. For $g(b)$ in (4.11) is certainly continuous at $b = 0$.

We note in passing that lemma 4.3 shows that the process of analytic continuation familiar from complex function theory does not give anything useful in the non-arch. context. There are, however, more sophisticated notions of analytic continuation which are useful, as we shall see in Chapter 12, §3.

The next result is a striking contrast with the situation in $\mathbb{R}$ and $\mathbb{C}$. It is the key of some of the more surprising applications of local fields.

<u>THEOREM 4.1</u> (Strassmann). <u>Let</u> k <u>be complete with respect to the non-arch. valuation</u> $|\ |$ <u>and let</u>

$$f(X) = \sum_0^\infty f_n X^n. \tag{4.12}$$

Suppose that $f_n \to 0$ (so $f(X)$ converges in Ω) but that not all f_n are 0. Then there is at most a finite number of $b \in \Omega$ such that $f(b) = 0$.

More precisely, there are at most N such b where N is defined by

$$|f_N| = \max_n |f_n| \tag{4.13}$$

$$|f_n| < |f_N| \quad (\text{all } n > N). \tag{4.14}$$

Proof. We use induction on N.

Suppose, first, that $N = 0$ but $f(b) = 0$ for some $b \in \Omega$. Then

$$f_o = - \sum_{n\geq 1} f_n b^n.$$

This is a contradiction since (4.14) and $N = 0$ implies that

$$\left|\sum_{n\geq 1} f_n b^n\right| \leq \max_{n\geq 1}\left|f_n b^n\right| \leq \max_{n\geq 1}\left|f_n\right| < \left|f_o\right|.$$

Now suppose that $N > 0$ and $f(b) = 0$, $b \in \Omega$. Let $c \in \Omega$. Then

$$\begin{aligned} f(c) = f(c) - f(b) &= \sum_{n\geq 1} f_n(c^n - b^n) \\ &= (c - b) \sum_{n\geq 1} \sum_{j<n} f_n c^j b^{n-1-j}. \end{aligned}$$

By Lemma 4.1 we may rearrange in powers of c, so

$$f(c) = (c - b) g(c)$$

where

$$g(X) = \sum g_j X^j$$

and

$$g_j = \sum_{r\geq 0} f_{j+1+r} b^r.$$

It is readily verified that (4.13), (4.14) imply that

$$|g_j| \leq |f_N| \quad \text{(all } j\text{)}$$

$$|g_{N-1}| = |f_N|$$

$$|g_j| < |f_N| \quad (j > N - 1).$$

Hence $g(X)$ satisfies hypotheses of the Theorem but with $N - 1$ instead of N. By the induction hypothesis $g(X)$ has at most $N - 1$ zeros $c \in \Omega$. But $f(c) = 0$ implies either $c = b$ or $g(c) = 0$. Hence $f(X)$ has at most N zeros, as required.

COROLLARY 1. *Suppose that* $f(X)$, $g(X)$ *both converge in* Ω *and that* $f(b) = g(b)$ *for infinitely many* $b \in \Omega$. *Then* $f(X)$, $g(X)$ *have the same coefficients*.

Proof. For $f(X) - g(X)$ has infinitely many zeros $b \in \Omega$.

A consequence of Strassmann's theorem is that a periodic function (other than a constant) cannot be defined by a power series:

COROLLARY 2. *Suppose that* k *has characteristic* 0. *Let* $f(X)$ *be a power series converging in* Ω. *Suppose that* $f(X + d) = f(X)$ *for some* $d \in \Omega$. *Then* $f(X)$ *is a constant*.

Proof. $f(X) - f(0)$ has then infinitely many zeros md $(m \in \mathbb{Z})$ in Ω.

5 A USEFUL EXPANSION

There are analogues in non-arch. valued fields of most of the standard functions of analysis. They share many properties with their analogues in $\mathbb{R}$ or $\mathbb{C}$ but there are also striking differences (as Theorem 4.1, Corollary 2 makes plain). We shall not go into this here in general (cf. Chapter 12, §2) but will prove the existence of one expansion which will be useful later. Before enunciating it we give a simple but useful estimate:

LEMMA 5.1.

$$|m!|_p = p^{-M}$$

where

$$M = [m/p] + [m/p^2] + [m/p^3] + \ldots \quad .$$

Proof. For $j \geqslant 1$ let $s(j)$ of the integers $1, \ldots, m$ be divisible by p^j but not by $j + 1$. Then

$$M = \sum_j j\,s(j)$$
$$= \sum_i t(i)$$

where

$$t(i) = s(i) + s(i + 1) + s(i + 2) \ldots \quad .$$

Here $t(i)$ is the number of the integers $1, \ldots, m$ which are divisible by p^i. Hence $t(i) = [m/p^i]$.

We shall usually use the

COROLLARY. $|m!| > p^{-m/(p-1)}$

Proof.

$$M < m/p + m/p^2 + \ldots = m/(p - 1).$$

LEMMA 5.2. Let $b \in Q_p$ and suppose that

$$\left.\begin{array}{ll} |b| \leqslant 2^{-2} & (p = 2) \\ |b| \leqslant p^{-1} & (\text{otherwise}). \end{array}\right\} \qquad (5.1)$$

where $|\ |$ is the p-adic valuation. Then there is a power series

$$\Phi_b(X) = \sum_0^\infty \gamma_n X^n, \qquad (5.2)$$

where

$$\gamma_n \in \mathbb{Q}_p, \qquad \gamma_n \to 0 \tag{5.3}$$

such that

$$(1 + b)^r = \Phi_b(r) \tag{5.4}$$

for all $r \in \mathbb{Z}$.

Proof. Suppose, first, that $r \geq 0$. Then

$$(1 + b)^r = \sum_{s=0}^{\infty} \binom{r}{s} b^s. \tag{5.5}$$

Here $\binom{r}{s} = 0$ for $s > r$ but we ignore this, and write (5.5) as

$$(1 + b)^r = \sum_{s=0}^{\infty} r(r - 1) \ldots (r - s + 1)(b^s/s!). \tag{5.6}$$

Now

$$|b^s/s!| \to 0$$

by (5.1) and Lemma 5.1, Corollary. By the rearrangement Lemma 4.1 we may therefore rearrange (5.6) in powers of r to obtain

$$(1 + b)^r = \sum_{n=0}^{\infty} \gamma_n r^n \tag{5.7}$$

where the $\gamma_n \in \mathbb{Q}_p$ are independent of r and satisfy (5.3). This proves the Lemma for $r \geq 0$.

We note first on putting $r = p^m$ $(m = 1,2,\ldots)$ in (5.7) that

$$\lim_m (1 + b)^{p^m} = 1. \tag{5.8}$$

Now let $r < 0$. Then $p^m + r > 0$ for large enough m, and so (5.7) applies with $p^m + r$ instead of r:

$$(1 + b)^{p^m+r} = \sum_n \gamma_n (p^m + r)^n. \tag{5.9}$$

Now let $m \to \infty$ (in the usual sense) so $p^m \to 0$. The left hand side tends to $(1 + b)^r$ by (5.8). The right hand side of (5.9) tends to $\sum \gamma_n r^n$ because a power series is continuous in its domain of convergence by Lemma 4.3, Corollary. Hence (5.7) hold also for $r < 0$ and the Lemma is proved.

Note The Lemma (and the proof) extend to any complete field $k \supset \mathbb{Q}_p$ with valuation extending the p-adic valuation. It is then appropriate to replace (5.1) by

$$|b| < p^{-1/(p-1)} \tag{5.10}$$

(both for $p = 2$ and $p > 2$).

6 AN APPLICATION TO RECURRENT SEQUENCES

We shall discuss the general theory of recurrent sequences in the next Chapter. Here we consider a couple of special cases. This will motivate the later discussion and exemplify some general points.

LEMMA 6.1. (Nagell) *Let u_n be defined by $u_o = 0$, $u_1 = 1$ and*

$$u_n = u_{n-1} - 2u_{n-2} \quad (n \geq 2). \tag{6.1}$$

Then $u_n = \pm 1$ only for $n = 1, 2, 3, 5$ and 13.

Proof. The first few values of u_n are

n	0	1	2	3	4	5	6	7	8	9
u_n	0	1	1	-1	3	-1	5	7	-3	-17

Solving the recurrence relation by sixth-form mathematics, we have

$$u_n = \frac{\alpha^n - \beta^n}{\alpha - \beta}, \tag{6.2}$$

where α, β are the roots of

$$F(X) = X^2 - X + 2. \tag{6.3}$$

Indeed one checks readily that the right hand side of (6.2) satisfies (6.1) and as the first two values u_o, u_1 are correct, then so are all the others. The roots of (6.3) are

$$\alpha = \tfrac{1}{2}(1 + \sqrt{-7}), \qquad \beta = \tfrac{1}{2}(1 - \sqrt{-7})$$

which, of course, at school we regard as complex numbers. It is an easy school exercise then to show that the sequence $\{u_n\}$ changes sign infinitely often.

We can, however, work instead in any p-adic field $\mathbb{Q}_p$ for which $F(X)$ splits. This is the case for $\mathbb{Q}_{11}$ by Lemma 3.1 Corollary 2 since $D = -7$ (by (3.16)) and $f(5) = 22 \equiv 0 \ (11)$. Using the machinery of Hensel's Lemma, we find that there is a root $\alpha \in \mathbb{Z}_{11}$ with

$$\alpha \equiv 16 \quad (11^2). \tag{6.4}$$

The other root is

$$\beta = 1 - \alpha \equiv 106 \quad (11^2). \tag{6.5}$$

Our first thought is to expand u_n as a power series in n and apply Strassmann's theorem. This does not work directly because α, β do not satisfy the conditions of Lemma 5.2. However by the "little Fermat theorem"

$$\left.\begin{aligned} A &= \alpha^{10} \equiv 1 \quad (\text{mod } 11) \\ B &= \beta^{10} \equiv 1 \quad (\text{mod } 11) \end{aligned}\right\} \tag{6.6}$$

and so Lemma 5.2 does apply to A, B.

We therefore write

$$n = r + 10s \qquad 0 \leq r \leq 9,$$

so

$$u_{r+10s} = \frac{\alpha^r A^s - \beta^r B^s}{\alpha - \beta} \tag{6.7}$$

We note that

$$u_{r+10s} \equiv u_r \quad (11), \tag{6.8}$$

and so the only r which we need consider are $r = 1, 2, 3, 5$.

r	$\alpha^r \bmod 11^2$	$\beta^r \bmod 11^2$
1	16	106
2	14	104
3	103	13
5	111	21
10	100	78

We now write

$$\alpha^{10} = A = 1 + a, \quad \beta^{10} = B = 1 + b, \tag{6.9}$$

so

$$a \equiv 99 \quad (11^2), \qquad b \equiv 77 \quad (11^2) \tag{6.10}$$

and develop

$$(\alpha - \beta)\,(u_{r+10s} \mp 1) = \alpha^r(1 + a)^s - \beta^r(1 + b)^s \mp (\alpha - \beta) \tag{6.11}$$

as a power series

$$c_o + c_1 s + c_2 s^2 + \ldots \tag{6.12}$$

using Lemma 5.2. Here the upper sign is the correct one for $r = 1,2$ and the lower for $r = 3, 5$. In every case $c_o = 0$. It is easy to see that

$$c_j \equiv 0 \quad (11^2) \quad (\text{all } j \geq 2). \tag{6.13}$$

For $r = 1, 2$ and 5 the values of α^r, β^r listed above show that

$$\begin{aligned} c_1 &\equiv \alpha^r a - \beta^r b && (11^2) \\ &\not\equiv 0 && (11^2). \end{aligned} \tag{6.14}$$

Hence the series (6.12) has at most one zero $s \in \Omega$. Since in every case $s = 0$ is a zero, there can be no others.

For $r = 3$, however, we have $c_1 \equiv 0 \quad (11^2)$ and so we must estimate the c_j more precisely. We have

$$2 \cdot 11^{-2} c_2 \equiv \alpha^3 (a/_{11})^2 - \beta^3 (b/_{11})^2 \equiv 6 \quad (11)$$

and so

$$c_2 \not\equiv 0 \quad (11^3).$$

Since $c_j \equiv 0 \quad (11^3) \quad (j \geqslant 3)$, as is easy to see, Strassmann's theorem tells us that (6.12) can vanish for at most two values of s. Since $u_3 = u_{13} = -1$, we conclude that there can be no others.

COROLLARY. The only solutions of

$$x^2 + 7 = 2^m \quad (x, m \in \mathbb{Z}) \tag{6.15}$$

have $m = 3, 4, 5, 7, 15$.

Proof. Clearly x is odd, say $x = 2y - 1 \quad (y \in \mathbb{Z})$ and then

$$y^2 - y + 2 = 2^{m-2} \tag{6.16}$$

The ring $\mathbb{Z}[\alpha]$, where $\alpha^2 - \alpha + 2 = 0$, has a Euclidean algorithm and so is a unique factorization domain [cf. Theorem 5.1, Cor. 2 of Chapter 10]. On considering factorization of both sides of (6.16), we get

$$y \pm \alpha = \pm \alpha^{m-2} \tag{6.17}$$

(for some choices of signs). Then

$$y \pm \beta = \pm \beta^{m-2} \tag{6.18}$$

for the conjugate root β. Hence

$$(\alpha - \beta) = \pm (\alpha^{m-2} - \beta^{m-2}) \tag{6.19}$$

which with $n = m - 2$ is the problem discussed above.

Before going on, we make some philosophical remarks about the proof of Lemma 6.1. It was clear at an early stage in the proof that it would give an upper bound for the number of solutions n of $u_n = \pm 1$ but it appeared only to be a piece of good luck that the bound was actually attained. This is indeed a correct impression. Instead of working in $\mathbb{Q}_{11}$ we could have taken $\mathbb{Q}_{23}$ since $F(X)$ given by (6.3) has roots

$$\alpha \equiv 33 \quad (23^2), \qquad \beta \equiv 497 \quad (23^2)$$

in $\mathbb{Q}_{23}$. Further

$$u_{12} = 45 \equiv -1 \quad (23).$$

On expanding $(\alpha - \beta)\, u_{12+22s}$ as a power series (6.2) we find

$$|c_o| = 23^{-1}, \qquad |c_1| = 23^{-1}, \qquad |c_j| \leq 23^{-2} \quad (j \geq 2).$$

Strassmann's theorem tells us that there is at most one s such that $u_{12+22s} = -1$. However by Lemma 6.1 we know that there are no such s. The fact is that the power series (6.2) has a zero $\theta \in \mathbb{Z}_{23}$ but θ is not a rational integer. By letting s approximate to θ in the 23-adic sense we can make

$$|u_{12+22s} + 1|_{23}$$

arbitrarily small. So long as we operate with the 23-adic information alone, there is no obvious way in which we can show that $u_{12+22s} \neq -1$ and indeed no-one has ever found one.

We shall revert to this theme when we discuss the "Weierstrass preparation theorem" (Chapter 6, §5).

<u>LEMMA 6.2</u>. (Mignotte) <u>Let</u> u_n <u>be given by</u> $u_o = u_1 = 0$, $u_2 = 1$ <u>and</u>

$$u_{n+3} = 2u_{n+2} - 4u_{n+1} + 4u_n \quad (n \geq 0)$$

<u>then</u> $u_n = 0$ <u>precisely for</u> $n = 0, 1, 4, 6, 13, 52$.

<u>Note</u> u_n is greater than 10^{11} in absolute value for $n = 48, 49, 50, 51$.

<u>Proof</u>. (sketch) Here the auxiliary polynomial is

$$F(X) = X^3 - 2X^2 + 4X - 4$$

The smallest prime for which it splits completely is 47, so we work in $\mathbb{Q}_{47}$. The roots of $F(X)$ are

$$\alpha \equiv 1398, \qquad \beta \equiv 550, \qquad \gamma \equiv 263 \qquad (47^2)$$

We have

$$u_n = A\alpha^n + B\beta^n + C\gamma^n \qquad \text{(all } n\text{)}$$

where

$$A \equiv 319, \qquad B \equiv 578, \qquad C \equiv 1312 \qquad (47^2)$$

Also

$$\alpha^{46} = 1 + a, \qquad \beta^{46} = 1 + b, \qquad \gamma^{46} = 1 + c$$

where

$$a \equiv 1457 = 31.47, \quad b \equiv 1316 = 28.47, \quad c \equiv 1363 = 29.47 \quad (47^2)$$

(cf. Appendix F). Put $n = r + 46s$. One checks that $u_n \equiv 0 \ (47)$ precisely when $r \equiv 0, 1, 4, 6$ or $13 \ (46)$.

The proof is then very similar to that of Lemma 6.1. For $r = 0, 1, 4, 13$ the Strassmann bound is 1 and there is a solution with $s = 0$. For $r = 6$ the Strassmann bound is 2 and there are solutions with $s = 0$ and $s = 1$.

<u>Notes</u>

§2 Rather surprisingly, Theorem 2.1 is almost best possible. The $\beta(q)$ $(q \neq 2)$ are best possible but $\beta(2)$ can be replaced by a somewhat smaller $\beta'(2)$ by the following modification of the argument. The finite group $G \subset GL_n(\mathbb{Q})$ leaves invariant a non-singular definite quadratic form in n variables with coefficients in $\mathbb{Q}$, namely $\sum_A F(A\underline{X}) = G(\underline{X})$, where A runs through G and F is any positive definite form. The improvement comes from arguing about the corresponding orthogonal group as we have done for $GL_n(\mathbb{F})$. [This corrects the statement on p51 of Cassels (1978). The required result on the orthogonal group is Example 7 to Chapter 6 on p101]. For groups G showing that Theorem 2.1 as amended is best possible, see Burnside (1911) pp479-484 or Bourbaki (1972) pp272-274, and for a more general result, see Schur (1905).

§3 For an interesting result about the number of solutions of a set of congruences mod p^m as m varies, see Denef (1984).

§4 We shall return to analysis in non-arch. fields in Chapter 12. For an account of ultrametric analysis at undergraduate level, see Schikhof (1985). Theorem 4.1 is proved in Strassmann (1928), perhaps the earliest paper on ultrametric analysis.

§6 Lemma 6.1 was first proved by Nagell (1960): he works in $\mathbb{Q}_7$ and uses Strassmann's theorem but his approach is slightly different. There is an account in Mordell (1969). Lemma 6.2 is proved in Mignotte (1973/4), see also Loxton and van der Poorten (1977).

Exercises

1. Find $a \in \mathbb{Z}$ such that

$$|5a + 1|_3 < 3^{-6}.$$

2. For what $a \in \mathbb{Z}$ is $5x^2 = a$ soluble in $\mathbb{Z}_5$? in $\mathbb{Q}_5$?

3. Find $a \in Z$ such that

$$|a^2 + 6|_5 < 5^{-4}.$$

4. Show that each of the following functions has a zero in $\mathbb{Z}_p$

for every p:

(i) $(X^2 - 2)(X^2 - 17)(X^2 - 34)$,

(ii) $(X^3 - 37)(X^2 + 3)$.

[Note For (ii) you will require the law of quadratic reciprocity cf. Chapter 10, Exercise 13.]

5. Show that

$$F(X) = 5X^3 - 7X^2 + 3X + 6$$

has a root $\alpha \in \mathbb{Z}_7$ with $|\alpha - 1|_7 < 1$. Find an $a \in \mathbb{Z}$ such that $|\alpha - a| \leqslant 7^{-4}$.

6. [Hard]. For $p = 2,3,5,7$ determine how many roots there are in $\mathbb{Q}_p$ of

$$F(X) = X^3 + 25X^2 + X - 9.$$

7. If $c \in \mathbb{Z}_p$ satisfies $|c|_p < 1$, show that

$$(1 + c)^{-1} = 1 - c + c^3 - c^4 + \ldots \quad .$$

Hence or otherwise find $a \in \mathbb{Z}$ such that

$$|4a - 1|_5 \leqslant 5^{-10}.$$

8. (Binomial expansion in $\mathbb{Q}_p$). Define the function $\binom{T}{n}$ of the indeterminate T and natural number n by $\binom{T}{0} = 1$ and

$$\binom{T}{n} = \frac{T(T - 1) \ldots (T - n + 1)}{n!}$$

(i) For indeterminates S,T show that

$$\binom{S + T}{n} = \sum_{j=o}^{n} \binom{S}{j}\binom{T}{n-j}.$$

[Hint Show first that it holds when S,T take values in the natural numbers. Compare Chapter 5, Lemma 2.1].

(ii) If $t \in \mathbb{Z}_p$ show that

$$\binom{t}{n} \in \mathbb{Z}_p .$$

[Hint t can be approximated arbitrarily closely by natural numbers.]

(iii) For $b,t \in \mathbb{Q}_p$ define

$$G(b,t) = 1 + \binom{t}{1}b + \binom{t}{2}b^2 + \ldots + \binom{t}{n}b^n + \ldots .$$

Show that the series converges whenever $t \in \mathbb{Z}_p$, $|b|_p < 1$. If also $s \in \mathbb{Z}_p$ show that

$$G(b,s)\, G(b,t) = G(b,\, s + t).$$

(iv) Let $u,v \in \mathbb{Z}$, and suppose that $v > 0$, $p \nmid v$. Show that

$$\{G(b,u/v)\}^v = (1 + b)^u.$$

(v) Hence find $e \in \mathbb{Q}$ such that

$$|e^2 - 11|_5 \leq 5^{-5}.$$

(vi) How should the above be modified when $t \in \mathbb{Q}_p$ but not $t \in \mathbb{Z}_p$? Show, in particular, that the series defining $G(\frac{1}{2}, -8)$ converges in $\mathbb{Q}_2$ and that

$$\{G(\tfrac{1}{2}, -8)\}^2 = -7.$$

9. Define the sequence $u(n)$ by $u(0) = 1$, $u(1) = 2$ and

$$u(n) = 3u(n - 1) - 5u(u - 2).$$

Show that $u(n) = 1$ only for $n = 0, 2, 6$.

[Hint Work in $\mathbb{Q}_3$].

10. Let $u(0) = 0$, $u(1) = 1$ and

$$u(n) = 3u(n - 1) - 7u(n - 2).$$

Find the smallest $m > 0$ such that

$$u(m) \equiv 0. \quad (5^4)$$

Show that there are $m > 0$ for which $u(m)$ is divisible by an arbitrarily high power of 5.

11. Let $r,s,a,b \in \mathbb{Z}_p$ for some $p \geq 5$ and suppose that

$$|r| = |s| = |a| = |b| = |a - b| = 1.$$

Show that

$$u(n) = r(1 + pa)^n + s(1 + pb)^n$$

can take no value more than twice $(n \in \mathbb{Z})$.

12. Let $u(0) = u(1) = 1$ and

$$u(n + 2) = 5u(n + 1) - 11u(n). \quad (n \geq 0).$$

By working in $\mathbb{Q}_5$, or otherwise, show that $u_n = 1$ only for $n = 0,1$. Is it possible that $u(n) = 0$?

13. (cf. §2).

(i) Let $A \in GL_n(\mathbb{Z}_2)$ and suppose that

$$A \equiv I \ (2^2), \quad A \neq I.$$

Show that A is of infinite order.

(ii) Let $A \in GL_n(\mathbb{Z}_2)$ and suppose that

$$A \equiv I \ (2). \qquad (*)$$

Show that either $A^2 = I$ or A is of infinite order.

(iii) Let H be a finite group of matrices $A \in GL_n(\mathbb{Z}_2)$ satisfying (*). Show that H is abelian and that its order divides 2^n.

(iv) Let G be a finite subgroup of $GL_n(\mathbb{Z}_2)$. Show that its order divides

$$2^n(2^n - 2^{n-1})(2^n - 2^{n-2}) \ldots (2^n - 1).$$

14. (i) Determine the domain $\mathcal{D}$ of convergence in $\mathbb{Q}_p$ of

$$E(X) = 1 + \frac{X}{1!} + \frac{X^2}{2!} + \ldots + \frac{X^n}{n!} + \ldots \quad .$$

(ii) For $a,b \in \mathcal{D}$ show that $E(a + b) = E(a)\,E(b)$.

(iii) Find the domain $\mathcal{E}$ of convergence of

$$L(X) = X + \frac{X^2}{2} + \frac{X^3}{3} + \ldots + \frac{X^n}{n} + \ldots \quad .$$

(iv) If a is small enough, show that

$$E(L(a)) = (1 - a)^{-1}.$$

15. Justify the steps in the following procedure for finding the number of roots in $\mathbb{Z}_p$ of $F(X) \in \mathbb{Z}_p[X]$.

(i) If the discriminant D is zero, then F and F' have a common factor. We are thus reduced to considering $D \neq 0$.

(ii) If $|D|_p = p^{-m}$ compute $F(a)$ for $0 \leq a < p^{2m+1}$. Let S be the set of such a for which $F(a) \equiv 0 \ (p^{2m+1})$.

(iii) For $a,b \in S$ write $a \sim b$ if

$$b \equiv a \quad (p^{2m+1} \, |F'(a)|_p).$$

This is an equivalence relation and the number of roots of $F(X)$ in $\mathbb{Z}_p$ is the number of equivalence classes.

16. Use the procedure of the previous exercise to verify your answer to exercise 6.

[Hint The discriminant D of $f_o + f_1X + f_2X^2 + f_3X^3$ is

$$D = f_1{}^2f_2{}^2 - 4f_of_2{}^3 - 4f_1{}^3f_3 - 27f_o{}^2f_3{}^2 + 18f_of_1f_2f_3.$$

Better use a programmable calculator!].

17. Show that the equation

$$(X^2 - 5)(X^2 - 41)(X^2 - 205) = 0$$

has solutions everywhere locally but not globally.

18. Show that the equation

$$3X^4 + 4Y^4 - 19Z^4 = 0$$

is soluble everywhere locally but not globally.

[Hint If there is a global solution (x,y,z), then without loss of generality x,y,z are integers without common factor. The equation can be written

$$(3x^2 + 8y^2 - 19z^2)(3x^2 - 8y^2 - 19z^2) = 57(x^2 - z^2)^2.$$

Show that

$$3x^2 - 8y^2 - 19z^2 < 0$$

and that

$$3x^2 + 8y^2 - 19z^2 \not\equiv 0 \quad (3)$$

$$\not\equiv 0 \ (19).$$

Show further that the three bracketed expressions cannot have an odd prime divisor in common. Deduce that there is one of the two following cases:

	Case I	Case II
$19z^2 - 8y^2 - 3x^2 =$	u^2	$2u^2$
$19z^2 + 8y^2 - 3x^2 =$	$57v^2$	$114v^2$
$z^2 - x^2 =$	uv	$2uv.$

Show that Case I is impossible in $\mathbb{Q}_3$ and that Case II is impossible in $\mathbb{Q}_2$.

There is a crib in Cassels (1985). The example is due to Bremner, Lewis and Morton (1984), who use a different technique.]

The following exercise is not concerned especially with local fields but helps to motivate the proof of Lemma 3 bis. 3 and the hint to the preceding Exercise.

19. Let k be any field of characteristic $\neq 2$ and let $F(X,Y,Z) \in k[X,Y,Z]$ be a non-singular quadratic form. Suppose that there is a non-trivial solution of $F = 0$ in k. Show that there are linear forms $L,M,N \in k[X,Y,Z]$ and $d \in k$ such that

$$F = LN + dM^2.$$

[Hint There are two alternative approaches.

(i) algebraic. After a non-singular linear transformation of X,Y,Z we may suppose that $F(1,0,0) = 0$, that is

$$F(X,Y,Z) = bY^2 + cZ^2 + 2eYZ + 2fZX + 2gXY.$$

After a further linear transformation on Y,Z only we may suppose that

$f = 0$. We can then take $L = Y$, $M = Z$.

(ii) geometric. Suppose that $F(\underline{a}) = 0$. Let $L = 0$ be the tangent at $\underline{a}$ to $F(\underline{X}) = 0$ and let $M = 0$ be the equation of any other line through $\underline{a}$. Then $F = 0$ meets $M = 0$ in a point $\underline{b}$ distinct from $\underline{a}$. Let $N = 0$ be the tangent at $\underline{b}$. Then $F = hLN + dM^2$ for some $d,h \in k$; and h can be absorbed in N.]

20. Let p be prime and let

$$m = a_o + a_1 p + \dots + a_J p^J \quad (0 \leq a_j < p).$$

Suppose that

$$m! = p^M N$$

where $p \nmid N$. Show that

$$(p - 1)M = m - \sum a_j$$

and

$$N \equiv (-1)^M \Pi(a_j!) \quad (p).$$

[Note Compare Lemma 5.1].

21. A set S in a topological space is <u>clopen</u> if it is both open and closed.

For $a \in \mathbb{Z}_p$ and non-negative rational integer N define $T(a,N) \subset \mathbb{Z}_p$ by

$$T(a,N) = \{b : b \equiv a \quad (p^N)\}.$$

Show, further, that any clopen $S \subset \mathbb{Z}_p$ is the <u>disjoint</u> union of a <u>finite</u> number of $T(a,N)$.

[Hint Show that S is compact].

22. A function f on a topological space is locally constant if every y has a neighbourhood $U(y)$ such that $f(x) = f(y)$ for all $x \in U(y)$. Give an example of a $\mathbb{Q}_p$-valued function on $\mathbb{Z}_p$ which is both continuous and locally constant but is not constant.

If the $\mathbb{Q}_p$-valued function $f(x)$ on $\mathbb{Z}_p$ is both continuous and locally constant, show that for any b the set of x with $f(x) = b$ is clopen. Deduce that f takes only finitely many values.

Show that any continuous $\mathbb{Q}_p$-valued function on $\mathbb{Z}_p$ is the uniform limit of continuous, locally constant functions.

23. Let k be complete with respect to a non-arch. valuation. Suppose that $g_j = g_j(T) \in k[[T]]$ $(0 \leq j \leq J)$ are not all 0 and that $f \in k[[T]]$ satisfies

$$\sum g_j f^j = 0.$$

If all the g_j converge in some neighbourhood of the origin, show that f converges in a (possibly smaller) neighbourhood.

Does the above statement hold also for archimedean valuations?

[Hint Chapter 2, Exercise 2].

24. Let $p \neq 2$. Show that $16 = b^8$ for some $b \in \mathbb{Q}_p$.

[Hint Show that at least one of $X^2 + 1$, $X^2 + 2$, $X^2 - 2$ has a root in $\mathbb{Q}_p$].

25. (i) Let k be a field of characteristic p complete with respect to the valuation $|\ |$ and suppose that the residue class field is finite with q elements. Show that k contains the field $\mathbb{F}_q$ of q elements.

(ii) If, further, $|\ |$ is discrete, show that $k = \mathbb{F}_q((\pi))$ in the notation of Chapter 2, §5, where π is any prime element.

[Note Compare Chapter 8, §2].

CHAPTER FIVE: EMBEDDING THEOREM

1 INTRODUCTION

In this chapter we prove the following theorem which often facilitates the use of p-adic methods:

THEOREM 1.1 *Let* K *be a finitely-generated extension of the rational field* $\mathbb{Q}$ *and let* C *be a finite set of non-zero elements of* K. *Then there exist infinitely many primes* p *for which there is an embedding*

$$\alpha: K \hookrightarrow \mathbb{Q}_p \tag{1.1}$$

and such that

$$|\alpha c| = 1 \quad (\text{all } c \in C). \tag{1.2}$$

Here, of course, $|\ | = |\ |_p$ is the p-adic valuation. We prove the theorem in the next two sections. In sections 4 and 5 we apply it to obtain a theorem of Selberg about matrix groups and a theorem of Mahler and Lech about recurrence relations.

2 THREE LEMMAS

LEMMA 2.1. Let

$$f_j(X_1,\dots,X_n) \in \mathbb{Z}[X_1,\dots,X_n] \quad (1 \leqslant j \leqslant J)$$

$$\neq 0 \tag{2.1}$$

be a finite set of polynomials in the indeterminates $X_1,\dots,X_n$. *Then there are* $a_1,\dots,a_n \in \mathbb{Z}$ *such that simultaneously*

$$f_j(a_1,\ldots,a_n) \neq 0 \quad (1 \leq j \leq J) \tag{2.2}$$

Proof. (i) $n = 1$. Pick a_1 distinct from the finitely many roots of the $f_j(X_1)$.

(ii) $n > 1$. By induction we may pick $a_2,\ldots,a_n$ such that $f_j(X_1,a_2,\ldots,a_n) \neq 0 \quad (1 \leq j \leq J)$. Now pick a_1.

LEMMA 2.2. Let

$$g(X) \in \mathbb{Z}[X] \tag{2.3}$$

be a non-constant polynomial in the single indeterminate X. *Then there are infinitely many primes* p *for which there is a solution* $b = b(p) \in \mathbb{Z}$ *of the congruence*

$$g(b) \equiv 0 \quad (p) \tag{2.4}$$

Proof. Let

$$g(X) = g_o + g_1X + \ldots + g_nX^n.$$

If $g_o = 0$ we can take $b = 0$ for all primes p: so we may suppose that

$$g_o \neq 0.$$

Suppose, if possible, that (2.4) has a solution b only for the primes p in a finite (possibly empty) set P. Let $c \in \mathbb{Z}$ be divisible by all the $p \in P$. Then

$$g(g_oc) = g_or$$

where

$$r = r(c) = g_ng_o^{n-1}c^n + g_{n-1}g_o^{n-2}c^{n-1} + \ldots + 1$$

$$\equiv 1 \quad (\text{all } p \in P).$$

Since $g(X)$ is not constant, by hypothesis, we may choose c so that $r \neq \pm 1$. Let p^* be a prime dividing r, so $p^* \notin P$, and put $b^* = g_o c$. Then

$$g(b^*) \equiv 0 \quad (p^*),$$

contrary to the definition of P.

<u>LEMMA 2.3</u>. <u>For any prime</u> p <u>the field</u> $\mathbb{Q}_p$ <u>has infinite transcendence degree over</u> $\mathbb{Q}$.

<u>Proof</u>. $\mathbb{Q}_p$ is uncountable (by Chapter 4, Lemma 1.4). On the other hand, the algebraic closure of any finitely generated extension $\mathbb{Q}(a_1,\ldots,a_n)$ of $\mathbb{Q}$ is easily seen to be countable. The argument shows, in fact, that the transcendence degree is uncountable.

3 <u>PROOF OF THE THEOREM</u>

We note, first, that, by taking a larger set for C if necessary, we may suppose that $c^{-1} \in C$ whenever $c \in C$. Then instead of (1.2) it will be enough to ensure that

$$|\alpha c| \leq 1 \quad (\text{all } c \in C). \tag{3.1}$$

Let $x_1,\ldots,x_m$ $(m \geq 0)$ be a transcendence base for $K/\mathbb{Q}$. Then $x_1,\ldots,x_m$ are independent transcendentals and

$$K = \mathbb{Q}(y,x_1,\ldots,x_m), \tag{3.2}$$

where y is algebraic over $\mathbb{Q}(x_1,\ldots,x_m)$. We may thus put each $c \in C$ in the shape

$$c = U_c(y,x_1,\ldots,x_n)/V_c(x_1,\ldots,x_n), \tag{3.3}$$

where

$$U_c(Y,X_1,\ldots,X_m) \in \mathbb{Z}[Y,X_1,\ldots,X_m] \tag{3.4}$$

$$V_c(X_1,\ldots,X_m) \in \mathbb{Z}[X_1,\ldots,X_m] \tag{3.5}$$
$$\neq 0.$$

Here $X_1,\ldots,X_m$, Y are indeterminates, while small letters (minuscules) such as $x_1,y,a,\ldots$ denote elements of the fields under discussion.

We can select an irreducible equation $G(Y) = 0$ for y over $\mathbb{Q}(x_1,\ldots,x_m)$ of the shape

$$G(Y) = H(Y,x_1,\ldots,x_m), \tag{3.6}$$

where

$$H(Y,X_1,\ldots,X_m) \in \mathbb{Z}[Y,X_1,\ldots,X_m]. \tag{3.7}$$

If s is the degree of H in Y we denote the coefficient of Y^s by $H_o(X_1,\ldots,X_m)$, so

$$H_o(X_1,\ldots,X_m) \in \mathbb{Z}[X_1,\ldots,X_m] \neq 0. \tag{3.8}$$

The discriminant of $G(Y)$ [cf. Appendix A] is of the shape $\Delta(x_1,\ldots,x_m)$, where

$$\Delta(X_1,\ldots,X_m) \in \mathbb{Z}[X_1,\ldots,X_m] \neq 0. \tag{3.9}$$

By Lemma 2.1 we can pick $a_1,\ldots,a_m \in \mathbb{Z}$ such that

$$\Delta(a_1,\ldots,a_m) \neq 0 \tag{3.10}$$

$$H_o(a_1,\ldots,a_m) \neq 0 \tag{3.11}$$

$$V_c(a_1,\ldots,a_m) \neq 0 \quad (\text{all } c \in C). \tag{3.12}$$

By (3.11) and Lemma 3.2 there are infinitely many primes p for which there is a solution $b \in \mathbb{Z}$ of

$$H(b,a_1,\ldots,a_m) \equiv 0 \quad (p). \tag{3.13}$$

On excluding finitely many primes p from consideration, we may also suppose, by (3.10) and (3.12) that

$$\Delta(a_1,\ldots,a_m) \not\equiv 0 \quad (p) \tag{3.14}$$

and

$$V_c(a_1,\ldots,a_m) \not\equiv 0 \quad (p) \quad (\text{all } c \in C). \tag{3.15}$$

By Lemma 3 we can select m independent transcendentals $\theta_1,\ldots,\theta_m \in \mathbb{Q}_p$. On replacing θ_j by $p^t\theta_j$ with large positive integral t if necessary, we may suppose that

$$|\theta_j| < 1 \quad (1 \leqslant j \leqslant m). \tag{3.16}$$

Then

$$\xi_j = a_j + \theta_j \quad (1 \leqslant j \leqslant m) \tag{3.17}$$

is a set of independent transcendentals and

$$|\xi_j - a_j| < 1 \quad (1 \leqslant j \leqslant m), \tag{3.18}$$

where $|\ | = |\ |_p$.

By (3.13) and (3.18) we have

$$|H(b,\xi_1,\ldots,\xi_m)| < 1. \tag{3.19}$$

Hence and by (3.14) Hensel's Lemma (Chapter 4, Lemma 3.1, Corollary 2) provides us with an $\eta \in \mathbb{Z}_b$ such that

$$H(\eta,\xi_1,\ldots,\xi_m) = 0. \tag{3.20}$$

Then

$$U_c(\eta,\xi_1,\ldots,\xi_n) \in \mathbb{Z}_p \ , \quad V_c(\xi_1,\ldots,\xi_n) \in \mathbb{Z}_p \tag{3.21}$$

by (3.4), (3.5). Indeed

$$|V_c(\xi_1,\ldots,\xi_n)| = 1 \tag{3.22}$$

by (3.15) and (3.18). [cf. Exercise 1].

We now define α by

$$\alpha x_j = \xi_j \quad (1 \leq j \leq m); \quad \alpha y = \eta. \tag{3.23}$$

This clearly does give an embedding of $K = \mathbb{Q}(y,x_1,\ldots,x_m)$ into $\mathbb{Q}_p$. Further,

$$|\alpha c| = |U_c(\eta,\xi_1,\ldots,\xi_m)|/|V_c(\xi_1,\ldots,\xi_m)|$$

$$\leq 1 . \tag{3.24}$$

This confirms (3.1) and so completes the proof.

4 APPLICATION. A THEOREM OF SELBERG

In this section we use Theorem 1.1 to prove a Theorem of Selberg about matrix groups. As in Chapter 4, §2 the group of invertible $n \times n$ matrices with elements in a ring R is denoted by $GL_n(R)$.

THEOREM 4.1 (Selberg). *Let k be a field of characteristic 0 and suppose that the group $G \subset GL_n(k)$ is finitely generated. Then G contains a normal torsion-free subgroup H of finite index.*

Note 1 The field k need not be finitely generated: the theorem applies, for example, to $k = \mathbb{C}$.

Note 2 It follows from the proof that H has other properties regarded as pleasant by group theorists, e.g. that it is residually nilpotent.

Proof. Let C be the set of non-zero elements of A and A^{-1} when A runs through a set of generators of G and let K be the field generated by C over $\mathbb{Q}$. Then $G \subset GL_n(K)$. Let $p \neq 2$ be a prime given by Theorem 1.1 and let α be the relevant embedding. Then $\alpha G \subset GL_n(\mathbb{Z}_p)$. Let $H \subset G$ consist of the matrices B such that $\alpha B \equiv I \pmod p$. Then H is clearly normal of finite index and it is torsion-free by Chapter 4, Lemma 2.1.

5 APPLICATION. THE THEOREM OF MAHLER AND LECH

Let k be a field. We shall be concerned with sequences $\{u(n)\}$ $(n \geq 0)$ of elements of k which satisfy a recurrence relation

$$u(n+\ell) = c_{\ell-1}\, u(n+\ell-1) + \ldots + c_o u(n). \tag{5.1}$$

Here $c_o, \ldots, c_{\ell-1} \in k$ are independent of n. The elements $u(0), \ldots, u(\ell-1)$ may be chosen arbitrarily and then (5.1) determines $u(\ell)$, $u(\ell+1), \ldots$ in order.

There is an alternative description. Consider the formal power series

$$\Phi(T) = u(0) + u(1)T + \ldots + u(n)T^n + \ldots \tag{5.2}$$

and put

$$\Gamma(T) = 1 - c_{\ell-1}T - c_{\ell-2}T^2 - \ldots - c_o T^{\ell}. \tag{5.3}$$

Then, clearly,

$$\Gamma(T)\Phi(T) = \Delta(T) \quad \text{(say)} \tag{5.4}$$

is a polynomial in T of degree at most $\ell - 1$. Hence $\Phi(T)$ is the formal expansion of the rational function $\Delta(T)/\Gamma(T)$. Conversely, the coefficients of the formal expansion of a rational function satisfy a recurrence relation, at least from some stage on. We shall, however, continue to express our results in terms of recurrence relations.

THEOREM 5.1 (Mahler-Lech). *Let the sequence $\{u(n)\}$ of elements of a field k of characteristic 0 satisfy a recurrence relation and let $c \in k$ be given. Then*

either $u(n) = c$ *for at most finitely many* n

or $u(n) = c$ *for all the* n *in some arithmetic progression.*

Note k need not be of finite transcendence degree. The theorem applies, for example, to $k = \mathbb{C}$.

Proof. We have already met special cases of the theorem in Chapter 4, §6. We follow the same paradigm. If $c_o = 0$ in (5.1), then $\{u(n)\}$ satisfies a recurrence relation of shorter length than ℓ. We may therefore suppose that

$$c_o \neq 0. \tag{5.5}$$

Let $\mathbb{K}$ be the splitting field over k of

$$F(X) = X^\ell - c_{\ell-1} X^{\ell-1} - \ldots - c_o. \tag{5.6}$$

Let the distinct roots be $\theta_1,\ldots,\theta_r$ with respective multiplicities $m_1,\ldots,m_r$, so $\sum m_j = \ell$. Then from schoolboy mathematics we know that

$$u(n) = \sum_j P_j(n)\, \theta_j^n , \tag{5.7}$$

where $P_j(T) \in K[T]$ has degree at most $m_j - 1$. [In terms of (5.4) this corresponds to the decomposition of $\Delta(T)/\Gamma(T)$ in partial fractions]. Let K be the field generated over $\mathbb{Q}$ by the θ_j and the coefficients of the $P_j(T)$. By (5.5) we have $\theta_j \neq 0$ $(1 \leq j \leq r)$ and so, by the embedding theorem 1.1, there is a prime $p \neq 2$ and an embedding

$$\alpha: K \hookrightarrow \mathbb{Q}_p \tag{5.8}$$

such that

$$|\alpha\theta_j| = 1 \quad (1 \leq j \leq r). \tag{5.9}$$

To avoid excessive notation we shall regard k as a subfield of $\mathbb{Q}_p$ and so suppress mention of the mapping α.

By (5.9) and the small Fermat Theorem we have $\theta_j^{p-1} \equiv 1 \ (p)$; that is

$$\theta_j^{p-1} = 1 + \lambda_j, \qquad |\lambda_j| \leq p^{-1} \tag{5.10}$$

For fixed integer r $(0 \leq r \leq p - 2)$ we may thus, by Chapter 4,

Lemma 5.2, develop

$$u(r + (p - 1)s) - c = - c + \sum_j P_j(r + (p - 1)s)\ \theta_j^r(1 + \lambda_j)^s$$

$$= \phi_r(s) \text{ (say)} \tag{5.11}$$

as a power series in s convergent for all $s \in \mathbb{Z}_p$. If $\phi_r(s)$ vanishes identically for some r, then we have the second alternative in the enunciation of the theorem. Otherwise, by Strassmann's theorem (Theorem 4.1 of Chapter 4) each $\phi_r(s)$ vanishes for at most finitely many $s \in \mathbb{Z}_p$: this gives the first alternative of the enunciation.

Note Clearly the Theorem can be generalized further, e.g. by replacing c with a polynomial in n.

Notes

§1 Theorem 1.1 is largely implicit in Lech (1954) but was first explicitly enunciated in Cassels (1976).

§5 For an algebraic number field Theorem 5.1 is due to Mahler (1935). The general case is due to Lech (1954). For recurrences of order 2 there is a universal upper bound to the number of times a value can be taken, see Beukers (1980).

Exercises

1. Let $\mathfrak{o}$ be the ring of integers in the field k with non-arch. valuation $|\ |$. Let $F(X_1,\ldots,X_n) \in \mathfrak{o}[X_1,\ldots,X_n]$ and $a_1,\ldots,a_n \in \mathfrak{o}$. Suppose that $F(a_1,\ldots,a_n) = c \neq 0$ and that $b_1,\ldots,b_n \in \mathfrak{o}$ satisfy $|b_j - a_j| < |c|$. Show that $|F(b_1,\ldots,b_n)| = |c|$.

[Hint Introduce variables $Y_j = X_j - a_j$.]

2. Let $k = \mathbb{F}(t)$ where $\mathbb{F}$ is the field of p elements (p prime) and t is transcendental over $\mathbb{F}$. Show that

$$u(n) = (t + 1)^n - t^n$$

satisfies a recurrence relation and that $u(n) = 1$ precisely when n is a power of p.

Deduce that Theorem 5.1 cannot be extended to fields of prime characteristic.

CHAPTER SIX: TRANSCENDENTAL EXTENSIONS. FACTORIZATION

1 INTRODUCTION

Let $|\ |$ be a non-arch. valuation on a field k. We introduce a family of extensions $\|\ \|$ of $|\ |$ to $k(X)$, where X is transcendental over k. In §2 we use them to prove a classical lemma of Gauss about factorization in $\mathbb{Z}[X_1,\dots,X_n]$ and a criterion of Eisenstein for the irreducibility of polynomials. In the rest of the chapter, k is complete. We show that the set of values $|f_j|$ of the coefficients of

$$f(X) = f_o + f_1X + \dots + f_nX^n \in k[X] \tag{1.1}$$

give a great deal of information about the factorization of $f(X)$ in $k[X]$. This information will be crucial when, in later chapters, we discuss the extensions of valuations on a field k to algebraic extensions of k. In the final section we consider formal power series and prove an analogue of Weierstrass' preparation theorem. This provides another approach to Strassmann's theorem (Theorem 4.1 of Chapter 4) and gives further insights.

LEMMA 1.1. *Let* $|\ |$ *be a non-arch. valuation on the field* k *and let* $c > 0$ *be arbitrary. For* $f(X) \in k[X]$ *given by* (1.1) *put*

$$\|f\| = \|f\|_c = \max_j c^j|f_j|. \tag{1.2}$$

For $h(X) \in k(X)$ *put*

$$\|h\| = \|f\| / \|g\| \tag{1.3}$$

where $h(X) = f(X)/g(X)$ *and* $f(X),\ g(X) \in k[X]$. *Then* $\|\ \|$ *is a* valuation on $k(X)$ which coincides with $|\ |$ on k.

Proof. Let $f(X), g(X) \in k[X]$. Clearly (1.2) implies that

$$\|f + g\| \leq \max\{ \|f\| , \|g\| \} \tag{1.4}$$

and

$$\|fg\| \leq \|f\| \ \|g\| . \tag{1.5}$$

Our first objective is to show that there is equality in (1.5). There is an integer I such that

$$\|f_I X^I\| = \|f\| ; \ \|f_i X^i\| < \|f\| \qquad (i < I). \tag{1.6}$$

If

$$g(X) = g_o + g_1 X + \ldots + g_m X^m, \tag{1.7}$$

we similarly define J by

$$\|g_J X^J\| = \|g\| \ ; \ \|g_j X^j\| < \|g\| \quad (j < J). \tag{1.8}$$

The coefficient of X^{I+J} in fg is

$$\sum_{i+j=I+J} f_i g_j \quad . \tag{1.9}$$

We subdivide cases.

(i) $i < I$. Then $\|f_i X^i\| < \|f\|$, that is $|f_i| < c^{-i} \|f\|$. Further $\|g_j X^j\| \leq \|g\|$, that is $|g_j| \leq c^{-j} \|g\|$. Hence

$$|f_i g_j| < c^{-I-J} \|f\| \ \|g\| . \tag{1.10}$$

(ii) $j < J$. Then (1.10) holds similarly

(iii) $i = I$, $j = J$. Here $|f_I| = c^{-I} \|f\|$, $|g_J\| = c^{-J} \|g\|$, and so

$$|f_I\, g_I| = c^{-I-J}\, \|f\|\, \|g\| \tag{1.11}$$

Hence

$$\Big|\sum_{i+j=I+J} f_i\, g_j\Big| = c^{-I-J}\, \|f\|\, \|g\| \tag{1.12}$$

and so

$$\|fg\| \geq \|f\|\, \|g\| \tag{1.13}$$

by the definition of $\|\ \|$. This with (1.5) gives

$$\|fg\| = \|f\|\, \|g\|, \tag{1.14}$$

as required.

Now let $h(X) \in k(X)$, say

$$h(X) = f(X)/g(X) = F(X)/G(X), \tag{1.15}$$

where $f,g,F,G \in k[X]$. Then

$$f(X)\, G(X) = F(X)\, g(X),$$

so

$$\|f\|\, \|G\| = \|F\|\, \|g\|. \tag{1.16}$$

Hence the definition (1.3) of $\|h\|$ is independent of the choice of f,g. It now follows immediately from (1.4) and (1.14) that $\|\ \|$ is a (non-arch.) valuation on $k(X)$.

COROLLARY. *Let $X_1,\ldots,X_n$ be independent transcendentals over k and let $c_1 > 0,\ldots,c_n > 0$. For*

$$f(X_1,\ldots,X_n) = \sum f(i_1,\ldots,i_n)\, X_1^{i_1} \ldots X_n^{i_n} \tag{1.17}$$

$$(f(i_1,\ldots,i_n) \in k)$$

put

$$||f|| = ||f||_{c_1,\ldots,c_n} = \max c_1^{i_1} \ldots c_n^{i_n} |f(i_1,\ldots,i_n)|.$$

Then $|| \; ||$ extends uniquely to $k(X_1,\ldots,X_n)$ and is a valuation.

Proof. Since $k(X_1,\ldots,X_n) = k(X_1,\ldots,X_{n-1})(X_n)$, this follows directly by induction on n.

2 GAUSS' LEMMA AND EISENSTEIN IRREDUCIBILITY

The valuation $|| \; ||$ just introduced give easy proofs of some classical results. As usual we denote the ring of valuation integers of k by $\mathfrak{o}$.

LEMMA 2.1. ("Gauss"). Suppose that

$$f(X_1,\ldots,X_n) \in \mathfrak{o}[X_1,\ldots,X_n]$$

is the product of two non-constant elements of $k[X_1,\ldots,X_n]$. Then it is the product of two non-constant elements of $\mathfrak{o}[X_1,\ldots,X_n]$.

Proof. We use the valuation $|| \; ||$ on $k(X_1,\ldots,X_n)$ given by Lemma 1.1, Corollary with $c_1 = \ldots = c_n = 1$. Then $\mathfrak{o}[X_1,\ldots,X_n]$ is just the set of elements of $k[X_1,\ldots,X_n]$ which are valuation-integers. Further, $| \; |$ and $|| \; ||$ have the same value-group.

Suppose that

$$f = gh \qquad g,h \in k[X_1,\ldots,X_n].$$

There is a $b \in k$ such that $|b| = ||g||$. On taking $b^{-1}g$ for g and bh for h we may assume that

$$||g|| = 1.$$

Then

$$1 \geqslant ||f|| = ||g|| \, ||h|| = ||h||.$$

Hence, $g,h \in \mathfrak{o}[X_1,\ldots,X_n]$, as required.

<u>COROLLARY</u>. <u>If</u> f <u>is irreducible in</u> $\mathfrak{o}[X_1,\ldots,X_n]$, <u>then it is irreducible in</u> $k[X_1,\ldots,X_n]$.

We deduce what Gauss actually proved.

<u>LEMMA 2.2</u>. (Gauss). <u>Suppose that</u> $f(X_1,\ldots,X_n) \in \mathbb{Z}[X_1,\ldots,X_n]$ <u>is the product of two non-constant elements of</u> $\mathbb{Q}[X_1,\ldots,X_n]$. <u>Then it is the product of two non-constant elements of</u> $\mathbb{Z}[X_1,\ldots,X_n]$.

<u>Proof</u>. Suppose that

$$f = gh \qquad g,h \in \mathbb{Q}[X_1,\ldots,X_n].$$

Then

$$g,h \in \mathbb{Z}_p[X_1,\ldots,X_n]$$

except for the primes p in a finite set S. If S is empty, we are done. Otherwise, for each $p \in S$ there is, by the proof of the preceding lemma, a power $p^{m(p)}$ such that

$$p^{m(p)} g \in \mathbb{Z}_p[X_1,\ldots,X_n], \quad p^{-m(p)} h \in \mathbb{Z}_p[X_1,\ldots,X_n].$$

Put

$$r = \prod_{p \in S} p^{m(p)}.$$

Then

$$rg \in \mathbb{Z}_p[X_1,\ldots,X_n], \quad r^{-1}h \in \mathbb{Z}_p[X_1,\ldots,X_n]$$

for all primes p. Hence

$$rg,\ r^{-1}h \in \mathbb{Z}[X_1,\ldots,X_n],$$

as required.

THEOREM 2.1 ("Eisenstein"). *Let the valuation* $|\;|$ *on* k *be discrete with prime element* π. *Suppose that*

$$f(X) = f_o + f_1 X + \ldots + f_n X^n \tag{2.1}$$

has

$$|f_n| = 1; \quad |f_j| < 1 \quad (j < n); \quad |f_o| = |\pi|. \tag{2.2}$$

Then $f(X)$ *is irreducible in* $k[X]$.

Proof. By Lemma 2.1 if $f(X)$ is reducible in $k[X]$ then it is reducible in $\mathfrak{o}[X]$, say

$$f(X) = g(X)h(X), \tag{2.3}$$

where

$$g(X) = g_o + \ldots + g_r X^r \tag{2.4}$$

$$h(X) = h_o + \ldots + h_s X^s \tag{2.5}$$

and $r + s = n$. Let us denote by a bar $(\bar{\ })$ the map from $\mathfrak{o}$ onto the residue class field $\mathfrak{o}/\mathfrak{p}$ and also the induced map from $\mathfrak{o}[X]$ to $\mathfrak{o}/\mathfrak{p}[X]$. Then $\bar{f}(X) = \bar{f}_n X^n$ and so $\bar{g}(X) = \bar{g}_r X^r$, $\bar{h}(X) = \bar{h}_s X^s$ by (2.3). In particular, $|g_o| < 1$, $|h_o| < 1$ so $|g_o| \leq |\pi|$, $|h_o| \leq |\pi|$. Thus $|f_o| = |g_o h_o| \leq |\pi|^2$, contrary to (2.2).

Note We shall refer to (2.2) as the *Eisenstein criterion*. A polynomial which satisfies it is an *Eisenstein polynomial*.

COROLLARY 1. The polynomial

$$\phi(X) = X^{p-1} + X^{p-2} \ldots + 1 = (X^p - 1)/(X - 1) \tag{2.6}$$

is irreducible in $\mathbb{Q}_p[X]$.

Proof.

$$\phi(Y + 1) = Y^{p-1} + \binom{p}{1} Y^{p-2} + \ldots + \binom{p}{2} Y + \binom{p}{1} \tag{2.7}$$

is an Eisenstein polynomial.

More generally we have

COROLLARY 2. For any $n \geq 1$ the polynomial

$$\psi(X) = (X^{p^n} - 1)/(X^{p^{n-1}} - 1) = \phi(X^{p^{n-1}}) \tag{2.8}$$

is irreducible in $\mathbb{Q}_p[X]$.

Proof. Again we put $X = Y + 1$, say

$$\psi(Y + 1) = \theta(Y).$$

By (2.8) we have $\theta(0) = \phi(1) = p$. Further,

$$\{(Y + 1)^{p^{n-1}} - 1\} \theta(Y) = \{(Y + 1)^{p^n} - 1\}.$$

On mapping the coefficients into the residue class field, as in the proof of the theorem, the two terms in { } map into $Y^{p^{n-1}}$ and Y^{p^n}. Hence $\bar{\theta}(Y) = Y^{p^n - p^{n-1}}$, and so θ is an Eisenstein polynomial.

3 NEWTON POLYGON

In the rest of this chapter, k is complete with respect to $|\ |$.

Let

$$f(X) = f_o + f_1 X + \ldots + f_n X^n \in k[X]. \tag{3.1}$$

We shall suppose that n is the precise degree of f, and it is also convenient to assume that $f(X)$ is not divisible by X, so

$$f_o \neq 0, \quad f_n \neq 0. \tag{3.2}$$

To obtain the so-called Newton polygon $\Pi = \Pi(f)$ of f we plot in $\mathbb{R}^2$ the pairs

$$P(j) = (j, \log|f_j|) \quad (f_j \neq 0). \tag{3.3}$$

Then Π is most simply described as the upper boundary of the convex cover of the $P(j)$. It thus consists of a set of line segments σ_s for $1 \leq s < r$ (say), where σ_s joins $P(m_{s-1})$, $P(m_s)$ and

$$0 = m_o < m_1 < \ldots < m_r = n. \tag{3.4}$$

The slope of σ_s is

$$\gamma_s \text{ (say)} = \frac{\log|f_{m_s}| - \log|f_{m_{s-1}}|}{m_s - m_{s-1}} \tag{3.5}$$

and

$$\gamma_1 > \gamma_2 > \ldots > \gamma_r \tag{3.6}$$

Every $P(j)$ lies either on or below Π.

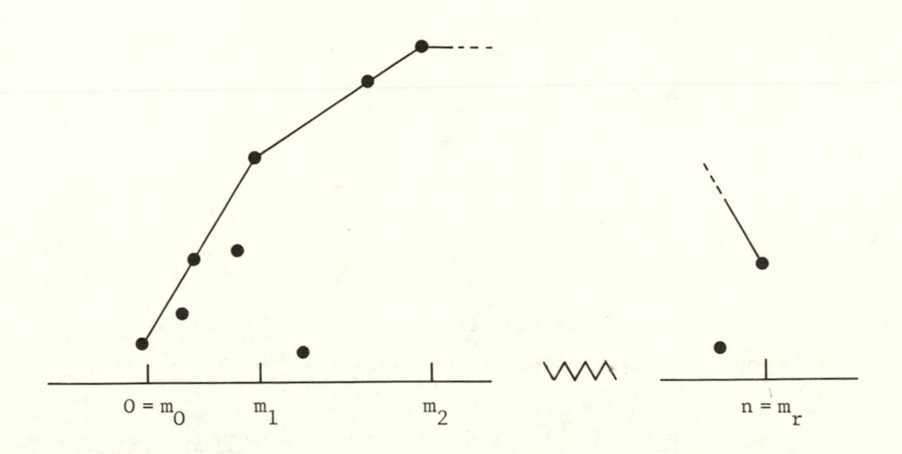

We shall say that f is of <u>type</u>

$$(\ell_1,\gamma_1;\ \ell_2,\gamma_2;\ \ldots,\ \ell_r,\ \gamma_r), \tag{3.7}$$

where

$$\ell_1 = m_1, \quad \ell_s = m_s - m_{s-1} \quad (s > 1). \tag{3.8}$$

If $r = 1$, we say that f is _pure_ [The terms "type" and "pure" are not standard].

THEOREM 3.1 ("Newton"). _Suppose that_ k _is complete and that_ $f(X) \in k[X]$ _is of type_ (3.7). _Then_

$$f(X) = g_1(X) \ldots g_r(X), \tag{3.9}$$

where $g_s(X)$ _is pure of type_ (ℓ_s,γ_s) $(1 \leq s \leq r)$.

Note It is not asserted that the $g_s(X)$ are irreducible.

The proof of Theorem 3.1 will require several lemmas. We note, first, that the Newton polygon is closely associated with the norms $\|\ \ \|_c$ defined by (1.2). If

$$\log c = -\gamma_s, \tag{3.10}$$

then

$$\|f_j X^j\|_c = \|f\| \qquad (j = m_{s-1}, m_s) \tag{3.11}$$

and

$$\|f(X) - \sum_{m_{s-1} \leq j \leq m_s} f_j X^j\| < \|f\|. \tag{3.12}$$

If, however, $\log c$ is distinct from the γ_s, then $\|f_j X^j\| = \|f\|$ for precisely one value of j [which is necessarily one of the m_s].

LEMMA 3.1. _Suppose that_ $f(X), g(X) \in k[X]$ _are pure with the same slope_ γ. _Then_ $f(X)\,g(X)$ _is also pure of slope_ γ.

Proof. Let $\log c = -\gamma$. Then in the notation (3.1) we have

$$\|f\| = \|f_o\| = \|f_n X^n\|$$

and, similarly

$$\|g\| = \|g_o\| = \|g_N X^N\|$$

in an obvious notation, where N is the degree of g. Hence

$$\|fg\| = \|f_o g_o\| = \|f_n g_N X^{n+N}\|,$$

which implies that fg is pure of slope γ.

LEMMA 3.2. Suppose that f is of type (3.7) and that g is pure of type (N,γ), where $\gamma < \gamma_r$. Then fg is of type

$$(\ell_1,\gamma_1;\ \ell_2,\gamma_2;\ \ldots;\ \ell_r,\gamma_r;\ N,\gamma). \tag{3.13}$$

Proof. Let c be given by (3.10). Then

$$\|g(X) - g_o\|_c < \|g\|_c$$

since $\gamma < \gamma_s$. Hence and by (3.12)

$$\|f(X)g(X) - g_o \sum_{m_{s-1}\leq j\leq m_s} f_j X^j\|_c < \|fg\|_c \quad . \quad (\log c = -\gamma_s)$$

Similarly, if we put $\log c = -\gamma$ we have

$$\|f(X)g(X) - f_n X^n g(X)\|_c < \|fg\|_c \quad (\log c = -\gamma).$$

These inequalities together with (3.11) and the purity of g fully determine the Newton polygon of fg and confirm that it is of type (3.13).

LEMMA 3.3. Let $\| \ \| = \| \ \|_c$ for some c. Let $R(X) \in k[X]$ and suppose that

$$G(X) = G_o + \dots + G_N X^N \in k[X] \tag{3.14}$$

has

$$\|G_N X^N\| = \|G\| . \tag{3.15}$$

Let $L(X)$, $M(X)$ be respectively the quotient and the remainder on dividing R by G:

$$R(X) = L(X)\, G(X) + M(X); \quad \deg M(X) < N. \tag{3.16}$$

Then

$$\|L\|\ \|G\| \leqslant \|R\| \tag{3.17}$$

and

$$\|M\| \leqslant \|R\| . \tag{3.18}$$

Proof. Let R have degree n so L has degree $n - N$. The coefficients $L_{n-N}, L_{n-N-1}, \dots, L_o$ of L are determined in order by the equations

$$G_N\, L_{n-N-j} + G_{N-1}\, L_{n-N-j+1} + \dots + G_{N-j}\, L_{n-N} = R_{n-j}$$

where R_{n-j} is the coefficient of X^{n-j} in $R(X)$. On using (3.15) it follows readily by induction on j that

$$\|L_{n-N-j}\, X^{n-N-j}\|\ \|G\| \leqslant \|R\| .$$

Hence (3.17) holds and then (3.18) follows at once from (3.16) and (3.17).

LEMMA 3.4. Let $\|\ \| = \|\ \|_c$ for some c and let f be given by (3.1). Suppose that there is some N with $0 < N < n$ such that

$$\|f_N X^N\| = \|f\| , \quad \|f_j X^j\| < \|f\| \qquad (j > N). \tag{3.19}$$

Then $f = gh$ *where* $g,h \in k[X]$ *have degrees* N, $n - N$ *respectively*.

Proof. There is a $\Delta < 1$ such that

$$\|f(X) - \sum_j^N f_j X^j\| = \Delta\|F\| . \tag{3.20}$$

We consider $G,H \in k[X]$ such that

$$\deg G = N, \quad \deg H \leqslant n - N \tag{3.21}$$

and

$$\|f - G\| \leqslant \Delta\|f\| , \quad \|H - 1\| \leqslant \Delta . \tag{3.22}$$

Define δ by

$$\|f - GH\| = \delta \|f\| , \tag{3.23}$$

so

$$\delta \leqslant \Delta. \tag{3.24}$$

One such choice is $G^{(o)} = \sum_j^N f_j X^j$, $H^{(o)} = 1$, $\delta = \Delta$. We shall show in the spirit of the proof of Hensel's Lemma that if G,H are given and $\delta > 0$ then we can find G^*, H^* satisfying (3.21), (3.22) and for which $\delta^* \leqslant \Delta\delta$.

It follows from (3.19) and (3.22) that G satisfies the condition (3.15) of Lemma 3.3. We apply that Lemma with $R = f - GH$ and obtain $L,M \in k[X]$ such that

$$f - GH = LG + M,$$

$$\deg L \leqslant n - N, \quad \deg M < N,$$

and

$$\|L\| \leqslant \delta, \quad \|M\| \leqslant \delta \|f\| .$$

Put

$$G^* = G + M, \quad H^* = H + L.$$

Then

$$\delta^* \|f\| = \|f - G^*H^*\| = \|(H - 1)M + ML\|$$

$$\leq \max\{\|H - 1\| \|M\|, \|M\| \|L\|\}$$

$$\leq \Delta\delta \|f\|.$$

Clearly G^*, H^* satisfy (3.21), (3.22). If $\delta^* > 0$ we can repeat the process. Clearly the sequence of polynomials G, H tend to polynomials g, h such that $f = gh$.

COROLLARY 1. *Suppose that* $f(X) \in k[X]$ *is irreducible. Then it is pure.*

Proof. For if f is not pure we can find a c and an N satisfying the conditions of the Lemma. For example one can take $-\log c$ to be the slope of the line segment joining P_o and P_n in the notation (3.3).

For a later we enunciate

COROLLARY 2. *We can suppose without loss of generality that* $h(0) = 1, \ \|h - 1\| < 1.$

Proof. For we can replace $h(X)$ by $\{h(0)\}^{-1} h(X)$.

Proof of Theorem 3.1. Let $f(X) = \prod_\lambda h_\lambda(X)$ be an expression of $f(X)$ as a product of irreducibles. By Lemma 3.4, Corollary 1, the $h_\lambda(X)$ are pure. If more than one of the $h_\lambda(X)$ have the same slope δ, then their product is also pure of slope δ. In this way we get an expression of $f(X)$ as the product of polynomials $g_\mu(X)$ $(1 \leq \mu \leq M)$ where g_μ is pure of type (q_μ, δ_μ) (say) and $\delta_1 > \delta_2 > \ldots > \delta_M$. By Lemma 3.2 and an obvious induction the type of $\prod g_\mu(X)$ is

$$(q_1, \delta_1; \ q_2, \delta_2; \ \ldots \ ; \ q_M, \delta_M).$$

This must be the type of $f(X)$, so $M = r$ and $q_s = \ell_s$, $\delta_s = \gamma_s$ $(1 \leq s \leq r)$. This concludes the proof.

4 FACTORIZATION OF PURE POLYNOMIALS

In this section we give a method for obtaining a factorization of $f(X) \in k[X]$ if we have a "sufficiently good" approximate factorization. It applies, in particular, to pure $f(X)$ of zero slope. It will be clear that it can be modified to deal with any pure polynomial. We shall not be making use of the results of this section.

LEMMA 4.1. Let

$$f(X) = f_o + f_1 X + \dots + f_n X^n \in \Omega[X] \tag{4.1}$$

with

$$|f_n| = 1. \tag{4.2}$$

Let $G(X), H(X) \in \Omega[X]$ *have degrees* r,s *respectively and suppose that*

$$\delta = \|f(X) - G(X)H(X)\| \tag{4.3}$$

satisfies

$$\delta < |R(G,H)|^2. \tag{4.4}$$

Here $\| \ \|$ *has* $c = 1$ *in the notation of* (1.2) *and* $R(G,H)$ *is the resultant of* G *and* H. (cf. Appendix A). *Then*

$$f(X) = g(X)h(X) \tag{4.5}$$

for some $g,h \in \Omega[X]$ *of degrees* r,s.

Proof. We may suppose without loss of generality that $f_n = 1$ and that the coefficients of X^r, X^s in f,g respectively are also 1. We shall find

$$G^*(X) = G(X) + \gamma(X), \quad H^*(X) = H(X) + \chi(X) \tag{4.6}$$

where $\gamma(X)$, $\chi(X)$ have degrees $\leq r - 1$, $\leq s - 1$ respectively and such that G^*H^* is a better approximate factorization. We have

$$G^*H^* - f = (GH - f) + G\chi + H\gamma + \gamma\chi \tag{4.7}$$

and will choose γ, χ such that

$$G\chi + H\gamma = f - GH. \tag{4.8}$$

The right hand side is of degree $\leq n - 1$ by our convention about the terms in X^n, X^r, X^s of f, g, h respectively. Hence we have $n = r + s$ linear equations for the $r + s$ unknown coefficients of γ, χ. The determinant of the corresponding homogeneous forms is $R(G,H)$ by the definition of resultant [Appendix A]. Further, the coefficients of the unknowns (i.e. the coefficients of G,H) are all in Ω. Hence all cofactors are in Ω. On solving for the unknowns determinantally, we have

$$\|\gamma\| \ , \ \|\chi\| \leq \delta/|R(G,H)| \tag{4.9}$$

on using the estimate (4.3).

Now (4.7), (4.8), (4.9) give

$$\begin{aligned} \delta^* \text{ (say) } &= \|f - G^*H^*\| \\ &= \|\gamma\chi\| \\ &\leq \delta^2/|R(G,H)|^2 \\ &< \delta \end{aligned} \tag{4.10}$$

by (4.4).

Further,

$$\|\gamma\| \ , \ \|\chi\| < |R(G,H)| \tag{4.11}$$

by (4.4) and (4.9). From this it easily follows that

$$|R(G^*,H^*)| = |R(G,H)| \tag{4.12}$$

(cf. Chapter 5, Exercise 1). The rest of the proof of the Lemma follows the by now familiar routine.

COROLLARY. *The conclusion of the Lemma holds if* (4.3) *is replaced by*

$$\delta < D(f), \tag{4.13}$$

where $D(f)$ *is the discriminant of* f.

Proof. We have

$$|D(GH)| = |D(f)|$$

as at the end of the proof of the Lemma. Here

$$D(GH) = D(G)\, D(H)\, \{R(G,H)\}^2$$

(Appendix A, Lemma 5). Now $D(G), D(H) \in \mathfrak{o}$ since $G,H \in \mathfrak{o}[X]$; and so

$$|D(f)| = |D(GH)| \leq |R(G,H)|^2.$$

Hence (4.13) implies (4.4) and the conditions of the Lemma are satisfied.

We note that Lemma 4.1 Corollary gives a procedure to decide whether or not f given by (4.1), (4.2) is reducible whenever the residue class field is finite, in particular when $\mathfrak{o} = \mathbb{Z}_p$. If $D(f) = 0$ then either f is inseparable or $f'(X) \neq 0$ and f,f' have a common factor: so in either case we are reduced to a situation of lower degree. Otherwise, let $|D(f)| = p^{-t}$. Then it is clearly enough to consider $G(X)$, $H(X)$ modulo p^t and there are only a finite number of cases to consider.

5 WEIERSTRASS PREPARATION THEOREM

We shall prove for power series an analogue of the Weierstrass preparation theorem:

THEOREM 5.1 ("Weierstrass"). Let

$$f(X) = \sum_{o}^{\infty} f_n X^n, \tag{5.1}$$

where

$$f_n \to 0. \tag{5.2}$$

Let N be defined by

$$|f_N| = \max|f_n|; \quad |f_n| < |f_N| \quad (\text{all } n > N). \tag{5.3}$$

Then there is a polynomial

$$g(X) = g_o + \ldots + g_N X^N \tag{5.4}$$

and a power series

$$h(X) = 1 + h_1X + h_2X^2 + \ldots \tag{5.5}$$

with

$$h_n \to 0 \tag{5.6}$$

such that

$$f(X) = g(X)\, h(X). \tag{5.7}$$

Further,

$$|g_N| = \max|g_n| \tag{5.8}$$

and

$$|h_n| < 1. \quad (n > 0). \tag{5.9}$$

Before proceeding to a proof we note that this implies

Strassmann's theorem (Chapter 4, Theorem 4.1). If $b \in \Omega$ then $h(b) \neq 0$ and so the zeros of $f(X)$ in Ω are precisely the zeros of $g(X)$ in Ω. It is, of course, quite possible for $g(X)$ to have fewer than N zeros in Ω (or, what is the same thing by (5.8), in k): when this happens the Strassmann bound fails to be attained.

The theorem we are about to prove is a generalization of Lemma 3.4 and its corollaries. The proof will be on similar lines but there are extra complications.

We denote by W the ring of power series $f(X)$ satisfying 5.2 (i.e. which are convergent in Ω) and write

$$\|f\| = \max |f_n|. \tag{5.10}$$

This makes W into a metric space with an ultrametric. Further,

$$\|f + g\| \leq \max \{\|f\|, \|g\|\} \tag{5.11}$$

and

$$\|fg\| = \|f\| \, \|g\| \tag{5.12}$$

for $f, g \in W$. (cf. Lemma 1.1).

<u>LEMMA 5.1</u>. W <u>is complete with respect to</u> $\| \ \|$.

<u>Proof</u>. Let $\{f^{(j)}(X)\}$ be a fundamental sequence of elements

$$f^{(j)}(X) = \sum_n f_n^{(j)} X^n \tag{5.13}$$

of W. For each n the sequence $\{f_n^{(j)}\}$ is fundamental, and so has a limit f_n^* by the completeness of k. We have to show that

$$f^*(X) = \sum_n f_n^* X^n \tag{5.14}$$

is in W. Let $\varepsilon > 0$ be given. Then there is some $J = J(\varepsilon)$ such that $\|f^{(J)} - f^{(j)}\| < \varepsilon$ for all $j \geq J$. Hence

$$|f_n^{(J)} - f_n^*| < \varepsilon \quad \text{(all } n\text{)}. \tag{5.15}$$

Since $f^{(J)}(X) \in W$, there is an $m = m(\varepsilon)$ such that

$$|f_n^{(J)}| < \varepsilon \quad (\text{all } n \geq m). \tag{5.16}$$

Hence

$$|f_n^*| < \varepsilon \quad (\text{all } n \geq m). \tag{5.17}$$

But ε is arbitrary, so $\lim_n f_n^* = 0$. Hence $f^* \in W$, and we are done.

LEMMA 5.2. *Let* $R(X) \in W$ *and let*

$$G(X) = G_o + \ldots + G_N X^N \in k[X] \tag{5.18}$$

have

$$|G_N| = \max |G_n|. \tag{5.19}$$

Then there is an $L(X) \in W$ *and* $M(X) \in k[X]$ *of degree* $\leq N - 1$ *such that*

$$R(X) = L(X)G(X) + M(X). \tag{5.20}$$

Further,

$$||L|| \, ||G|| \leq ||R|| \tag{5.21}$$

$$||M|| \leq ||R|| . \tag{5.22}$$

Note This is a generalization of Lemma 3.3.

Proof. Let $R^{(j)}(X) \in k[X]$ be a sequence of polynomials tending to $R(X)$. By Lemma (3.3) here are $L^{(j)}(X) \in k[X]$ and $M^{(j)}(X) \in k[X]$ of degree $\leq N - 1$ such that

$$R^{(j)}(X) = L^{(j)}(X)G(X) + M^{(j)}(X). \tag{5.23}$$

and so

$$\{R^{(i)}(X) - R^{(j)}(X)\} = \{L^{(i)}(X) - L^{(j)}(X)\} G(X) + M^{(i)}(X) - M^{(j)}(X). \tag{5.24}$$

The estimates (3.17), (3.18) with $R^{(i)} - R^{(j)}$ instead of R shows that $\{L^{(j)}\}$ and $\{M^{(j)}\}$ are fundamental sequences. By Lemma 5.1 they have limits L, M. Clearly L, M satisfy the conclusions of the Lemma.

Proof of Theorem 5.1. As already remarked, Theorem 5.1 is a generalization of Lemma 3.4 and its Corollary 2. The proof of Theorem 5.1 is similar to that of Lemma 3.4, except that one invokes Lemma 5.2 instead of Lemma 3.3.

Notes

§§3,4 A very powerful algorithm for factorizing $F(T) \in \mathbb{Z}[T]$ depends on factorizing it in $\mathbb{Z}_p[T]$ for an appropriate p first, see A.K. Lenstra et al. (1982) and the article by A.K. Lenstra in H.W. Lenstra & Tijdeman (1982). But see Chapter 9, Exercise 16.

§5. We shall encounter Lemma 5.1 again in Chapter 12, §3.

Exercises

1. For $f(X_1,\ldots,X_n) \in \mathbb{Z}[X_1,\ldots,X_n]$ define the content $c(f)$ to be the greatest common divisor of the coefficients. If also $g(X_1,\ldots,X_n) \in \mathbb{Z}[X_1,\ldots,X_n]$ show that

$$c(fg) = c(f)\, c(g).$$

2. Assumptions and notation as in Strassmann's theorem (Theorem 4.1 of Chapter 4). Let $\operatorname{Char} k = 0$. Denote the successive (formal) derivatives of $f(X)$ by $f'(X), f''(X), \ldots, f^{(r)}(X), \ldots$. If $b \in \Omega$ is a zero of $f(X)$, show that not all $f^{(r)}(b)$ vanish. Define the multiplicity of b as a zero to be the least r for which $f^{(r)}(b) \neq 0$. Show that the sum of the multiplicities of the zeros $b \in \Omega$ is at most N.

[Hint Use "Weierstrass" preparation theorem].

3. Let $b \in \mathbb{Z}_p$, $|b| = 1$ and put $f(X) = X^4 - b$.

(i) if $p \equiv 3\ (4)$ show that $f(X)$ is always reducible.

(ii) if $p \equiv 1\ (4)$ show that $f(X)$ is reducible precisely when $b \in (\mathbb{Q}_p^*)^2$.

(iii) What is the situation for $p = 2$?

4. Let $f(X) \in \mathfrak{o}[X]$ and let $\bar{f}(X) \in \rho[X]$, where $\rho = \mathfrak{o}/\mathfrak{p}$ be obtained by mapping the coefficients of $\mathfrak{o}$ into $\mathfrak{o}/\mathfrak{p}$. Suppose that $f(X)$ and $\bar{f}(X)$ have the same degree (i.e. the highest coefficient of f is a unit) and that $\bar{f}(X)$ is irreducible in $\rho[X]$. Show that $f(X)$ is irreducible in $k[X]$.

5. Let $u(0) = 1$, $u(1) = 2$ and

$$u(n+2) = 5u(n+1) - 9u(n) \quad (n \geq 2).$$

(i) Show that $u(2m)$ and $u(2m+1)$ can both be expressed as power series in m with coefficients in $\mathbb{Q}_5$ convergent for $m \in \mathbb{Z}_5$.

(ii) Find all solutions of $u(n) = \pm 1$.

(iii) Show that the bounds for the numbers of solutions of $u(n) = 1$ obtained by considering $u(r + 4s)$, $r = 0,1,2,3$ (according to the paradigm of Chapter 5, §5) in $\mathbb{Q}_5$ are not attained.

(iv) Show that the two zeros of the power series in s for $r = 0$ are $s = 0$ and $s = \frac{1}{2} \in \mathbb{Z}_5$, and explain the other additional zeros.

6. Show that W, as defined in §5, is not locally compact.

7. Suppose that

$$f(X) = 1 + \sum_{j=1}^{\infty} f_n X^n$$

converges for all $X \in k$ where k is a complete with respect to a non-arch. valuation.

(i) Show that for any C there are only finitely many zeros x of $f(X)$ in $|x| \leqslant C$.

(ii) Suppose, further, that k is algebraically closed. Let $B_1, \ldots, B_m, \ldots$ be the zeros of $f(X)$ arranged in increasing order of value (multiple zeros being taken multiply). Show that

$$f(X) = \lim_{M \to \infty} \prod_{m=1}^{M} (1 - B_m^{-1} X),$$

where the limit is in the topology of W (cf. Lemma 5.1).

CHAPTER SEVEN: ALGEBRAIC EXTENSIONS (COMPLETE FIELDS)

1 INTRODUCTION

Let $k \subset K$ be fields. We shall say that K is a *finite algebraic extension* of k if the relative degree $[K:k]$ is finite. We shall show that if k is complete with respect to a valuation $|\ |$, there is precisely one extension of $|\ |$ to K and we shall study the situation further. If $|\ |$ is archimedean, we saw in Chapter 3 that the only case with $K \neq k$ is $k = \mathbb{R}$, $K = \mathbb{C}$ and we proved Theorem 1.1 below for this. We may therefore suppose that $|\ |$ is non-archimedean.

If $[K:k] < \infty$ and $A \in K$, we denote by $N_{K/k}(A)$ the relative *norm* of A, that is the determinant of the map

$$B \to AB \quad (B \in K),$$

of K into itself, where K is considered as a k-vector space. Then $N_{K/k}$ gives a homomorphism of the multiplicative group K^* of non-zero elements of K into k^*. Further,

$$N_{K/k}(a) = a^n \quad (a \in k),$$

where $n = [K:k]$. (See Appendix B).

Our first objective is

THEOREM 1.1 Let k be complete with respect to $|\ |$ and let K be an extension of relative degree

$$[K:k] = n. \tag{1.1}$$

Then there is precisely one extension $\|\ \|$ of $|\ |$ to K. It is given by

$$\|A\| = |N_{K/k}(A)|^{1/n} \quad (A \in K). \tag{1.2}$$

Further, K is complete with respect to $\| \; \|$.

In §2 we show that there is at most one extension $\| \; \|$ of $| \; |$ and that K is complete with it if it exists. In §3 we finish the proof of Theorem 1.1 by showing that (1.2) does in fact define a valuation.

In the rest of the Chapter we shall discuss, under appropriate additional conditions, the residue class fields and the value groups of k and K. Two kinds of extension K/k have an especial significance. For the first, unramified extensions, the value group remains the same: They can be described entirely in terms of the residue class field extension. For the second, completely ramified extensions, the residue class fields are the same: They can be described in terms of Eisenstein polynomials. For a finite algebraic extension K/k there is a field L with $k \subset L \subset K$ such that L/k is unramified and K/L is completely ramified.

2 UNIQUENESS

To prove the uniqueness part of Theorem 1.1 we must introduce the notion of norm on a finite dimensional vector space V over a valued field k. Since it causes no extra trouble and has some interest we shall (in this section only) permit $| \; |$ to be archimedean.

DEFINITION 2.1. Let V be a vector-space over the field k and let $| \; |$ be a valuation on k satisfying the triangle inequality. A real valued function $\| \; \|$ on V is called a norm if

(i) $\|\underline{a}\| \geq 0$ for all $\underline{a} \in V$, with equality and for $\underline{a} = \underline{0}$.

(ii) $\|\underline{a} + \underline{b}\| \leq \|\underline{a}\| + \|\underline{b}\|$ for all $\underline{a}, \underline{b} \in V$.

(iii) $\|c\underline{a}\| = |c| \, \|\underline{a}\|$ for $c \in k$, $\underline{a} \in V$.

Note 1 There is an unfortunate clash in nomenclature as we are using the word "norm" also in a different sense.

Note 2 In an obvious way, $\| \;\|$ induces a metric and so a topology in V.

DEFINITION 2.2. Two norms $\| \;\|_1$ and $\| \;\|_2$ are said to be equivalent if there are $C_1, C_2 \in \mathbb{R}$ such that

$$\|\underline{a}\|_1 \leq C_2 \|\underline{a}\|_2 ; \quad \|\underline{a}\|_2 \leq C_1 \|\underline{a}\|_1 \quad (\text{all } \underline{a} \in V) \tag{2.1}$$

Note It is readily verified that this is an equivalence relation. Equivalent norms induce the same topology.

LEMMA 2.1. Suppose that k is complete with respect to $| \;|$. Then any two norms on the same finite-dimensional k-vector space V are equivalent. Further, V is complete under the induced metrics.

Proof. Let $\underline{e}_1, \ldots, \underline{e}_n$ be any k-basis for V. Put

$$\underline{a} = a_1\underline{e}_1 + \ldots + a_n\underline{e}_n \quad (a_j \in k) \tag{2.2}$$

and

$$\|\underline{a}\|_o = \max |a_j| . \tag{2.3}$$

Clearly $\| \;\|_o$ is a norm and V is complete with respect to it.

It is enough to show that any norm $\| \;\|$ on V is equivalent to $\| \;\|_o$. One of the equalities (2.1) is easy:

$$\begin{aligned} \|a\| &= \|\sum a_j\underline{e}_j\| \\ &\leq \sum |a_j| \, \|\underline{e}_j\| \\ &\leq C_o \|\underline{a}\|_o , \end{aligned} \tag{2.4}$$

where

$$C_o = \sum_j \|\underline{e}_j\| .$$

It remains to show that there is some C such that

$$\|\underline{a}\|_o \leq C\|\underline{a}\| \quad (\text{all } \underline{a} \in V). \tag{2.5}$$

If not, for every $\varepsilon > 0$ there is a $\underline{b} = \underline{b}(\varepsilon) \in V$ such that

$$\|\underline{b}\| < \varepsilon\|\underline{b}\|_o. \tag{2.6}$$

On recalling (2.3) and permuting $\underline{e}_1,\ldots,\underline{e}_n$ if necessary, we may suppose without loss of generality that there is a $\underline{b} = \underline{b}(\varepsilon)$ satisfying (2.6) and

$$\|\underline{b}\|_o = |b_n|.$$

On replacing $\underline{b}$ by $b_n^{-1}\underline{b}$ we have $\underline{b} = \underline{c} + \underline{e}_n$, where $\underline{c}$ is in the subspace W spanned by $\underline{e}_1,\ldots,\underline{e}_{n-1}$. To sum up, if (2.5) is false, we can find a sequence $\underline{c}^{(m)}$ $(m = 1,2,\ldots)$ of elements of W such that

$$\|\underline{c}^{(m)} + \underline{e}_n\| \to 0 \quad (m \to \infty).$$

By (ii) of the definition of a norm we have

$$\|\underline{c}^{(\ell)} - \underline{c}^{(m)}\| \to 0 \quad (\ell, m \to \infty).$$

We now use induction on the dimension n of V. Since W has dimension $n - 1$, it is complete under $\|\ \|$. There is therefore a $\underline{c}^* \in W$ such that

$$\|\underline{c}^{(m)} - \underline{c}^*\| \to 0 \quad (m \to \infty).$$

Now

$$\|\underline{c}^* + \underline{e}_n\| = \lim_{m\to\infty} \|\underline{c}^{(m)} + \underline{e}_n\|$$

$$= 0.$$

This contradicts (i) of Definition 2.1. The contradiction shows that (2.5) holds, and so $\|\ \|$ and $\|\ \|_o$ are equivalent.

COROLLARY 1. (Uniqueness in Theorem 1.1). *Let* k *be complete with respect to* $|\ |$, *and let* K *be a finite algebraic extension of* k. *Then there is at most one extension* $\|\ \|$ *of* $|\ |$ *to* K.

Proof. The function $\|\ \|$ on K, regarded as a finite-dimensional k-vector space, satisfies Definition 2.1 and so is a norm. By the Lemma, any two valuations $\|\ \|_1$, $\|\ \|_2$ extending $|\ |$ are equivalent as norms and so induce the same topology on K. By Lemma 3.2 of Chapter 2 they are thus equivalent as valuations. Since they take the same values on k, they must be identical.

COROLLARY 2. *Let* k *be complete with respect to* $|\ |$ *and suppose that* $\|\ \|$ *is an extension to the finite algebraic extension* K *of* k. *Then* K *is complete with respect to* $\|\ \|$.

Proof. The last sentence of the enunciation of the Lemma.

3 EXISTENCE

We first complete the proof of Theorem 1.1. After the two corollaries to Lemma 2.1 it remains to show that $\|\ \|$ given by (1.2) is a valuation on K and that it extends $|\ |$. For properties of norms, characteristic polynomials and minimal polynomials we refer to Appendix B.

Let $a \in k$. Then $N_{K/k}(a) = a^n$. Hence $\|a\| = |a|$.

Let $A, B \in K$. Then $N_{K/k}(AB) = N_{K/k}(A)\, N_{K/k}(B)$, so $\|AB\| = \|A\|\ \|B\|$. In particular, if $A \neq 0$ we have $\|A\|\ \|A^{-1}\| = \|1\| = 1$, so $\|A\| > 0$. We have thus checked parts (i), (ii) of the definition of a valuation (Definition 1.1 of Chapter 2).

It remains to show that

$$\|A\| \leq 1 \tag{3.1}$$

implies $\|1 + A\| \leq C$ for some C, and our proof gives directly $C = 1$. Let

$$F_A(T) = F(T) = T^n + F_{n-1}\, T^{n-1} + \dots + F_o \in k[T] \tag{3.2}$$

be the characteristic polynomial of A. Then

$$|F_o| = |\pm N_{K/k}(A)| \leqslant 1 \tag{3.3}$$

by (3.1). By Lemma 3 of Appendix B we have

$$F(T) = \{f(T)\}^r \tag{3.4}$$

for some $r > 0$, where $f(T)$ is the minimal polynomial for A over k. Since $f(X)$ is irreducible, it is pure (Chapter 4, Theorem 3.1 Corollary 1) and so F is pure (Chapter 4, Lemma 3.1). In particular, $F(T) \in \mathfrak{o}[T]$ by (3.3). Now $N_{K/k}(1 + A) = (-1)^n F(-1)$, so

$$\|1 + A\| = |F(-1)|^{1/n} \leqslant 1,$$

as required.

This concludes the proof of Theorem 1.1. Since $\| \ \|$ is unique, there will usually be no confusion in writing $| \ |$ instead of $\| \ \|$.

COROLLARY 1. (to Theorem 1.1) _There is a unique extension of_ $| \ |$ _to the algebraic closure_ $\bar{k}$ _of_ k.

Note We shall see later (Chapter 8, §3) that $\bar{k}$ is not, in general, complete with respect to the extension.

Proof. A standard application of Zorn's Lemma (or transfinite induction).

COROLLARY 2. Let $A, A' \in K$ _be conjugate over_ k. _Then_ $\|A\| = \|A'\|$.

Proof. They have the same minimum polynomial, so the same norm. Alternatively, we can suppose that K is normal over k. Then $A' = \sigma A$ for some $\sigma \in \mathrm{Gal}(K/k)$. Define $\| \ \|_\sigma$ on K by $\|B\|_\sigma = \|\sigma B\|$. Then $\| \ \|_\sigma$ is an extension of $| \ |$, so $\| \ \|_\sigma = \| \ \|$.

COROLLARY 3. (Krasner) _Let_ A _and_ $A' \neq A$ _be conjugate over_ k _and let_ $a \in k$. _Then_ $\|a - A\| \geqslant \|A - A'\|$.

Proof. For otherwise $\|a - A'\| = \|A - A'\| > \|a - A\|$, contrary to the previous Corollary (with $a - A$ for A).

4 RESIDUE CLASS FIELDS

In this section k and $K \supset k$ are fields with relative degree

$$[K:k] = n \tag{4.1}$$

and $|\ |$ is a valuation on K with respect to which k (and so K) is complete. The ring of integers and maximal ideal of k are $\mathfrak{o}, \mathfrak{p}$ and the corresponding things for K are $\mathfrak{O}, \mathfrak{P}$. We denote the residue class fields by ρ, P:

$$\rho = \mathfrak{o}/\mathfrak{p} \quad ; \qquad P = \mathfrak{O}/\mathfrak{P} \tag{4.2}$$

LEMMA 4.1. There is a natural injection

$$\rho \hookrightarrow P. \tag{4.3}$$

Further,

$$f \text{ (say)} = [P:\rho] \leq n = [K:k]. \tag{4.4}$$

Note The notation f is standard in this context.

Proof. An element $b \in \mathfrak{o}$ is in $\mathfrak{P}$ if and only if it is in $\mathfrak{p}$. Hence the inclusion $\mathfrak{o} \hookrightarrow \mathfrak{O}$ induces an injection (4.3).

Now let $A_1, \ldots, A_{n+1} \in \mathfrak{O}$. We shall show that the residue classes $\bar{A}_1, \ldots, \bar{A}_{n+1} \in P$ are linearly dependent over ρ. By (4.1) there are $a_1, \ldots, a_{n+1}$ not all 0 such that

$$\sum a_j A_j = 0. \tag{4.5}$$

We may suppose, without loss of generality, that

$$\max|a_j| = 1. \tag{4.6}$$

Then $a_j \in \mathfrak{o}$ $(1 \leq j \leq n + 1)$ and not every residue class $\bar{a}_j \in \rho$ is 0.

Now $\sum \bar{a}_j \bar{A}_j = 0$ by (4.5), and we have proved (4.4).

DEFINITION 4.1. If $f = n$ we say that the extension K/k is *unramified*.

DEFINITION 4.2. If $f = 1$ we say that K/k is *completely ramified*.

LEMMA 4.2. *Let* L *be a field,* $k \subset L \subset K$. *Then the degrees of the corresponding residue class field extensions satisfy*

$$f(K/k) = f(K/L)\, f(L/k). \tag{4.7}$$

Proof. Clear.

To avoid "pathological" behaviour, we must make some restrictions on the residue class fields, and so recall some definitions and facts. Let ρ be a field and let $\phi(T) \in \rho[T]$ be a polynomial in the indeterminate T. We say that $\phi(T)$ is *inseparable* if $\phi'(T) = 0$. An example of an inseparable polynomial is $T^p - b$ where $b \in \rho$ and p is the characteristic of ρ. If ϕ is not inseparable then it is *separable*. An element α of some overfield which is algebraic over ρ is separable by definition if its minimum polynomial $\phi(T)$ is separable. Clearly then $\phi'(\alpha) \neq 0$. A finite algebraic extension P/ρ is separable by definition if every $\alpha \in P$ is separable. It can then be shown (e.g. van der Waerden (1949)) that P is obtainable from ρ by a single adjunction: $P = \rho(\beta)$ for some β. Finally, the field ρ is *perfect* if every finite algebraic extension of ρ is separable. A necessary and sufficient condition for ρ to be perfect is that either (i) it has characteristic 0 or (ii) it has characteristic p and every element of ρ is a p-th power. Indeed if

$$\phi(T) = \sum_j a_j\, T^{pj}$$

is inseparable and $a_j = b_j^p$, then

$$\phi(T) = \left(\sum_j b_j\, T^j\right)^p$$

is reducible. In particular (and this is the case we shall be mainly concerned with), any finite field is perfect.

THEOREM 4.1 *Let* K,k *be as before, with* P, ρ *given by* (4.2). *Let* $\alpha \in P$ *be separable over* ρ. *Then there is an* $A \in \alpha$ *such that*

$$[k(A) : k] = [\rho(\alpha) : \rho]. \tag{4.8}$$

Further, the field $k(A)$ *depends only on* α.

Note In terms of Definition 4.1, the equality (4.8) states that $k(A)/k$ is unramified.

Proof. Let $\phi(T) \in \rho[T]$ be the minimum polynomial for α over ρ, so $\phi'(\alpha) \neq 0$ by the separability hypothesis. Let $\Phi(T) \in \mathfrak{o}[T]$ be any *lift* of $\phi(T)$, that is (i) ϕ and Φ have the same degree and (ii) the coefficients of ϕ are the residue classes of those of Φ. Let $A_o \in \mathfrak{O}$ any element of the residue class α. Then

$$|\Phi(A_o)| < 1, \quad |\Phi'(A_o)| = 1.$$

By Hensel's Lemma (Chapter 4, Lemma 3.1) with $k(A_o)$ as groundfield there is some

$$A \in k(A_o) \subset K \tag{4.9}$$

such that

$$\Phi(A) = 0; \quad |A - A_o| < 1. \tag{4.10}$$

Then $A \in \alpha$ and (4.8) holds. If we suppose, further, that $[k(A_o) : k] = [\rho(\alpha) : \rho]$ then (4.9) gives $k(A) = k(A_o)$. This proves the last sentence of the enunciation.

COROLLARY 1. *Suppose that* P/ρ *is separable. Then there is a bijection between the fields* $M \subset K$ *which are unramified over* k *and the fields* μ *with* $\rho \subset \mu \subset P$. *The field* $\mu = \mu(M)$ *corresponding to* M *is*

$$(M \cap \mathfrak{O}) \bmod \mathfrak{P}.$$

Proof. By the facts about separability stated before the enunciation, every μ is of the form $\mu = \rho(\alpha)$ for some $\alpha \in P$.

COROLLARY 2. *Suppose that* P/ρ *is separable. There is a field* $L, k \subset L \subset K$ *such that* L/k *is unramified and such that every* $M \subset K$ *which is unramified over* k *is contained in* L. *Further,* K/L *is completely ramified.*

Proof. L corresponds to P in previous Corollary.

COROLLARY 3. *Suppose that* ρ *is perfect. Then the residue class field of the algebraic closure* $\bar{k}$ *of* k *is the algebraic closure of* ρ. *There is a subfield* k_u *of* $\bar{k}$ *such that a finite algebraic extension* K/k *is unramified precisely when* $K \subset k_u$.

Note 1 c.f. Theorem 1.1, Corollary 1.

Note 2 The Lemma shows that the unramified extensions of k can be described entirely in terms of ρ.

Proof. Let $\phi(T) \in \rho[T]$ be irreducible and $\Phi(T)$ any lift to $k[T]$. Then $\bar{k}$ contains all the roots of $\Phi(T)$, so its residue class field contain all the roots of $\phi(T)$. Hence the residue class field of $\bar{k}$ is the algebraic closure of ρ. The rest follows from Corollary 2 and Zorn's Lemma.

5 RAMIFICATION

We now consider the relation between the value groups G_k and G_K for a finite algebraic extension K/k under the hypothesis that G_k is discrete.

LEMMA 5.1. *Suppose that* $| \ |$ *is discrete on* k. *Then it is discrete on* K.

Proof. Follows at once from (1.2).

DEFINITION 5.1. The index

$$e = [G_K : G_k] \tag{5.1}$$

is the <u>ramification index</u>. The notation e is standard.

It follows at once from the proof of Lemma 5.1 that e divides $n = [K : k]$. There is a more precise statement below (Theorem 5.1).

<u>LEMMA 5.2</u>. <u>Let</u> L <u>be a field</u>, $k \subset L \subset K$. <u>Then in an obvious notation</u>

$$e(K/k) = e(K/L)e(L/k). \tag{5.2}$$

<u>Proof</u>. Clear.

Before proceeding, we require a lemma which might well have been put in Chapter 4.

We recall that an abelian group $\mathfrak{M}$ (written additively) is an $\mathfrak{o}$-module if for every $a \in \mathfrak{o}$, $A \in \mathfrak{M}$ there is given an element $aA \in \mathfrak{M}$ satisfying the axioms:

$$1A = A$$

$$a(A + B) = aA + aB$$

$$(a + b)A = aA + bA$$

$$(ab)A = a(bA).$$

It is <u>torsion-free</u> if $aA = 0$ implies that either $a = 0$ or $A = 0$. The module $\mathfrak{M}$ is <u>finitely generated</u> if there are $E_1,\ldots,E_n \in \mathfrak{M}$ such that every $A \in \mathfrak{M}$ can be written as $a_1E_1 + \ldots + a_nE_n$ $(a_j \in \mathfrak{o})$. The set $\{E_1,\ldots,E_n\}$ of generators is a <u>basis</u>, if $a_1E_1 + \ldots + a_nE_n = 0$ implies $a_1 = \ldots = a_n = 0$. In the above definitions $\mathfrak{o}$ is any ring with a 1.

<u>LEMMA 5.3</u>. <u>Let</u> $\mathfrak{o}$ <u>be the ring of integers of a (not necessarily</u>

complete) field k with respect to a valuation $| \ |$. Then every torsion-free finitely generated $\mathfrak{o}$-module $\mathfrak{M}$ has a basis.

Proof. Let $\{E_1,\ldots,E_n\}$ be a set of generators. If they are not a basis, there are $a_1,\ldots,a_n \in \mathfrak{o}$ not all 0 such that

$$a_1E_1 + \ldots + a_nE_n = 0.$$

Without loss of generality

$$|a_n| = \max|a_j| \ ; \quad a_j = a_nb_j, \ b_j \in \mathfrak{o}.$$

Hence

$$a_n(b_1E_1 + \ldots + b_{n-1}E_{n-1} + E_n) = 0.$$

Since $\mathfrak{M}$ is torsion-free, we have

$$E_n = - b_1E_1 - \quad - b_{n-1}E_{n-1}$$

and so $\{E_1,\ldots,E_{n-1}\}$ is a set of generators. If it is not a basis, we repeat the argument.

We shall need the following lemma later in a rather different context. This dual purpose explains the somewhat barock formulation:

LEMMA 5.4. Let $k \subset K$ be fields and let $| \ |$ be a valuation on K. Suppose that

(i) k is complete with respect to $| \ |$

(ii) $| \ |$ is discrete on both k and K. Let e be defined by (5.1)

(iii) The residue class field extension P/ρ is of finite relative degree $[P : \rho] = f$.

Then the extension K/k is of finite relative degree

$$[K : k] = ef. \tag{5.3}$$

Let Π _be a prime element for_ K _and (in the notation (4.2)) let_ $B_1,\dots,B_f$ _be any lift to_ $\mathcal{O}$ _of a basis of_ P/ρ. _Then_

$$\mathcal{B} = \{B_i\,\Pi^j : 1 \leq i \leq f,\quad 0 \leq j \leq e-1\} \tag{5.4}$$

_is an_ $\mathfrak{o}$-_basis of_ $\mathcal{O}$.

Proof. By the definition of e we have

$$|\Pi|^e = |\pi|, \tag{5.5}$$

where π is a prime element of k.

We show first that $\mathcal{B}$ is linearly independent over k. If not, we should have

$$\sum_{i,j} a_{ij}\, B_i\, \Pi^j = 0, \tag{5.6}$$

where $a_{ij} \in k$ are not all zero. Without loss of generality

$$\max|a_{ij}| = 1,$$

and so there are I,J such that

$$|a_{ij}| \leq |\pi| \quad (1 \leq i \leq f,\ j < J); \qquad |a_{IJ}| = 1.$$

Then

$$|\sum_i a_{iJ}\, B_i| = 1$$

by the definition of the B_i. Hence

$$|\sum_i a_{ij}\, B_i\, \Pi^j| \quad \begin{cases} \leq |\pi| = |\Pi|^e & (j < J) \\ = |\Pi|^J & (j = J) \\ \leq |\Pi|^{J+1} & (j > J). \end{cases}$$

This contradicts (5.6). Hence $\mathcal{B}$ is linearly independent over k, and _a fortiori_ over $\mathfrak{o}$.

We now show that $\mathcal{B}$ is a set of generators of $\mathfrak{O}$. The proof is an elaboration of that of Lemma 1.4 of Chapter 4. Let $A \in \mathfrak{O}$. By the definition of the B_i, there are $a_{io} \in \mathfrak{o}$ such that

$$A - \sum_i a_{io} B_i = \Pi A_1 \quad \text{(say)} \in \Pi\mathfrak{O}.$$

We repeat the process with A_1 and so on until we obtain $a_{ij} \in \mathfrak{o}$ such that

$$A - \sum_{j=o}^{e-1} \sum_i a_{ij} B_i \Pi^j = \Pi^e A_e \in \Pi^e \mathfrak{O}.$$

By (5.5) we have

$$\Pi^e A_e = \pi A^{(1)}$$

for some $A^{(1)} \in \mathfrak{O}$. We now start again with $A^{(1)}$ instead of A. In this way we find in succession linear combinations C_r of $\mathcal{B}$ with coefficients in $\mathfrak{O}$ such that

$$A - C_o - \pi C_1 \dots - \pi^s C_s \in \pi^{s+1} \mathfrak{O}$$

for every s. On letting $s \to \infty$ and using the completeness of k, we express A as a linear combination of $\mathcal{B}$ with coefficients in $\mathfrak{O}$, as required. We have thus shown that $\mathcal{B}$ is an $\mathfrak{o}$-basis for $\mathfrak{O}$ and _a fortiori_ a k-basis for K. This completes the proof.

THEOREM 5.1 _Let_ k _be complete with respect to the discrete valuation_ $|\ |$ _and let_ K _be an extension with finite relative degree_

$$n = [K : k].$$

Then

$$n = ef.$$

Proof. Follows at once from the previous lemma and Theorem 1.1.

<u>COROLLARY</u>. K/k <u>is unramified (Definition 4.1) precisely when</u> $e = 1$, <u>and is completely ramified (Definition 4.2) precisely when</u> $e = n$.

<u>Note</u> Definitions 4.1, 4.2 are valid whether or not $|\ |$ is discrete, and so in circumstances not covered by the above Corollary.

6 DISCRIMINANTS

In the case of valued fields there is a notion of discriminant of a finite algebraic extension K/k which is more precise than the usual notion, which we shall here call <u>field-discriminant</u>. We first review the theory of the latter.

Let $K \supset k$ be fields with

$$[K : k] = n < \infty \tag{6.1}$$

We recall that the <u>trace</u> $S_{K/k}(A)$ of an element $A \in K$ is defined to be the trace of the k-linear map

$$B \to AB \qquad (B \in K)$$

of K into itself. The trace is a k-linear map of K into k. For a more detailed discussion cf. Appendix B.

Let $A_1,\ldots,A_n$ be any k-basis of K. We write

$$D(A_1,\ldots,A_n) = \det(S(A_iA_j))_{i,j} \tag{6.2}$$

where $S = S_{K/k}$ is the trace. Any other basis $B_1,\ldots,B_n$ is of the shape

$$B_i = \sum_j t_{ij} A_j, \tag{6.3}$$

where

$$t_{ij} \in k, \quad T \text{ (say)} = \det(t_{ij}) \neq 0. \tag{6.4}$$

Clearly,

$$D(B_1,\ldots,B_n) = T^2 D(A_1,\ldots,A_n) \tag{6.5}$$

by the k-linearity of the trace.

Now suppose that K/k is separable and let N be a finite normal extension of k which contains K. There are then n embeddings

$$\sigma_\ell : K \to N \quad (1 \leq \ell \leq n) \tag{6.6}$$

of K in N which are the identity on k. If $K = k(C)$, these are given by

$$C \longmapsto C^{(\ell)} \tag{6.7}$$

where $C = C^{(1)}, C^{(2)}, \ldots, C^{(n)}$ are the conjugates of C over k. For any basis $A_1, \ldots, A_n$ of K/k we put

$$\Delta(A_1, \ldots, A_n) = \det(\sigma_\ell A_j)_{\ell,j}. \tag{6.8}$$

Since the order of the σ_ℓ is arbitrary, Δ is defined only up to sign. In any case,

$$\begin{aligned} \{\Delta(A_1, \ldots, A_n)\}^2 &= \det(\sum_\ell \sigma_\ell A_i A_j)_{i,j} \\ &= \det(S(A_i A_j))_{i,j} \\ &= D(A_1, \ldots, A_n). \end{aligned} \tag{6.9}$$

In particular, we have

$$\begin{aligned} \Delta(1, C, \ldots, C^{n-1}) &= \prod_{i>j} (\sigma_i C - \sigma_j C) \\ &\neq 0. \end{aligned} \tag{6.10}$$

Hence and by (6.4), (6.5)

$$D(A_1, \ldots, A_n) \neq 0 \tag{6.11}$$

for all K/k bases $A_1, \ldots, A_n$. From (6.5) and (6.11) we have

LEMMA 6.1. Let K/k be separable. Then the class of $k^*/(k^*)^2$ given by $D(A_1,\dots,A_n)$ is the same for all K/k bases $A_1,\dots,A_n$.

Note We recall that k^* is, by definition, the multiplicative group of non-zero elements of k.

DEFINITION 6.1. The element of $k^*/(k^*)^2$ just defined is the field-discriminant.

Now suppose that k is complete with respect to a discrete valuation $|\ |$. Then we can consider $\mathfrak{o}$-bases $A_1,\dots,A_n$ of $\mathfrak{O}$. If $B_1,\dots,B_n$ is another such basis then in (6.3) we have $t_{ij} \in \mathfrak{o}$ and so $T \in \mathfrak{o}$. The inverse transformation to (6.3) has determinant T^{-1}, so $T^{-1} \in \mathfrak{o}$. Hence $|T| = 1$ or, in other words, $T \in U$, where U is the group of units of k. Hence we have

LEMMA 6.2. Suppose that k is complete with respect to the discrete valuation $|\ |$ and that K/k is separable. Then $D(A_1,\dots,A_n)$ for all $\mathfrak{o}$-bases $A_1,\dots,A_n$ of $\mathfrak{O}$ lies in the same non-zero class of $\mathfrak{o}$ modulo U^2.

DEFINITION 6.2. The class of $\mathfrak{o}$ modulo U^2 just defined is the discriminant of K/k and will be denoted $D_{K/k}$.

Note In particular $|D(A_1,\dots,A_n)|$ is the same for all $\mathfrak{o}$-bases $A_1,\dots,A_n$. We shall denote it by $|D_{K/k}|$.

We can now enunciate the key result.

THEOREM 6.1 Suppose that k is complete with respect to the discrete valuation $|\ |$. Suppose that K/k is separable and also that the corresponding residue class P/ρ is separable. Then $|D_{K/k}| = 1$ precisely when K/k is unramified.

Proof. (i) Suppose that K/k is ramified, so we have a basis

$$\mathcal{B} = \{B_i \Pi^j : 1 \leq i \leq f,\ \ 0 \leq j \leq e - 1\} \qquad (6.11.\ \text{bis})$$

of $\mathcal{O}$ with $e > 1$. The valuation $|\ |$ extends to the normal extension N of k containing K and in the notation (6.6) we have

$$|\sigma_\ell(B_i\,\Pi^j)| = |B_i\,\Pi^j| = |\Pi|^j \tag{6.12}$$

by Theorem 1.1, Corollary 2. Hence a whole column of the matrix in (6.8) which defines $\Delta(\mathcal{B})$ has value < 1. Thus $|\Delta(\mathcal{B})| < 1$ and

$$|D_{K/k}| = |\Delta(\mathcal{B})|^2 < 1. \tag{6.13}$$

(ii) Suppose that K/k is unramified.

Denote the map from $\mathcal{O}$ to $P = \mathcal{O}/\mathfrak{P}$ by a bar ($\bar{\ }$). We shall show below (Lemma 6.3) that

$$\overline{S_{K/k}(A)} = S_{P/\rho}(\bar{A}) \tag{6.14}$$

for all $A \in \mathcal{O}$. Using a suffix K/k or P/ρ to denote the field extension under consideration, it follows from the definition (6.2) that

$$\overline{D_{K/k}(B_1,\ldots,B_n)} = D_{P/\rho}(\bar{B}_1,\ldots,\bar{B}_n). \tag{6.15}$$

The right hand side is $\neq 0$ by Lemma 6.1 applied to the extension P/ρ. But then (6.15) gives

$$|D_{K/k}(B_1,\ldots,B_n)| = 1,$$

as required.

<u>COROLLARY</u>.

$$|D_{K/k}| \leq |\pi|^{(e-1)f} \tag{6.16}$$

<u>Proof</u>. By (6.12) f of the columns defining $\Delta(\mathcal{B})$ are divisible by Π^j for $j = 1, 2, \ldots, e-1$. Hence

$$|D(\mathcal{B})| = |\Delta(\mathcal{B})|^2 \leq |\Pi|^{2(1+2+\ldots+(e-1))f}$$

$$= |\Pi|^{e(e-1)f}$$

$$= |\pi|^{(e-1)f}.$$

It remains to justify (6.14). We prove more generally

LEMMA 6.3. *Suppose that* K/k *is unramified and* P/ρ *is separable. For any* $A \in \mathfrak{O}$ *the characteristic equation of* $\bar{A} \in P$ *is obtained from that of* A *by applying the map* $\mathfrak{o} \to \mathfrak{o}/\mathfrak{p}$ *to the coefficients.*

Proof. Since $B_1,\ldots,B_n$ is a basis for K/k, the characteristic equation of A is obtained by eliminating $B_1,\ldots,B_n$ from the equations

$$AB_i = \sum_j a_{ij} B_j \quad (a_{ij} \in k).$$

(Appendix B). Since $B_1,\ldots,B_n$ is an $\mathfrak{o}$-basis for $\mathfrak{O}$, we have $a_{ij} \in \mathfrak{o}$ and can map into the residue classfields:

$$\bar{A}\bar{B}_i = \sum_j \bar{a}_{ij} \bar{B}_j.$$

But $\bar{B}_1,\ldots,\bar{B}_n$ is a basis of P/ρ and the result follows.

It is convenient to prove here a very special result which we shall want to quote later.

LEMMA 6.4. *Let* k *be a finite extension of* $\mathbb{Q}_2$. *Then*

$$D_{k/\mathbb{Q}_2} \equiv 1 \text{ or } 0 \quad (4) \tag{6.17}$$

Note By definition, $D_{k/\mathbb{Q}_2}$ is an element of $\mathbb{Z}_2$ modulo U^2 (multiplicatively), where U is the group of 2-adic units. Since U^2 is precisely the set of $v \in \mathbb{Z}_2$ satisfying $v \equiv 1$ (8), the congruence (6.17) makes sense.

Proof. Suppose, first, that $k/\mathbb{Q}_2$ is ramified. Then $D_{k/\mathbb{Q}_2} \equiv 0$ (4) by (6.16) except, possibly, when $e = 2$, $f = 1$. Then k is a quadratic extension of $\mathbb{Q}_2$. Since it is ramified, it must be one of

$$\mathbb{Q}_2(\sqrt{\pm 2}), \quad \mathbb{Q}_2(\sqrt{\pm 6}), \quad \mathbb{Q}_2(\sqrt{-1}), \quad \mathbb{Q}_2(\sqrt{-5})$$

by Chapter 4, Lemma 3.3., Corollary. It is readily verified that (6.17) holds in these cases also (cf. Exercise 5).

Suppose, now, that $k/\mathbb{Q}_2$ is unramified, and let $\mathcal{B}$ be a $\mathbb{Z}_2$-basis of the integers of k. Then

$$D(\mathcal{B}) = \Delta(\mathcal{B})^2 \equiv 1 \quad (2),$$

where $\Delta(\mathcal{B}) \in k$. Hence either $\Delta(\mathcal{B}) \in \mathbb{Q}_2$ or $\mathbb{Q}_2(\Delta(\mathcal{B}))$ is an <u>unramified</u> quadratic extension. In the first case, we have $D(\mathcal{B}) \equiv 1 \quad (8)$ and in the second we have $\mathbb{Q}_2(\Delta(\mathcal{B})) = \mathbb{Q}_2(\sqrt{5})$ and so $D(\mathcal{B}) \equiv 5 \quad (8)$.

7 COMPLETELY RAMIFIED EXTENSIONS

There is a very useful explicit description of totally ramified extensions of discrete valued fields. We recall that an <u>Eisenstein polynomial</u> is one satisfying the conditions of Theorem 2.1 of Chapter 6.

<u>THEOREM 7.1</u> *Let $|\ |$ be discrete on k. A necessary and sufficient condition that a finite algebraic extension K/k be completely ramified is that $K = k(B)$, where B is a root of an Eisenstein polynomial.*

<u>Proof</u>. (i) Suppose, first, that B is the root of an Eisenstein polynomial, say

$$f_o + f_1 B + \ldots + f_n B^n = 0,$$

where

$$|f_n| = 1; \qquad |f_j| < 1 \quad (j < n); \qquad |f_o| = \pi$$

Then $|B|^n = |\pi|$. Hence $e(k(B)/k) \geq n$.

(ii) Suppose that K/k is completely ramified, $[K:k] = n$ and let Π be a prime element of K. Then $1, \Pi, \ldots, \Pi^{n-1}$ are linearly independent over k because their values are in distinct cosets of the value group G_K modulo G_k. There must be an equation

$$\Pi^n + f_{n-1}\Pi^{n-1} + \ldots + f_o = 0. \quad (f_j \in k).$$

Here $|f_j| < 1$ because two of the summands must have the same value, and $|f_o| = |\Pi|^n = |\pi|$.

8 ACTION OF GALOIS

Let the field k be complete with respect to the discrete valuation $|\ |$ and let K be a normal extension of k with galois group Γ and finite relative degree

$$n = [K:k]. \tag{8.1}$$

The existence of the valuation severely restricts the possibilities for Γ and for its action on k.

By Theorem 1.1, Corollary 2 we have

$$|\sigma A| = |A| \quad (\text{all } \sigma \in \Gamma, A \in K). \tag{8.2}$$

In particular

$$\sigma\mathfrak{O} = \mathfrak{O}; \quad \sigma\mathfrak{P} = \mathfrak{P}. \tag{8.3}$$

Define subgroups Δ_j $(j = 0, 1, 2, \ldots)$ of Γ by

$$\sigma \in \Delta_j \iff |\sigma A - A| < |\Pi|^j \quad (\text{all } A \in \mathfrak{O}) \tag{8.4}$$

The group Δ_o is called the inertia group (German: Trägheitsgruppe), Δ_1 is the ramification group (German: Verzweigungsgruppe), and the Δ_j $(j > 1)$ are the higher ramification groups.

<u>LEMMA 8.1</u>. <u>The</u> Δ_j <u>are normal subgroup of</u> Γ. <u>Further</u>, $\Delta_j = \{1\}$ <u>for all large enough</u> j.

Proof. (i) For $\sigma, \tau \in \Gamma$ we have

$$|\tau^{-1}\sigma\tau A - A| = |\sigma A' - A'| \quad (A' = \tau A) \tag{8.5}$$

by (8.2) Hence $\tau^{-1}\sigma\tau \in \Delta_j$ precisely when $\sigma \in \Delta_j$.

(ii) Let $\sigma \neq 1$. Then there is some A (depending on σ) such that $\sigma A \neq A$. Hence $\sigma \notin \Delta_j$ as soon as $|\sigma A - A| \geq |\Pi|^j$.

THEOREM 8.1 Let K/k be normal. Suppose that P/ρ is separable. Then

(i) P/ρ is normal with galois group Γ/Δ_o.

(ii) The fixed field of Δ_o is the maximal unramified extension $L \subset K$ of k.

(iii) For $j \geq 1$ the group Δ_j is the set of $\sigma \in \Delta_o$ such that

$$|\sigma\Pi - \Pi| < |\Pi|^j.$$

Proof. Let L, $k \subset L \subset K$ be the maximal unramified extension of k given by Theorem 4.1, Corollary 2. Then σL $(\sigma \in \Gamma)$ is unramified over k by (8.2) so $\sigma L = L$: that is, L is a normal extension of k.

Let $P = \rho(\alpha)$ and let $A \in L$ be a lift of α to L. Since L/k is normal, there are f conjugates $A = A^{(1)}, A^{(2)}, \ldots, A^{(f)}$ of A in L and $\prod_j (X - A^{(j)}) \in \mathfrak{o}[X]$. Denoting the map $\mathfrak{O} \to P = \mathfrak{O}/\mathfrak{P}$ by a bar, we see that α is a root of $\prod(X - \bar{A}^{(j)}) \in \rho[X]$. Hence P/ρ is normal, the $\bar{A}^{(j)}$ are distinct, and they are the conjugates of α over ρ. Hence the galois groups of P/ρ and L/k are naturally isomorphic, the isomorphism $\alpha = \bar{A}^{(1)} \to \bar{A}^{(j)}$ of P/ρ correspond to $A = A^{(1)} \to A^{(j)}$ for L/k.

Since the $\sigma \in \Delta_o$ are by definition precisely the $\sigma \in \Gamma$ which induce the identity on P, it follows that the automorphism group of P/ρ is precisely Γ/Δ_o. By what is said above, this is also the galois group of L/k. This gives (i) and (ii) of the Theorem.

Finally, $\mathfrak{O} = \mathfrak{o}_L[\Pi]$ by Lemma 5.4 where $\mathfrak{o}_L$ is the integers of L. Since Δ_o leaves $\mathfrak{o}_L$ elementwise fixed, this gives (iii) of the enunciation.

COROLLARY 1. (i) Δ_o/Δ_1 is isomorphic to a subgroup of P^*, (ii) For $j > 0$, the quotient Δ_j/Δ_{j+1} is isomorphic to a subgroup of the additive group P^+ of P.

Proof. (i) Let $\sigma \in \Delta_o$. Then $\sigma\Pi$ is a prime element of K, so

$$\sigma\Pi = A_\sigma\Pi \tag{8.6}$$

where A_σ is a unit. Further, $\sigma \in \Delta_1$ precisely when $A_\sigma \equiv 1 \ (\mathfrak{P})$. Let also $\tau \in \Delta_o$. Then

$$(\tau\sigma)\Pi = (\tau A_\sigma)\tau\Pi$$

$$= (\tau A_\sigma)A_\tau\Pi. \tag{8.7}$$

Here $\tau A_\sigma \equiv A_\sigma(\mathfrak{P})$ since $\tau \in \Delta_o$. Hence the map

$$\sigma \to \bar{A}_\sigma \quad (= A_\sigma \bmod \mathfrak{P}) \tag{8.8}$$

is the required injection of Δ_o/Δ_1 into P^*.

(ii) Let $\sigma \in \Delta_j$, where $j > 0$. Then

$$\sigma\Pi = \Pi + A_\sigma \Pi^{j+1} \tag{8.9}$$

where $A_\sigma \in \mathfrak{O}$. The argument is now similar to that for (i).

In a later application we shall need

COROLLARY 2. Suppose that the galois group Γ is abelian. Then $\bar{A}_\sigma \in \rho$ where $\sigma \in \Delta_o$ and $\bar{A}_\sigma$ is given by (8.6) and (8.8). Hence Δ_o/Δ_1 is isomorphic to a subgroup of ρ^*

Proof. The definition (8.6) makes sense for all $\sigma \in \Gamma$, and (8.7) holds for all $\sigma,\tau \in \Gamma$: that is

$$A_{\tau\sigma} = (\tau A_\sigma)A_\tau. \tag{8.10}$$

If $\tau\sigma = \sigma\tau$, it follows that

$$(\tau A_\sigma)A_\tau = (\sigma A_\tau)A_\sigma.$$

Now restrict σ to $\sigma \in \Delta_0$. Then $\sigma A_\tau \equiv A_\tau(\mathfrak{P})$ and so (since A_τ is a unit)

$$\tau A_\sigma \equiv A_\sigma \quad (\mathfrak{P}) \quad (\text{all } \tau \in \Gamma).$$

But this implies $\bar{A}_\sigma \in \rho$ by (i) of Theorem 8.1.

Note Those conversant with group cohomology will recognize that (8.10) states that $\{A_\sigma\}$ is a cocycle.

Notes

§4 There is some discussion of the "pathological" cases mentioned before Theorem 4.1 in Artin (1951).

Witt (1937) has shown that every perfect field ρ of characteristic p is the residue class field of a field k of characteristic 0 complete with respect to a valuation having p for prime element. The construction is ingenious and quite explicit ("Witt vectors"). There are accounts in Hasse (1949) and Serre (1962).

§5 For a fuller discussion of ramification, see Serre (1962) and Serre's contribution to the Brighton book (Cassels and Fröhlich (1967)).

Exercises

1. Suppose that $\mathfrak{o}$ is compact (Lemma 1.5 of Chapter 4) and let $A_1,\ldots,A_n$ be a k-basis of the extension K of k. Show that there is a real $c > 0$, depending only on $A_1,\ldots,A_n$ such that

$$|N_{K/k}(a_1A_1 + \ldots + a_nA_n)| \geq c \max_j |a_j|^n$$

for all $a_1,\ldots,a_n \in k$.

[Hint The set of $(a_1,\ldots,a_n)$ with $\max|a_j| = 1$ is compact.]

2. Under the conditions of the preceding exercise show that $\|A\| = |N_{K/k}(A)|^{1/n}$ is a valuation on K by generalizing the proof of

Lemma 2.3 of Chapter 3.

[Crib. Chapter 2 of Cassels and Fröhlich (1967)].

3. Assumptions and notation as in Strassmann's theorem and Exercise 2 of Chapter 6. Show that there is some finite algebraic extension K/k such that the sum of the multiplicities of the zeros of $f(X)$ in $\mathfrak{O}$ is precisely N.

4. Let $A \in K$ have minimal polynomial $G(X) \in k[X]$ over k, where the highest coefficient of G is 1. Show that $A \in \mathfrak{O}$ if and only if $G(X) \in \mathfrak{o}[X]$.

5. (i) Show that there are precisely 7 extensions K of $\mathbb{Q}_2$ with $[K:Q_2] = 2$, namely $\mathbb{Q}_2(\delta)$ where $\delta^2 = d$ and $d = -1, \pm 2, \pm 5, \pm 10$.

(ii) In every case show that the extension of $|\ | = |\ |_2$ to K is given by

$$|a + b\delta| = |a^2 - db^2|^{\frac{1}{2}} \quad (a,b \in \mathbb{Q}_2)$$

(iii) Show that the only unramified $K/\mathbb{Q}_2$ is that with $d = 5$.

(iv) Show that $1, \frac{1}{2}(1 + \delta)$ is an $\mathbb{Z}_2$-basis for $\mathfrak{O}$ when $d = 5$ but that otherwise a basis is $1, \delta$.

(v) Compute $|D_{K/\mathbb{Q}_2}|$ and confirm that Theorem 6.1 is satisfied.

6. Let $K = k(A)$ for some field k where A is a root of the irreducible polynomial

$$F(X) = X^n + f_{n-1} X^{n-1} + \dots + f_o \in k[X].$$

In the notation of (6.2) show that $D(1,A,\dots,A^{n-1})$ is equal to the discriminant of F.

7. Obtain the following generalization of Lemma 6.3. Suppose that P/ρ is separable but $e \geq 1$. Let $A \in \mathfrak{O}$ have characteristic polynomial $F(X) \in k[X]$ over k and let $\bar{A} \in P$ have characteristic polynomial $\phi(X) \in \rho[X]$ over ρ. Then

$$\bar{F}(X) = \{\phi(X)\}^e.$$

8. Let K/k be unramified and suppose that the residue class field extension P/ρ is normal. Show that K/k is normal.

9. In the notation of §8, show that Δ_j/Δ_{2j} is abelian for all j.

10. (i) Let θ be a root of $G(X) = X^p - X - 1 \in \mathbb{Q}_p[X]$. Show that $\mathbb{Q}_p(\theta)$ is normal over $\mathbb{Q}_p$, and that it is an unramified extension of $\mathbb{Q}_p$ of degree p.

(ii) Let $a \in \mathbb{Q}_p$ be a unit and let ϕ be a root of $X^p - X - a = 0$. Show that $\mathbb{Q}_p(\phi) = \mathbb{Q}_p(\theta)$.

11. (i) Let $p \neq 2$. Show that $\mathbb{Q}_p(\zeta)$, where $\zeta^p = 1$, $\zeta \neq 1$, is completely ramified and that $\lambda = 1 - \zeta$ is a prime element.

(ii) Let $\sigma: \zeta \to \zeta^g$ be an automorphism of $\mathbb{Q}_p(\zeta)/\mathbb{Q}_p$. Show that

$$\sigma\lambda \equiv g\lambda \quad (\lambda^2).$$

(iii) Show that

$$\lambda^{-p+1} p \equiv -1 \quad (\lambda).$$

[Hint. $p = \Pi(1 - \zeta^j)$].

(iv) Let $\alpha \in \mathbb{Q}_p(\zeta)$ be a unit, $\alpha \equiv 1(\lambda)$. Show that $\zeta^m\alpha \equiv 1\ (\lambda^2)$ for some $m \in \mathbb{Z}$.

(v) Let $\beta \in \mathbb{Q}_p(\zeta)$ be a unit, $\beta \equiv 1\ (\lambda)$. Show that a necessary and sufficient condition that $\beta = \alpha^p$ for some $\alpha \in \mathbb{Q}_p(\zeta)$ is that $\beta \equiv 1 \ (\lambda^{p+1})$.

[Hint Necessity, cf. (iv). Sufficiency, apply Hensel's lemma to $(1 + \lambda Y)^p - \beta$, using (iii)].

(vi) Let $\beta \equiv 1 \ (\lambda^p)$, $\beta \not\equiv 1 \ (\lambda^{p+1})$. Show that $\mathbb{Q}_p(\zeta,\theta)$, where $\theta^p = \beta$, is an unramified extension of $\mathbb{Q}_p(\zeta)$ of degree p.

[Hint Previous hint and Exercise 10].

12. Let ζ be a primitive p^m-th root of 1. Show that $\{\zeta^j\}$ $(0 \leq j < p^m - p^{m-1})$ is a $\mathbb{Z}_p$-basis of the integers of $Q_p(\zeta)$.

[Hint Show first that $\{\lambda^j\}$ is a basis, where $\lambda = 1 - \zeta$].

13. (i) Show that $k = \mathbb{Q}_2(\delta)$ is completely ramified of degree 4 where $\delta^2 = 1 + \gamma$, $\gamma^2 = -2$.

(ii) Show that $1, \gamma, \delta, \gamma\delta$ is a $\mathbb{Z}_2$-basis for the integers $\mathfrak{o}$ of k.

13 bis. Determine a prime element and the ramification groups of $\mathbb{Q}_2(2^{\frac{1}{2}},3^{\frac{1}{2}})$.

14. Show that an extension k of $\mathbb{Q}_3$ of degree 6 is given by $k = \mathbb{Q}_3(\delta,\theta)$ where $\delta^3 = 3$, $\theta^2 = 2 - \delta$. Show that $e = 3$, $f = 2$. Find an unramified extension L of Q_3 of degree 2 with $L \subset k$.

15. Show that the quadratic extension $k = \mathbb{Q}_7(\theta)$ is completely ramified, where $\theta^2 = \theta - 2$. By using Strassmann's Theorem for k, obtain another proof of Lemma 6.1 of Chapter 4.

[Note This is the approach of Nagell (1960)]

16. Let k be any field of characteristic $\neq 3$ and suppose that $d = D^3 \in k$, $D \notin k$. Show that

$$N_{k(D)/k}(a + bD + cD^2) = a^3 + db^3 + d^2c^3 - 3abcd$$

for $a,b,c \in k$.

17. Let $K/\mathbb{Q}_p$ be unramified of degree n and let $A_1,\ldots,A_n$ be a $\mathbb{Z}_p$-basis of the integers of K.

(i) Show that

$$N_{K/\mathbb{Q}_p}(a_1A_1 + \ldots + a_nA_n) = G(a_1,\ldots,a_n) \quad (a_1,\ldots,a_n \in \mathbb{Q}_p)$$

where

$$G(X_1,\ldots,X_n) \in \mathbb{Z}_p[X_1,\ldots,X_n]$$

is homogeneous of degree n.

(ii) Show that $|G(a_1,\ldots,a_n)| < 1$ for $a_1,\ldots,a_n \in \mathbb{Z}_p$ implies $|a_j| < 1$ $(1 \leq j \leq n)$.

(iii) Define the form $H(X_{11},\ldots,X_{nn})$ in the n^2 indeterminates $X_{11},\ldots,X_{nn}$ by

$$H(X_{11},\ldots,X_{nn}) = \sum_{j=1}^{n} p^{j-1} H(X_{j1},\ldots,X_{jn}).$$

Show that $H(a_{ij}) = 0$ for $a_{ij} \in \mathbb{Q}_p$ implies $a_{ij} = 0$ (all i,j).

[Note It was long believed that a form of degree n in $> n^2$ variables always represented 0; and this was ascribed to E. Artin as a conjecture. It was disproved by Terjanian (1966) and is now known to be very false, see Arhipov and Karacuba (1981, 1982), Lewis and Montgomery (1983), Brownawell (1984). On the other hand, Ax and Kochen (1965) have shown that for given n there are only finitely many p for which the conjecture fails.].

18. Let K/k be a finite algebraic extension, where k is complete with respect to a discrete valuation. Let Π be a prime element of K.

(i) Let $A \in \mathfrak{O}$ and suppose that $A \bmod \mathfrak{P}$ generates the residue class field extension P/ρ in the notation of §4. Let $\phi(T) \in \rho[T]$ be the minimum polynomial of $A \bmod \mathfrak{P}$ and let $\Phi(T) \in \mathfrak{o}[T]$ be any lift of $\phi(T)$. If $|\Phi(A)| = |\Pi|$, show that $\mathfrak{O} = \mathfrak{o}[A]$.

(ii) If $|\Phi(A)| \neq |\Pi|$, show that $|\Phi(A + \Pi)| = |\Pi|$. Deduce that there is a $B \in \mathfrak{O}$ such that $1, B, \ldots, B^{n-1}$ is an $\mathfrak{o}$-basis of $\mathfrak{O}$, where $n = [K:k]$. (Here P/ρ is separable).

19. Suppose that $\mathfrak{O} = \mathfrak{o}[B]$, where B has minimal polynomial $F(X) \in \mathfrak{o}[X]$. Show that for $C \in K$ a necessary and sufficient condition that

$$S(AC) \in \mathfrak{o} \quad (\text{all } A \in \mathfrak{O})$$

is that

$$F'(B)C \in \mathfrak{O}.$$

[Note Here $S = S_{K/k}$ is the trace.

Hint Put

$$F'(B)C = G(B)$$

where $G(X) \in K[X]$ has lower degree than $F(X)$. Show that

$$\begin{aligned} G(X) &= \sum_j \frac{G(B_j)}{F'(B_j)} \frac{F(X)}{X - B_j}, \\ &= \sum_j C_j \sum_i F_i (X^{i-1} + \ldots + B_j^{i-1}) \end{aligned}$$

where the B_j are the roots of $F(X)$ in some overfield $\mathbb{K}$ of K. On equating coefficients of powers of X, determine the $S(B^i C)$ $(i = 0, 1, 2, \ldots)$ recursively in terms of the coefficients of $F(X)$ and $G(X)$.]

20. This exercise introduces the <u>different</u>, which is an invariant of a field extension related to, but more subtle than, the

discriminant. Define $d \geq 0$ by

$$|\Pi|^{-d} = \sup \{|C| : S(AC) \in \mathfrak{o} \text{ for all } A \in \mathfrak{O}\}. \qquad (£)$$

The different $\mathfrak{D} = \mathfrak{D}_{K/k}$ is the element of $\mathfrak{O}$ modulo units determined by Π^d. [In the language of divisors, which will be introduced in Chapter 10, one writes $\mathfrak{D} = \mathfrak{P}^d$.]

(i) Show that

$$|\Pi|^d = \sup_A |F_A'(A)|$$

where $F_A(X)$ is the characteristic polynomial of A over k and A runs through $\mathfrak{O}$.

(ii) Show that

$$|D_{K/k}| = |N_{K/k}\Pi|^d .$$

[Note Here $D_{K/k}$ is the discriminant. In the language of divisors this is usually written $D_{K/k} = N_{K/k}\,\mathfrak{D}_{K/k}$.

Hint Compare (6.10)].

(iii) Show that $d = 0$ if and only if K/k is unramified.

(iv) Let $k \subset L \subset K$. Show that

$$d_{K/k} = d_{K/L} + [K:L]\, d_{L/k}$$

where, e.g., $d_{K/L}$ refers to the extension K/L.

[Note In divisorial notation this is written $\mathfrak{D}_{K/k} = \mathfrak{D}_{K/L}\,\mathfrak{D}_{L/k}$.

Hint Use $S_{K/k}(A) = S_{L/k}(S_{K/L}(A))$. Consider the generalization of (£) in which "$S(AC) \in \mathfrak{o}$" is replaced by $|S(AC)| \leq |\pi|^e$, where e is fixed; and argue from the definition.]

CHAPTER EIGHT: $\mathfrak{p}$-ADIC FIELDS

1 INTRODUCTION

In this Chapter we study the complete valued fields which will arise when we look at algebraic number fields in Chapter 10. The first two sections are straightforward applications and amplifications of the results of the previous Chapter and prepare the way for Chapter 10. The remainder of the Chapter gives a couple of results which are important in the further development of the theory but which are not required further in this book.

We start by defining the fields we shall be considering:

DEFINITION 1.1. Let the field k be complete with respect to the (non-arch.) valuation $|\ |$. We say that k is a $\mathfrak{p}$-*adic field* if

(i) k has characteristic 0

(ii) $|\ |$ is discrete

(iii) the residue class field ρ is finite.

We can give at once an alternative characterization:

LEMMA 1.1. *The valued field* k *is a* $\mathfrak{p}$-*adic field if and only if it is a finite extension of* $\mathbb{Q}_p$ *for some* $\mathfrak{p}$.

Proof. (i) Suppose that k is a finite extension of $\mathbb{Q}_p$. Then it is a $\mathfrak{p}$-adic field by Lemma 4.1 and 5.1 of Chapter 7.

(ii) Let k be a $\mathfrak{p}$-adic field. Then $k \supset \mathbb{Q}$ by (i) of the definition. Since the residue class field is finite by (iii) of the definition, it has characteristic p for some prime p. Hence the valuation on k induces a valuation equivalent to the p-adic valuation.

Hence $\mathbb{Q}_p \subset k$, since k is complete. We are now in the situation described by Lemma 5.4 of Chapter 7 and conclude that $[k : \mathbb{Q}_p] < \infty$.

An alternative characterization is given by

LEMMA 1.2. *A field k of characteristic 0 complete with respect to a non-arch. valuation is a $\mathfrak{p}$-adic field if and only if its ring $\mathfrak{o}$ of integers is compact.*

Proof. Lemma 1.5 of Chapter 4.

A natural normalization of the valuation on a $\mathfrak{p}$-adic field k is the extension of the valuation $|\ |_p$ on the $\mathbb{Q}_p \subset k$. There is, however, another normalization which will turn out to be the appropriate one in Chapter 10 and which we introduce briefly here.

DEFINITION 1.1. Let q be the cardinality of the residue class field of the $\mathfrak{p}$-adic field k. The *renormalized valuation* $\|\ \|_k$ on k is determined by

$$\|\pi\|_k = q^{-1}, \tag{1.1}$$

where π is a prime element.

Note When $k = \mathbb{Q}_p$ this coincides with $|\ |_p$.

LEMMA 1.3. *Suppose that* $[k : \mathbb{Q}_p] = n$. *Then*

$$\|a\|_k = |a|^n, \tag{1.2}$$

where $|\ |$ *is the valuation which coincides with* $|\ |_p$ *on* $\mathbb{Q}_p$.

Proof. It is enough to verify (1.2) for one non-unit a, and we choose $a = p$. We have

$$\|p\|_k = \|\pi\|_k^e,$$

where e is the ramification of $k/\mathbb{Q}_p$. Further, $q = p^f$, where f is the degree of the residue class field extension. Hence

$$\|p\|_k = q^{-e} = p^{-ef} = p^{-n}.$$

<u>COROLLARY 1</u>.

$$\|a\|_k = |N_{k/\mathbb{Q}_p}(a)|_p.$$

<u>Proof</u>. Cf. Theorem 1.1 of Chapter 7.

We note that $\|a\|_k$ depends on k as well as on a:

<u>COROLLARY 2</u>. <u>Let</u> $a \in k \subset K$. <u>Then</u>

$$\|a\|_K = \|a\|_k^{[K:k]}.$$

<u>Proof</u>. Clear.

For convenience we mention here the appropriate renormalization of the ordinary absolute value $|\ |_\infty$. The only complete fields to consider are $\mathbb{R}$ and $\mathbb{C}$.

<u>DEFINITION 1.2</u>.

$$\|a\|_{\mathbb{R}} = |a|_\infty \quad \text{and} \quad \|a\|_{\mathbb{C}} = |a|_\infty^2.$$

The analogues of the two Corollaries above hold:

<u>LEMMA 1.4</u>.

(i) $\|A\|_{\mathbb{C}} = \|N_{\mathbb{C}/\mathbb{R}}(A)\|_{\mathbb{R}}$ for $A \in \mathbb{C}$

(ii) $\|a\|_{\mathbb{C}} = \|a\|_R^2$ <u>for</u> $a \in \mathbb{R} \subset \mathbb{C}$.

<u>Proof</u>. Clear.

In the rest of this Chapter we shall not, however, use the renormalized valuation. In §2 we discuss unramified extensions, as more precise statements can be made than follow from the results of Chapter 7. In §§3,4 we consider two results which, while important in some areas of application will not be used later: These two §§ can be omitted at a first reading.

2 UNRAMIFIED EXTENSIONS

For $\mathfrak{p}$-adic fields we can add a little to what was done in §4 of Chapter 7.

LEMMA 2.1. *For each* $n = 1,2,...$ *there is precisely one unramified extension* k *of* $\mathbb{Q}_p$ *with* $[k : \mathbb{Q}_p] = n$. *It is the splitting field of*

$$X^q - X, \quad q = p^n. \tag{2.1}$$

over $\mathbb{Q}_p$.

Proof. The residue class field of $\mathbb{Q}_p$ is the finite field $\mathbb{F}_p$ of p elements. By the theory of finite fields, for every n there is precisely one extension ρ of $\mathbb{F}_p$ of degree n. It has q elements and the multiplicative group ρ^* of non-zero elements is cyclic, so $\alpha^q = \alpha$ for all $\alpha \in \rho$ and ρ is the splitting field of (2.1) over $\mathbb{F}_p$.

By Theorem 4.1 Corollary 3, there is precisely one unramified field extension k of $\mathbb{Q}_p$ whose residue class field is ρ. Put

$$f(X) = X^q - X, \tag{2.2}$$

so

$$f'(X) = qX^{q-1} - 1$$

and

$$|f'(a)| = 1 \quad (\text{all} \quad a \in \mathfrak{o}).$$

Hence by Hensel's Lemma for every $\alpha \in \rho = \mathfrak{o}/\mathfrak{p}$ there is some $\hat{\alpha} \in \alpha \subset \mathfrak{o}$ such that $f(\hat{\alpha}) = 0$. Hence (2.1) is split by k. The splitting field of

(2.1) over $\mathbb{Q}_p$ cannot be smaller than k because its residue class field must contain at least q elements. This concludes the proof.

DEFINITION 2.1. The $\hat{\alpha} \in \alpha$ defined above is the Teichmüller representative of α.

We note that the map $\alpha \to \hat{\alpha}$ gives an injection of ρ into k which respects multiplication (but not, of course, addition).

COROLLARY 1. *Let k be a $\mathfrak{p}$-adic field and let the cardinality of its residue classfield be q. For every $n = 1,2,\ldots$ there is precisely one unramified extension K of k of relative degree n. It is the splitting field over k of*

$$X^Q - X, \quad Q = q^n. \tag{2.3}$$

The extension K/k is normal with cyclic galois group.

There is a generator σ of this group which induces the automorphism $\beta \mapsto \beta^q$ of the residue classfield P of K.

DEFINITION 2.2. The σ just defined is the Frobenius automorphism of K/k.

Proof. The residue class field P must be the field of cardinality Q. Hence K must contain the field L given by Lemma 2.1 but with Q instead of q. Hence K is the compositum of L and k. The field L is the splitting field of (2.3) over $\mathbb{Q}_p$, so K is the splitting field over k.

Every splitting field is normal. By the theory of finite fields, P/ρ is cyclic and a generating automorphism is $\beta \mapsto \beta^q$. Since K/k is unramified, its galois group is that of P/ρ. (Theorem 8.1 of Chapter 7).

COROLLARY 2. *The unramified closure k_u of the $\mathfrak{p}$-adic field k is obtained by adjoining the mth roots of unity for all m prime to the residue class field characteristic p.*

Note For k_u, see Theorem 4.1, Corollary 3 of Chapter 7.

Proof. By the previous Corollary k_u is obtained by adjoining the $(q^n - 1)$-th roots of unity for $n = 1,2,\ldots$. For every m prime to p there is an n such that $q^n - 1$ is divisible by m.

LEMMA 2.2. *Let k be a $\mathfrak{p}$-adic field, let q be the cardinality of ρ and let $b \in \mathfrak{o}$. Then*

$$\hat{b} = \lim_{t\to\infty} b^{q^t} \tag{2.4}$$

exists. Further, $\hat{b}$ is the Teichmüller representative of (the residue class of) b.

Proof. If $|b| < 1$, then $\hat{b} = 0$, and we are done. Otherwise

$$b^q = b + c,$$

where $|c| < 1$. Then

$$b^{q^2} = (b + c)^q = b^q + qb^{q-1}c + \ldots + c^q.$$

Hence

$$|b^{q^2} - b^q| \leq \max\{|q||c|, |c|^2\} < |c|.$$

Continuing in this way, we see that (2.4) converges. Clearly $\hat{b}^q = \hat{b}$, so $\hat{b}$ is the Teichmüller representative.

3 NON-COMPLETENESS OF $\bar{\mathbb{Q}}_p$

The results of this section will not be required later. It can be omitted at a first reading.

Since the algebraic closure $\bar{\mathbb{Q}}_p$ of $\mathbb{Q}_p$ is the union of all finite algebraic extensions of $\mathbb{Q}_p$, there is a unique extension of the p-adic valuation to $\bar{\mathbb{Q}}_p$.

LEMMA 3.1. *The algebraic closure $\bar{\mathbb{Q}}_p$ of $\mathbb{Q}_p$ is not complete. The completion Ω_p of $\bar{\mathbb{Q}}_p$ is algebraically closed.*

Note Ω_p is a natural "universal" field in which to conduct p-adic analysis in the same way as the complex field is a natural field in which to conduct archimedean analysis.

Proof. Put $b_1 = 1$. For $n = 2,3, \ldots$ let $b_n \in \bar{\mathbb{Q}}_p$ be a root of unity of order prime to p such that $b_{n-1} \in \mathbb{Q}_p(b_n)$ and

$$[\mathbb{Q}_p(b_n) : \mathbb{Q}_p(b_{n-1})] > n. \tag{3.1}$$

Put

$$c = \sum_n b_n p^n \in \Omega_p. \tag{3.2}$$

Suppose, if possible, that $c \in \bar{\mathbb{Q}}_p$ and let

$$t = [\mathbb{Q}_p(c) : \mathbb{Q}_p]. \tag{3.3}$$

Put

$$c_t = \sum_{n=o}^{t} b_n p^n, \tag{3.4}$$

so

$$|c_t - c| \leq p^{-t-1}.$$

Hence

$$|\sigma c_t - \sigma c| \leq p^{-t-1} \tag{3.5}$$

for all automorphisms σ of $\bar{\mathbb{Q}}_p/\mathbb{Q}_p$. By (3.1) there are $t + 1$ automorphisms $\sigma_1, \ldots, \sigma_{t+1}$ which leave $\mathbb{Q}_p(b_{t-1})$ invariant but such that the $\sigma_j b_t$ are distinct. But $\mathbb{Q}_p(b_t)$ is unramified, so by (3.4)

$$|\sigma_i c_t - \sigma_j c_t| \geq p^{-t} \quad (i \neq j)$$

since b_t is a root of unity, by hypothesis. It follows from (3.5) that the $\sigma_j c$ are distinct. This contradicts (3.3), and so shows that $c \notin \bar{\mathbb{Q}}_p$.

It remains to show that Ω_p is algebraically closed. If not, there is an irreducible

$$f(X) = X^n + f_{n-1} X^{n-1} + \ldots + f_o \in \Omega_p[X]$$

of degree $n > 1$. By a substitution $X \to p^m X$ $(m \in \mathbb{Z})$, we may suppose that

$$|f_j| \leq 1 \qquad (0 \leq j \leq n - 1).$$

Since we are in characteristic 0, irreducibility implies that the discriminant $D(f) \neq 0$. Let $g_j \in \bar{\mathbb{Q}}_p$ satisfy

$$|g_j - f_j| < \{D(f)\}^2 \quad (0 \leq j \leq n - 1).$$

There is a $b \in \bar{\mathbb{Q}}_p$ such that

$$b^n + g_{n-1} b^{n-1} + \ldots + g_o = 0.$$

Then

$$|f(b)| < \{D(f)\}^2.$$

Now f has a zero in Ω_p by a form of Hensel's Lemma (Lemma 3.1, Corollary 2 of Chapter 4). This contradicts the supposed irreducibility of $f(X)$.

4 "KRONECKER-WEBER" THEOREM

A field extension K/k is *cyclotomic* if $K = k(u)$ for some root of unity, u. It is *abelian* if it is normal with an abelian galois group. Clearly every extension K/k which is contained in a cyclotomic extension is abelian. A famous theorem of Kronecker and Weber says that for $k = \mathbb{Q}$ the converse holds: every abelian extension is contained in a cyclotomic one. Here we shall prove the corresponding fact for $\mathbb{Q}_p$. We shall not need this section later except when we shall show (Chapter 10, §12) that the classical Kronecker-Weber theorem follows easily from the p-adic one.

THEOREM 4.1 ("Kronecker-Weber"). Every abelian extension K of $\mathbb{Q}_p$ is contained in a cyclotomic one.

It is convenient to enunciate the stages in the proof as lemmas. We recall first that we know already (Lemma 2.1, Corollary 2) that the unramified extensions of $\mathbb{Q}_p$ are precisely the extensions generated by roots of unity of order prime to p. Hence we shall have to study the extensions generated by the p^m-th roots of unity $(m = 1,2,\ldots)$.

For $n > 2$ we denote by w_n a primitive n-th root of 1:

$$w_n^n = 1, \quad w_n^m \neq 1 \quad (m < n). \tag{4.1}$$

We require the notion of the compositum of two fields K_1/k, K_2/k. If K_1, K_2 are subfields of the same field K and K_1, K_2 are not both contained in a proper subfield of K, then K is said to be their compositum (written $K = K_1K_2$). (See also Exercise 10).

LEMMA 4.1. (i) Let $K \supset L \supset k$ and let K/k be abelian. Then so are K/L and L/k.

(ii) Let K_1/k, K_2/k be abelian and let $K_1K_2 = K$ be the compositum. Then K/k is abelian.

Proof. Straightforward application of galois theory.

LEMMA 4.2. Let $p \neq 2$. Then

(i) $\mathbb{Q}_p(w_p)$ is a completely ramified cyclic extension of degree $p - 1$.

(ii) $\mathbb{Q}_p(w_n)$ where $n = p^m$, $m > 1$ is the compositum of $\mathbb{Q}_p(w_p)$ and a completely ramified cyclic extension of degree p^{m-1}.

Note By a cyclic extension we mean an abelian extension with cyclic galois group.

Proof. (i) follows at once from Theorem 2.1, Corollary 1 of Chapter 6.

(ii) By Theorem 2.1, Corollary 2 of Chapter 6, $\mathbb{Q}_p(w_n)$ is completely ramified of degree $p^m - p^{m-1}$. Its automorphisms are

$$\sigma_a : w_n \to w_n^{\,a} \tag{4.2}$$

where $0 \leq a < p^m$ and $p \nmid a$. This group is cyclic, and so is uniquely expressible as the product of cyclic groups of order $p - 1$ and p^{m-1}. Now (ii) follows by Galois theory.

LEMMA 4.3. (i) $\mathbb{Q}_2(w_4) = \mathbb{Q}_2(i)$, *where* $i^2 = -1$, *is completely ramified.*

(ii) $\mathbb{Q}_2(w_n)$ *where* $n = 2^m$, $m > 2$ *is the compositum of* $\mathbb{Q}_2(i)$ *and a completely ramified cyclic extension of degree* 2^{m-2}.

Proof. The only difference with the preceding case is that the group of the σ_a given by (4.2) is no longer cyclic. It is the product of the cyclic group of order 2 generated by σ_{-1} and the cyclic group of order 2^{m-2} generated by σ_5.

METALEMMA 4.1. *To prove Theorem 4.1 it is enough to prove it in the following three cases*

(i) $[K : \mathbb{Q}_p] = \ell^s$ *where* $\ell \neq p$ *is prime and* $s \geq 1$.

(ii) $p \neq 2$ *and* $[K : \mathbb{Q}_p] = p$

(iii) $p = 2$ *and* $[K : \mathbb{Q}_p] = 4$ *with cyclic galois group and with* $i \in K$.

Proof. An abelian group is the product of groups of prime power order. Hence an abelian extension $K/\mathbb{Q}_p$ is the compositum of extensions $K_j/\mathbb{Q}_p$ of prime-power degree. If the K_j are contained in cyclotomic extensions, so is K. This reduces Theorem 4.1 to the case (i) of the enunciation and the case when $[K : \mathbb{Q}_p]$ is a power of p, which we now suppose.

Suppose, first, that $p \neq 2$ and let p^t be the exponent of the Galois group. After forming the compositum of K with the unramified extension L of $\mathbb{Q}_p$ of degree p^t and the completely ramified extension M of degree p^t given by Lemma 4.2, we may suppose that K contains both those fields. If now $[K : \mathbb{Q}_p] = p^{2t}$, then K is precisely the compositum of the two known extensions, and we are done.

Otherwise there is a subgroup Γ_1 of $\mathrm{Gal}(K/\mathbb{Q}_p)$ of index p which fixes LM elementwise. Let K_1 be the fixed field of Γ_1. Then

$[K_1 : \mathbb{Q}_p] = p$ and K_1 is not contained in LM. To prove the Theorem we must show that K_1 does not exist, and this is the objective of (ii) of the Metalemma.

Now let $p = 2$. Every quadratic extension of $\mathbb{Q}_2$ is contained in

$$N = \mathbb{Q}_2(i, \sqrt{2}, \sqrt{-3}). \tag{4.3}$$

This is a cyclotomic field since $\mathbb{Q}_2(w_8) = \mathbb{Q}_2(i, \sqrt{2})$ and $\mathbb{Q}_2(w_3) = \mathbb{Q}_2(\sqrt{-3})$. Now let $K/\mathbb{Q}_2$ be an abelian extension and let 2^t be the exponent of its galois group where, after what has been said above, we may suppose that $t \geq 2$. As in the p odd case, we may suppose without loss of generality that K contains the unramified extension L of degree 2^t and the completely ramified cyclic extension M of degree 2^t given by Lemma 4.3. We also suppose that $i \in K$. If $K = LM(i)$, there is nothing to prove, so we suppose that K is strictly larger. The galois group $\Gamma = \mathrm{Gal}(K/\mathbb{Q}_p)$ cannot have a quotient group of type $(2,2,2,2)$ because every quadratic extension of $\mathbb{Q}_2$ is in (4.3). Hence Γ must be of type $(2^t,2^t,2^s)$ with $s > 1$. Let Γ_1 be a subgroup of Γ of index 4 which fixes L and M. Then the fixed field K_1 of Γ_1 has the properties required by (iii) of the enunciation of the metalemma.

<u>LEMMA</u> 4.4. <u>The Theorem is true in case (i) of the metalemma.</u>

<u>Proof</u>. Let ℓ^r $(r \geq 0)$ be the exact power of ℓ dividing $p - 1$. By galois theory $\mathbb{Q}_p(w_p)$ contains a subfield M of degree ℓ^r, which is completely ramified because $\mathbb{Q}_p(w_p)$ is. We may suppose without loss of generality that $M \subset K$. Let f be the residue class degree of K, so K contains the unramified extension L of degree f (where f is, of course, a power of ℓ).

We now consider the inertia group Δ_o and the ramification groups $\Delta_1, \Delta_2, \ldots$ introduced in Theorem 8.1 of Chapter 7. By Corollary 1 of that theorem, the orders of the quotient groups $\Delta_1/\Delta_2, \Delta_2/\Delta_3, \ldots$ are powers of p. Since $p \neq \ell$ we have $\Delta_1 = \{1\}$. By Corollary 2 of the quoted theorem, Δ_o/Δ_1 is isomorphic to a subgroup of $\mathbb{F}_p^*$, where $\mathbb{F}_p$ is the field of p elements. Hence the order of Δ_o divides $(p - 1)$. Since $e =$ order Δ_o is a power of ℓ

we have $e \leq \ell^r$. Hence $[K : \mathbb{Q}_p] = ef \leq \ell^r f = [LM : \mathbb{Q}_p]$. Thus $K = LM$ and we are done.

The next Lemma is a piece of general commutative algebra.

LEMMA 4.5. *Let p be an odd prime, k a field of characteristic $\neq p$ and K/k a cyclic extension of degree p. Let ζ be a p-th root of unity and*

$$\sigma: \zeta \to \zeta^g \tag{4.4}$$

be a generator of the galois group of $k(\zeta)/k$. Then

$$K(\zeta) = k(\zeta, \theta) \tag{4.5}$$

where

$$\theta^p = A \text{ (say)} \in k(\zeta), \ \theta \notin k(\zeta), \tag{4.6}$$

and

$$\sigma A = B^p A^g \tag{4.7}$$

for some $B \in k(\zeta)$.

Proof. Extend the action of σ to $K(\zeta)$ by making it act trivially on K. Let τ be a generator of the galois group of K/k and let it act trivially on ζ.

By Kummer theory, (4.5), (4.6) hold for some $\theta \in K(\zeta)$ with

$$\tau\theta = \zeta\theta. \tag{4.8}$$

Since θ and $\sigma\theta$ generate the same extension of $k(\zeta)$, we have

$$\sigma\theta = B\theta^h \tag{4.9}$$

for some $B \in k(\zeta)$ and some h, $p \nmid h$.

Now

$$\sigma(\tau\theta) = \zeta^g B\theta^h$$

and

$$\tau(\sigma\theta) = B(\zeta\theta)^h.$$

Hence $\zeta^g = \zeta^h$ since $\sigma\tau = \tau\sigma$. Without loss of generality $g = h$. Then (4.7) follows from (4.9) and (4.6).

Note Conversely, given $A \in k(\zeta)$ satisfying (4.7) for some B, then there is a cyclic extension K/k. We do not prove this as we do not need it.

COROLLARY 1. *Let H be the set of $A \in k(\zeta)^*$ for which there is a B such that (4.7) holds. Then H is a group under multiplication. The Lemma gives an injection of the set of cyclic extensions into the set of cyclic subgroups of the quotient group $H/\{k(\zeta)^*\}^p$.*

Proof. For A_1 gives K precisely when $A_1 = B^p A^h$, $p \nmid h$, $B \in k(\xi)$.

LEMMA 4.6. *The Theorem holds in case (ii) of the metalemma.*

Proof. We use Lemma 4.5. with $k = \mathbb{Q}_p$ and retain the notation $\zeta = w_p$. Some of the properties of $\mathbb{Q}_p(\zeta)$ which we need are set out in Exercise 11 of Chapter 7. Put $\lambda = 1 - \zeta$, so λ is a prime element of $\mathbb{Q}_p(\zeta)$.

Let $A \in \mathbb{Q}_p(\zeta)$ satisfy (4.7) and put $A = \lambda^m C$, where C is a unit. On comparing valuations on both sides of (4.7) and using $g \not\equiv 1 \ (p)$, we get $p | m$. Hence without loss of generality A is a unit. On taking $b^{-p}A$ instead of A for suitable $b \in \mathbb{Z}_p$, we may suppose that $A \equiv 1 \ (\lambda)$.

Clearly $A \mapsto \zeta$ satisfies (4.7) with $B = 1$. On taking $\zeta^n A$ for A with suitable n we may suppose that

$$A \equiv 1 \quad (\lambda^2).$$

There is some $j \geq 2$ such that

$$A \equiv a\lambda^j \quad (\lambda^{j+1})$$

with $a \in \mathbb{Z}_p$, $p \nmid a$. If $j \geq p + 1$ then $A \in \{k(\zeta)^*\}^p$ by (v) of the quoted exercise. Hence $j \leq p$. Further, $B^p \equiv 1 \quad (\lambda^{p+1})$ from the same source. Hence

$$\sigma A \equiv A^g \equiv 1 + ga\lambda^j \quad (\lambda^{j+1}).$$

But $\sigma\lambda \equiv g\lambda \quad (\lambda^2)$, so

$$\sigma A \equiv 1 + a(\sigma\lambda)^j \equiv 1 + ag^j \lambda^j \quad (\lambda^{j+1}).$$

Hence $g^j \equiv g\ (\lambda)$ and $g^j \equiv g(p)$ since $g \in \mathbb{Z}$. But g is a primitive root and $2 \leq j \leq p$. Hence $j = p$.

We have now shown that the group $H/\{k(\zeta)^*\}^p$ of Lemma 4.5 Corollary 1 is generated by ζ and $1 + \lambda^p$. There are thus at most $p + 1$ cyclic extensions $K/\mathbb{Q}_p$. But the compositum LM contains $p + 1$ cyclic extensions, where L is unramified of degree p and M is the totally ramified extension of degree p given by Lemma 4.2 (ii) with $m = 2$. Hence there are no further K, which proves the theorem for case (ii) of the Metalemma.

<u>Note</u> That $A = 1 + a\lambda^p$ gives an unramified K is confirmed by (vi) of Chapter 7, Exercise 11.

We again need a bit of general commutative algebra.

<u>LEMMA</u> 4.7. <u>Let</u> k <u>be a field of characteristic</u> $\neq 2$ <u>and let</u> K/k <u>be cyclic of degree</u> 4. <u>Let</u> $\delta \in K$, $\delta \notin k$ <u>with</u> $\delta^2 = d \in k$. <u>Then</u> $d = a^2 + b^2$ <u>for some</u> $a, b \in k$.

<u>Proof</u>. $K = k(\theta)$ for some θ with $\theta^2 \in k(\delta)$. Let σ be a generator of the galois group, so $\sigma\delta = -\delta$ and $\sigma^2\theta = -\theta$. Hence

$$\sigma\{\theta(\sigma\theta)\} = -\theta(\sigma\theta)$$

and so

$$\theta(\sigma\theta) = s\delta \qquad s \in k^*. \tag{4.10}$$

Put $\theta^2 = u + v\delta$ $(u,v \in k)$, so $(\sigma\theta)^2 = u - v\delta$. On squaring (4.10) we get

$$u^2 - dv^2 = ds^2,$$

so d has the required form.

LEMMA 4.8. *The Theorem holds in case* (iii) *of the Metalemma*.

Proof. Suppose, if possible, that there is a cyclic $K/\mathbb{Q}_2$ of degree 4 with $i \in K$. By Lemma 4.7 we should have $-1 = a^2 + b^2$ with $a,b \in \mathbb{Q}_2$. But this is impossible.

We have now verified the Theorem in the three cases specified by the metalemma, and so the Theorem has been proved.

Notes

§4 The local "Kronecker-Weber" theorem gives an explicit description of the abelian extensions of $\mathbb{Q}_p$. Local classfield theory, as first conceived, set out to give an equally explicit description of the abelian extensions of a general $\mathfrak{p}$-adic field k. If K/k is an abelian extension, there is an isomorphism between the galois group $\mathrm{Gal}(K/k)$ and the quotient group $k^*/N(K^*)$ where $N(K^*)$ is the group of norms of non-zero elements of K (the Artin isomorphism). An alternative way of describing this is to say that there is an isomorphism between the galois group $\Gamma = \mathrm{Gal}(k^{ab}/k)$ of the maximal abelian extension k^{ab} of k (so Γ is equipped with a canonical topology) and the completion $\hat{k}$ of k^* with respect to an appropriate topology. When $k = \mathbb{Q}_p$ this isomorphism can be described as follows. By "Kronecker-Weber" $\mathbb{Q}_p^{ab}$ is the compositum of the two disjoint fields $\mathbb{Q}_p^{un}$ and $\mathbb{Q}_p^{ram}$, where $\mathbb{Q}_p^{un}$ is obtained by adjoining all the roots of unity of order prime to p while $\mathbb{Q}_p^{ram}$ contains all roots of unity whose order is a power of p. The extension $\mathbb{Q}_p^{un}/\mathbb{Q}_p$ is unramified. Its galois group Γ^{un} is generated "topologically" by the Frobenius automorphism $A \to A^p$. It follows that Γ^{un} is isomorphic to the completion $\hat{\mathbb{Z}}$ of the infinite cyclic group $\mathbb{Z}$. On the other hand, the galois group of $\mathbb{Q}_p(C)$, where C is a p^m-th root of unity, consists of the $C \to C^u$ where $u \in \mathbb{Z}$ is prime to p and u is considered mod p^m. Hence the galois group Γ^{ram} of $\mathbb{Q}_p^{ram}/\mathbb{Q}_p$ is canonically isomorphic to the group U_p of p-adic units. Now consider $\mathbb{Q}_p^*$. It is the product of the cyclic group $p^{\mathbb{Z}}$ generated by p and the group U_p. We have already seen that U_p is isomorphic to Γ^{ram}. We map $p^{\mathbb{Z}}$ into Γ^{un} by making p correspond to the Frobenius automorphism.

In its developed form, local classfield theory also deals with extensions K/k which are not abelian. There is a good account by Serre in the Brighton Book (Cassels and Fröhlich (1967)).

Although in some ways conceptually simpler than the global theory, local classfield theory was developed later. At first, in fact, it was developed via the global theory and it was only later that it was made autonomous. Now the argument runs in the opposite direction. For a discussion, and also the origin of the name "class field theory", see the notes to Chapter 10, §12.

Exercises

1. Prove the converse of Lemma 4.5.

2. Show that the value group of Ω_p (defined in §3) is $\{p^r\}$, where r runs through $\mathbb{Q}$.

3. Let M be a primitive p-th root of unity, where $p > 2$ is prime.

(i) Show that

$$X^p - Y^p = (X - Y) \prod_{j=1}^{p-1} (M^j X - M^{-j} Y)$$

for indeterminates X,Y.

(ii) Deduce that

$$\prod_{j=1}^{p-1} (M^j - M^{-j}) = p.$$

(iii) Put

$$D(M) = D = \prod_{j=1}^{\frac{1}{2}(p-1)} (M^j - M^{-j}).$$

Show that

$$D^2 = (-)^{\frac{1}{2}(p-1)} p.$$

4. (i) Notation as in Exercise 3. For $p \nmid b$ show

$$D(M^b)/D(M) = \left(\frac{b}{p}\right),$$

the quadratic residue symbol.

[Hint It takes only the values ±1, is multiplicative in b, and depends only on b mod p].

(ii) Let $q > 2$ be prime, $q \neq p$ and let N be a primitive q-th root of unity. Show that

$$\left(\frac{q}{p}\right) = \prod_{j=1}^{\frac{1}{2}(p-1)} \prod_{k=1}^{q-1} (M^j N^k - M^{-j} N^{-k}).$$

[Hint Exercise 3(i)].

(iii) On interchanging the rôles of p,q in (ii), show that

$$\left(\frac{q}{p}\right)\left(\frac{p}{q}\right) = (-1)^{\frac{1}{4}(p-1)(q-1)}.$$

[Note This is the law of <u>quadratic reciprocity</u>]

5. (i) Notation as in Exercise 3. Operate in the ring $\mathbb{Z}[M]$. Let $q > 2$ be prime, $q \neq p$. Show that

$$(M^j - M^{-j})^q \equiv M^{jq} - M^{-jq} \qquad (q)$$

and deduce that

$$D(M)^q \equiv \left(\frac{q}{p}\right) D(M) \qquad (q).$$

(ii) Comparing with Exercise 3 (iii) deduce that

$$\left(\frac{q}{p}\right) \equiv d^{\frac{1}{2}(q-1)}, \qquad (q)$$

where

$$d = (-1)^{\frac{1}{2}(p-1)}\ p.$$

(iii) Deduce the law of quadratic reciprocity.

[Note We could have operated with S (introduced in Exercise 7) instead of D].

6. Let $p > 1$ be prime. For $a \in \mathbb{Z}_p$, $p \nmid a$ let $\omega(a)$ be the Teichmüller representative of a mod p. For $j \in \mathbb{Z}$ put

$$\chi(a) = \chi_j(a) = \{\omega(a)\}^j$$

and extend the definition of χ to all of $\mathbb{Z}_p$ by the convention that $\chi(a) = 0$ for $p|a$.

Let M be a primitive p-th root of unity and put

$$T(\chi,M) = \sum_{a \bmod p} \chi(a)\ M^a$$

(i) For $b \in \mathbb{Z}$, $p \nmid b$ show that

$$T(\chi,M^b) = \bar{\chi}(b)\ T(\chi,M),$$

where

$$\bar{\chi}(b) = \{\chi(b)\}^{-1}.$$

(ii) Deduce that

$$\{T(\chi,M)\}^{\ell} \in \mathbb{Z}_p,$$

where ℓ is the order of χ (that is, the least $\ell > 0$ such that $\chi^{\ell} = \chi_o$).

(iii) For $\chi \neq \chi_o$ show that

$$T(\chi, M)\, T(\bar{\chi}, M) = \chi(-1)\, p.$$

[Hint The left hand side is

$$\sum_{a,b} \chi(a)\, \bar{\chi}(b)\, M^{a+b}.$$

Make the substitution $a \equiv bc \mod p$.

Note Compare Appendix D.]

(iv) Let $\Pi = M - 1$, so Π is a prime element for $\mathbb{Q}_p(M)$. For $0 < j < p - 1$ show that

$$T(\chi_j, M) \equiv \frac{-\Pi^{p-j-1}}{(p-j-1)!} \mod \Pi^{p-j}.$$

[Hint On expanding $M^a = (1 + \Pi)^a$, we have

$$T(\chi_j, M) = \sum_{k=0}^{p-1} \frac{\Pi^k}{k!}\, t_k,$$

where

$$t_k = \sum_{a=1}^{p-1} \chi_j(a)\, a(a-1) \ldots (a-k+1),$$

$$\equiv \sum_a a^j.\, a(a-1) \ldots (a-k+1) \quad (p).$$

Show that $t_k \equiv 0\ (p)$ for $k < p - j - 1$ but $t_{p-j-1} \equiv -1\ (p)$.]

7. Let p, M be as in Exercise 3 and put

$$S = \sum_{a=0}^{p-1} M^{a^2}.$$

Show that

$$S^2 = (-1)^{\frac{1}{2}(p-1)}\, p$$

and

$$S + \frac{\Pi^{\frac{1}{2}(p-1)}}{\{\frac{1}{2}(p-1)\}!} \equiv 0 \qquad (\Pi^{\frac{1}{2}(p+1)}).$$

[Hint $S = T(\chi_j, M)$ for $j = \frac{1}{2}(p-1)$.]

8. (i) Show that

$$D \equiv 2^{\frac{1}{2}(p-1)} [\{\frac{1}{2}(p-1)\}!] \, \Pi^{\frac{1}{2}(p-1)}, \quad (\Pi^{\frac{1}{2}(p+1)})$$

where D is as in Exercise 3 (iii) and Π is as in Exercise 6 (iv). Deduce from the result of the previous Exercise that

$$S = \left(\frac{-2}{p}\right) D \qquad (*)$$

where $(\frac{\cdot}{p})$ is the quadratic residue symbol $(= \chi_j$ for $j = \frac{1}{2}(p-1))$.

(ii) Embed $\mathbb{Q}(M)$ into the complexes $\mathbb{C}$ by putting

$$M = \exp(2\pi i/p).$$

Show that

$$D = i^{\frac{1}{2}(p-1)} \sqrt{p},$$

where $\sqrt{p}$ is the positive square root.

(iii) Deduce that

$$S = \begin{cases} \sqrt{p} & \text{if} \quad p \equiv 1 \quad (4) \\ i\sqrt{p} & \text{if} \quad p \equiv -1 \quad (4). \end{cases}$$

[Note This is Gauss' famous determination of the sign of Gauss' Sum. It took him 4 years to find a proof. For an interesting proof on quite different lines, see Schur (1921). For this and other proofs see also Landau (1927), vol. I. For cubic and quartic sums, see Matthews (1979, 1979a).]

9. Let K/k be an unramified extension of $\mathfrak{p}$-adic fields. Show that every unit of k is the norm of some unit of K.

10. This exercise supplements the notion of compositum introduced in §4.

(i) Let K_j/k be finite field extensions $(j = 1,2)$ and suppose that K_1/k is normal. Let σ_j be an injection of K_j into the algebraic closure $\bar{k}$ and let $K(\sigma_1,\sigma_2)$ be the smallest subfield of $\bar{k}$ containing $\sigma_1 K_1$ and $\sigma_2 K_2$. Show that $K(\sigma_1,\sigma_2)$ is unique up to isomorphism.

(ii) Let $K_1 = \mathbb{Q}(\theta)$, $K_2 = \mathbb{Q}(\phi)$ where

$$(\theta^2 - 1)^2 - 5 = 0 : (\phi^2 - 2)^2 - 5 = 0.$$

Show that there are embeddings σ_j, τ_j of K_j into $\bar{\mathbb{Q}}$ $(j = 1,2)$ such that (in the notation of (i)) $K(\sigma_1,\sigma_2)$ is not isomorphic to $K(\tau_1,\tau_2)$.

CHAPTER NINE: ALGEBRAIC EXTENSIONS (INCOMPLETE FIELDS)

1 INTRODUCTION

Let K/k be a finite algebraic extension and let $|\ |$ be a valuation on k. We do not suppose that k is complete and ask ourselves what extensions, if any, there are of $|\ |$ to K. We shall encounter this problem also when $|\ |$ is archimedean. For most of this chapter we can consider arch. and non-arch. valuations together. The situation we shall be considering is usually evoked by the keyword <u>semi-local</u>.

Suppose that the valuation $|\ |_1$ on K extends $|\ |$ and let K_1 be the completion of K with respect to it. Then K_1 contains the completion $\bar{k}$ of k with respect to $|\ |$. A basis $\{B_i\}$ of K/k clearly generates K_1 as a $\bar{k}$-vector space. There is, however, no reason to expect that the B_i, considered as elements of K_1, will be linearly independent over $\bar{k}$, and we can conclude only that $[K_1 : \bar{k}] \leq [K : k]$. Multiplication gives K_1 a natural structure as K-module.

We shall also require the <u>tensor product</u> $\bar{k} \otimes_k K$. This can be described concretely, if non-canonically, as follows. Let $B_1,\ldots,B_n$ be a basis for K/k. There are $c_{ij\ell} \in k$ such that

$$B_i B_j = \sum_\ell c_{ij\ell} B_\ell \tag{1.1}$$

Then $\bar{k} \otimes_k K$ is an n-dimensional $\bar{k}$-vector space with a basis which we identify with the B_i:

$$\bar{k} \otimes_k K = \{a_1B_1 + \ldots + a_nB_n : a_1,\ldots,a_n \in \bar{k}\} . \tag{1.2}$$

It has a ring structure, multiplication being defined by (1.1) and by $\bar{k}$-linearity. We identify K in $\bar{k} \otimes_k K$ with the linear combinations of the B_i with coefficients in k.

After these explanations we can enunciate the main result of this chapter.

THEOREM 1.1 *Let K/k be a separable extension with*

$$[K : k] = n < \infty \tag{1.3}$$

and let $|\ |$ be a valuation (arch. or non-arch.) on k. Then there are just finitely many extensions $|\ |_j$ $(1 \leq j \leq J)$ of $|\ |$ to K. Let $\bar{k}$ be the completion of k with respect to $|\ |$ and let K_j $(1 \leq j \leq J)$ be the completion of K with respect to $|\ |_j$. Then

$$\bar{k} \otimes_k K = \bigoplus_j K_j \quad \text{(direct)}. \tag{1.4}$$

Here both sides can be regarded as $\bar{k}$-vector spaces, as K-modules or as rings. In particular,

$$\sum_j [K_j : \bar{k}] = [K : k]. \tag{1.5}$$

By (1.4) we mean, of course, that every $C \in \bar{k} \otimes_k K$ can be expressed uniquely as

$$C = \sum_j C_j \qquad (C_j \in K_j).$$

If, similarly

$$D = \sum_j D_j,$$

then

$$C + D = \sum (C_j + D_j),$$

$$CD = \sum C_j D_j,$$

where $C_j + D_j$, $C_j D_j \in K_j$. Further,

$$aC = \sum_j aC_j$$

$$BC = \sum_j BC_j$$

for $a \in \bar{k}$, $B \in K$, where aC_j, $BC_j \in K_j$.

We prove the Theorem in §2 taking arch. and non-arch. valuations together. The proof gives explicit descriptions of the K_j and $|\ |_j$. In the same section we deduce some straightforward consequences about the norms and traces of elements of K for the extensions K/k and $K_j/\bar{k}$. In §3 we suppose $|\ |$ non-arch. and introduce a notion of discriminant for K/k, which we relate to the discriminants of the $K_j/\bar{k}$. In §4 we suppose K/k normal, and look at the action of galois: here again we can treat arch. and non-arch. valuations together.

2 PROOF OF THEOREM AND COROLLARIES

LEMMA 2.1. *Let* $K = k(A)$ *be a separable extension and let*

$$F(X) \in k[X] \tag{2.1}$$

be the minimum polynomial for A. *Let* $\bar{k}$ *be the completion of* k *with respect to an (arch. or non-arch.) valuation* $|\ |$. *Let*

$$F(X) = \phi_1(X) \ldots \phi_J(X) \tag{2.2}$$

be the decomposition of $F(X)$ *into irreducibles in* $\bar{k}[X]$. *Then the* ϕ_j *are distinct. Let*

$$K_j = \bar{k}(B_j), \tag{2.3}$$

where B_j *is a root of* $\phi_j(X)$. *Then there is an injection*

$$K = k(A) \hookrightarrow K_j = \bar{k}(B_j) \tag{2.4}$$

extending $k \hookrightarrow \bar{k}$ *under which* $A \to B_j$. *Denote by* $|\ |_j$ *the valuation on* K *induced by (2.4) and the unique valuation on* K_j *extending* $|\ |$.

Then the $|\ |_j$ $(1 \leq j \leq J)$ *are precisely the extensions of* $|\ |$ *to* K. *Further,* K_j *is the completion of* K *with respect to* $|\ |_j$.

Proof. This is (almost) shorter than the enunciation. Let $\|\ \|$ be any valuation of K extending $|\ |$ and let $\bar{K}$ be the completion with respect to it. Then $\bar{k} \subset \bar{K}$ and $A \in K \subset \bar{K}$. Further, $\bar{k}(A)$ is complete, by Theorem 1.1 of Chapter 7 if $|\ |$ is non-arch. and by Theorem 1.1 of Chapter 3 if $|\ |$ is arch. Hence $\bar{K} = \bar{k}(A)$. Let $\phi(X) \in \bar{k}[X]$ be the minimum polynomial for A over $\bar{k}$. Since $F(A) = 0$, we have $\phi(X) | F(X)$ and so ϕ is one of the ϕ_j in (2.2) and we have the situation described in the Lemma.

We now go in the opposite direction. Let B_j be as in the enunciation. Then $F(B_j) = 0$ by (2.2) and so the extensions $k(A) = K$ and $k(B_j) \subset \bar{k}(B_j) = K_j$ are isomorphic. We can thus identify K with a subfield of K_j and have the situation already discussed.

It remains to show that the ϕ_j are distinct. If not, $F(X)$ and $F'(X)$ would have a common factor in $\bar{k}[X]$. Since the common factor can be determined by the euclidean algorithm, there would be a common factor in $k[X]$: and this is impossible since F is irreducible and separable, by hypothesis. This proves the Lemma.

Proof of Theorem 1.1. In the notation of the preceding proof, we have the following obvious ring isomorphisms

$$\bar{k}[X]/F(X) \cong \bar{k} \otimes_k K, \tag{2.5}$$

$$\bar{k}[X]/\phi_j(X) \cong K_j, \tag{2.6}$$

where in both cases $X \mapsto A$. After Lemma 2.1, the truth of Theorem 1.1. follows from the following general result of commutative algebra:

LEMMA 2.2. Let k *be any field and let*

$$F(X) = \phi_1(X) \dots \phi_J(X), \tag{2.7}$$

where the $\phi_j(X) \in k[X]$ *are coprime in pairs. Then*

$$k[X]/F(X) \cong \bigoplus_j k[X]/\phi_j(X). \quad \text{(direct)} \tag{2.8}$$

<u>Note</u> In the application we have $\bar{k}$ instead of k.

<u>Proof</u>. The two sides of (2.8) have the same dimension as k-vector spaces. Let Θ be the map of the left hand side of (2.8) into the right hand side induced by the identity map on $k[X]$ and let $f(X) \bmod F(X)$ be in the kernel. Then $f(X) \equiv 0 \bmod \phi_j(X)$ for all j. Hence $f(X) \equiv 0 \bmod F(X)$, i.e. Θ is a monomorphism. Because of the equality of the dimensions, Θ is an isomorphism; which is the assertion of the Lemma.

<u>COROLLARY 1</u>. <u>Let</u> $A \in K$. <u>Then the trace and norm are given by</u>

$$S_{K/k}(A) = \sum_j S_{K_j/\bar{k}}(A) \tag{2.9}$$

<u>and</u>

$$N_{K/k}(A) = \prod_j N_{K_j/\bar{k}}(A). \tag{2.10}$$

<u>Proof</u>. By definition, $S_{K/k}$ and $N_{K/k}$ are respectively the trace and the determinant of the k-linear map induced on K by multiplication with A. The result now follows at once from (1.4).

<u>COROLLARY 2</u>.

$$\prod_j |A|_j^{n(j)} = |N_{K/k}(A)|, \tag{2.11}$$

<u>where</u> $n(j) = [K_j : \bar{k}]$.

<u>Proof</u>. Immediate from (2.10) and Theorem 1.1 of Chapter 7.

<u>COROLLARY 3</u>. <u>Suppose that either</u> $\bar{k}$ <u>is</u> $\mathfrak{p}$-<u>adic or it is one of</u> $\mathbb{R}$, $\mathbb{C}$. <u>Let</u> $\| \ \|$, <u>be the renormalization of</u> $| \ |$ <u>on</u> $\bar{k}$ <u>introduced in Chapter 8, §1 and let</u> $\| \ \|_j$ <u>be the renormalization of</u> $| \ |_j$ <u>on</u> K_j. <u>Then</u>

$$\prod_j \|A\|_j = \|N_{K/k}(A)\|.$$

<u>Proof</u>. Lemma 1.3 Corollary 2 or Lemma 1.4 of Chapter 8.

3 INTEGERS AND DISCRIMINANTS

Let $| \ |$ be non-arch. We shall call $A \in K$ a (semi-local) integer if $|A|_j \le 1$ for all j. The ring of such A will be denoted by $\mathfrak{O}$. Clearly

$$\mathfrak{O} = \bigcap_j \{K \cap \mathfrak{O}_j\}, \tag{3.1}$$

where $\mathfrak{O}_j$ is the ring of integers of the complete field K_j. Denote the integers of $k, \bar{k}$ by $\mathfrak{o}, \bar{\mathfrak{o}}$ respectively.

LEMMA 3.1.

$$\mathfrak{O} \otimes_{\mathfrak{o}} \bar{\mathfrak{o}} = \bigoplus_j \mathfrak{O}_j \quad \text{(direct)} \tag{3.2}$$

Proof. We use the identification (1.4) in which we identify $A \in K$ with $1 \otimes A \in \bar{k} \otimes_k K$. In terms of these identifications

$$\mathfrak{O} = K \cap \{\bigoplus_j \mathfrak{O}_j\}. \tag{3.3}$$

Let B_{ij} $(1 \le i \le n_j)$ be an $\bar{\mathfrak{o}}$-basis of $\mathfrak{O}_j$ $(1 \le j \le J)$. By Theorem 3.1 of Chapter 2, we can choose $C_{ij} \in K$ such that

$$\left.\begin{aligned} |C_{ij} - B_{ij}|_j &< 1 \\ |C_{ij}|_\ell &< 1 \qquad (\ell \ne j,\ 1 \le \ell \le J). \end{aligned}\right\} \tag{3.4}$$

Then $C_{ij} \in \bigoplus_j \mathfrak{O}_j$. Indeed the matrix $\mathfrak{M}$ expressing the C_{ij} in terms of the B_{ij} is congruent to the identity modulo the maximal ideal $\bar{\mathfrak{p}}$ of $\bar{\mathfrak{o}}$. It follows that the C_{ij} are an $\bar{\mathfrak{o}}$-basis of $\bigoplus_j \mathfrak{O}_j$. But $C_{ij} \in K$ so the C_{ij} are an $\mathfrak{o}$-basis of $\mathfrak{O}$ by (3.3). This proves (3.2).

We now extend the notion of discriminants introduced in §6 of Chapter 7 to the present situation.

DEFINITION 3.1. Let $A_1, \ldots, A_n$ be an $\mathfrak{o}$-basis of $\mathfrak{O}$. Then the element of $\mathfrak{o}/U^2$ given by

$$\det(S_{K/k}\, A_i A_j) \tag{3.5}$$

is the <u>(semi-local)</u> <u>discriminant</u> $D_{K/k}$.

<u>Note</u> We recall that U is the group of units of k. Clearly different bases give the same element of $\mathfrak{o}/U^2$.

<u>LEMMA 3.2.</u>

$$D_{K/k} \to \prod_j D_{K_j/\bar{k}}$$

<u>under the homomorphism</u> $\mathfrak{o}/U^2 \to \bar{\mathfrak{o}}/\bar{U}^2$ <u>induced by</u> $k \hookrightarrow \bar{k}$. <u>In particular,</u>

$$|D_{K/k}| = \prod_j |D_{K_j/\bar{k}}|.$$

<u>Proof</u>. We are interested in (3.5) only up to a factor in $\bar{U}^2$ and so may take for $A_1,\ldots,A_n$ an $\bar{\mathfrak{o}}$-basis of $\bar{\mathfrak{o}} \otimes_{\mathfrak{o}} \mathfrak{O}$. As in the proof of Lemma 3.1, we take for $\{A_1,\ldots,A_n\}$ the union of $\bar{\mathfrak{o}}$-bases $\{B_{ij}\}$ $(1 \leq i \leq n_j)$ of $\mathfrak{O}_j$. Then $B_{ij}\, B_{uv} = 0$ for $j \neq v$ and by (2.10) the matrix $(S_{K/k}\, A_iA_j)$ becomes a chain of submatrices along the diagonal. The determinant of the j-th submatrix is

$$\det(S_{K_j/\bar{k}}\, B_{uj}\, B_{vj})_{u,v},$$

which maps into $D_{K_j/\bar{k}} \in \bar{\mathfrak{o}}/\bar{U}^2$.

<u>COROLLARY. A necessary and sufficient condition that all the</u> $|\ |_j$ <u>be unramified is that</u> $|D_{K/k}| = 1$.

<u>Proof</u>. For $|D_{K_j/\bar{k}}| \leq 1$, with equality only when $K_j/\bar{k}$ is unramified, by Theorem 6.1 of Chapter 7.

The following criterion is often useful:

<u>LEMMA 3.3.</u> <u>Let</u> $K = k(B)$ <u>be an extension of degree</u> n <u>and suppose that</u> B <u>is a root of</u> $F(X)$, <u>where</u>

$$F(X) \in \mathfrak{o}[X] \tag{3.6}$$

<u>has top coefficient</u> 1. <u>Suppose, further, that</u>

$$|F'(B)|_j = 1 \qquad (3.7)$$

<u>for all extensions</u> $|\ |_j$ <u>of</u> $|\ |$ <u>to</u> k. <u>Then all the</u> $|\ |_j$ <u>are unramified and</u> $1, B, \ldots, B^{n-1}$ <u>is an</u> $\mathfrak{o}$<u>-basis of</u> $\mathfrak{O} = \bigcap_j \mathfrak{O}_j$.

<u>Note 1</u> This is useful even in the special situation of Chapter 7, when k is complete and there is only one $|\ |_j$.

<u>Note 2</u> We shall apply the Lemma when F is not the minimum polynomial of B.

<u>Proof</u>. It follows at once from (3.6) that

$$B \in \mathfrak{O}. \qquad (3.8)$$

Let $G(X) \in k[X]$ be the minimum polynomial for B (with top coefficient 1), so

$$F(X) = G(X)H(X) \qquad (3.9)$$

for some $H(X) \in k[X]$. By the "Gauss' Lemma" 2.1 of Chapter 6, we have

$$G(X),\ H(X) \in \mathfrak{o}[X]. \qquad (3.10)$$

Now

$$F'(B) = G'(B)\ H(B),$$

where $|H(B)|_j \leq 1$ for all j by (3.8) and (3.10). Hence

$$|G'(B)|_j = 1$$

for all j. Thus the conditions of the theorem are satisfied with G instead of B. It is therefore enough to prove the Lemma under the additional assumption that F is the minimum polynomial of B, which we now suppose.

Let K be the splitting field of F over k, let

$B_1,\dots,B_n \in \mathbb{K}$ be the roots, and let $\|\ \|$ be any extension of $|\ |$ to $\mathbb{K}$. The discriminant of the set $1, B, \dots, B^{n-1}$ of elements of $\mathfrak{O}$ is

$$D(1, B, \dots, B^{n-1}) = \prod_{i<j} (B_i - B_j)^2$$

$$= \pm \prod_j F'(B_j). \tag{3.11}$$

Now

$$\|F'(B_j)\| = |F'(B)|_\ell \tag{3.12}$$

for the valuation $|\ |_\ell$ with $\ell = \ell(j)$ induced by $\|\ \|$ on $k(B)$ by the injection $B \to B_j$. Hence

$$|D(1, B, \dots, B^{n-1})| = \|D(1, B, \dots, B^{n-1})\|$$

$$= 1 \tag{3.13}$$

by (3.7), (3.11), (3.12).

Now let $A_1,\dots,A_n$ be an $\mathfrak{o}$-basis of $\mathfrak{O}$, say

$$B^{j-1} = \sum_i t_{ji} A_i \quad (1 \leq j \leq n),$$

with $t_{ji} \in \mathfrak{o}$. Then

$$|D(1, B, \dots, B^{n-1})| = |T|^2 \, |D(A_1, \dots, A_n)|$$

$$= |T|^2 \, |D_{K/k}|$$

where $T = \det(t_{ji})$. By (3.13) we have

$$|T| = 1, \quad |D_{K/k}| = 1.$$

Hence $1, B, \dots, B^{n-1}$ is a basis and the $|\ |_j$ are unramified by Lemma 3.2, Corollary.

4 APPLICATION TO CYCLOTOMIC FIELDS

We shall apply the results of the preceding section to cyclotomic fields, that is fields obtained by adjoining roots of unity to the rational field $\mathbb{Q}$; but must first develop some of their basic properties.

We denote by $\mathbb{Q}^{(m)}$ the splitting field of

$$X^m - 1 \tag{4.1}$$

over $\mathbb{Q}$. Since obviously $\mathbb{Q}^{(2m)} = \mathbb{Q}^{(m)}$ for m odd, we shall make the convention that

$$\text{either } 2 \nmid m \quad \text{or} \quad 2^2 | m. \tag{4.2}$$

The roots of unity of order precisely m are the roots of the polynomial

$$F_m(X) = \prod_{d|m} (X^d - 1)^{\mu(m/d)} \in \mathbb{Z}[X], \tag{4.3}$$

which has degree

$$\phi(m) = m \prod_{\substack{q|m \\ q \text{ prime}}} (1 - q^{-1}) \tag{4.4}$$

where μ is the Moebius function and ϕ is Euler's totient function. Hence there are $\phi(m)$ roots of unity M of order precisely m, and clearly

$$\mathbb{Q}^{(m)} = \mathbb{Q}(M) \tag{4.5}$$

for any one of them. The non-trivial fact which we shall require is that $F_m(X)$ is irreducible in $\mathbb{Q}[X]$, or, what is the same thing, that $\mathbb{Q}^{(m)}/\mathbb{Q}$ has degree $\phi(m)$.

LEMMA 4.1.

(i) $\mathbb{Q}^{(m)}/\mathbb{Q}$ has degree $\phi(m)$.

(ii) A prime q is ramified in $\mathbb{Q}^{(m)}$ precisely when $q|m$ (with the convention (4.2) if $q = 2$).

Proof. Suppose, first that $q \nmid m$. Then the q-adic valuation is unramified in $\mathbb{Q}^{(m)}$ by Lemma 3.3 with $F(X) = X^m - 1$ and $B = M$.

Now suppose that $m = q^\alpha$ (with $\alpha \geq 2$ if $q = 2$). Then the degree of $\mathbb{Q}^{(m)}$ is $q^\alpha - q^{\alpha-1} = \phi(q^\alpha)$ and q is completely ramified, by Corollaries 1, 2 to Theorem 2.1 of Chapter 6 and Theorem 7.1 of Chapter 7.

Finally, suppose that

$$m = q^\alpha \ell$$

where $q \nmid \ell$. Clearly $\mathbb{Q}^{(m)}$ is the compositum of the two fields $\mathbb{Q}^{(q^\alpha)}$ and $\mathbb{Q}^{(\ell)}$ (which gives (ii) for $q|m$). Let

$$I = \mathbb{Q}^{(q^\alpha)} \cap \mathbb{Q}^{(\ell)}.$$

Then q is completely ramified in I (since $I \subset \mathbb{Q}^{(q^\alpha)}$) but is also unramified (since $I \subset \mathbb{Q}^{(\ell)}$). The only possibility is that

$$I = \mathbb{Q}.$$

Since $\mathbb{Q}^{(m)}$ is a normal (Galois) extension of $\mathbb{Q}$ it follows that the degree of $\mathbb{Q}^{(m)}$ is the product of the degrees of $\mathbb{Q}^{(q^\alpha)}$ and $\mathbb{Q}^{(\ell)}$. This proves (i) by induction on the number of primes dividing m.

We now consider the semilocal situation when $k = \mathbb{Q}$, $|\ | = |\ |_p$ and $K = \mathbb{Q}^{(m)}$ for some m and some prime p.

LEMMA 4.2. Let $\mathfrak{O}$ be the set of elements of $\mathbb{Q}^{(m)}$ which are integral for all valuations extending the p-adic valuation. Then a $\mathbb{Z}_p$-basis of $\mathfrak{O}$ is given by $1, M, \ldots, M^{\phi-1}$ where M is any primitive m-th root of unity and $\phi = \phi(m)$.

Proof. As we saw in the proof of the previous lemma, this follows immediately from Lemma 3.6 when $p \nmid m$.

Now suppose that $m = p^{\alpha}\ell$, $p \nmid \ell$ and let $L = 1 - N$ for some primitive p^{α}-th root of unity N. Then any $A \in Q^{(m)}$ is uniquely of the form

$$A = \sum_{j=o}^{\psi-1} L^j A_j \tag{4.6}$$

where $\psi = \phi(p^{\alpha}) = p^{\alpha} - p^{\alpha-1}$ and $A_j \in \mathbb{Q}^{(\ell)}$. Further, $A \in \mathfrak{O}$ precisely when all the A_j are in $\mathfrak{O} \cap \mathbb{Q}^{(\ell)}$. By the unramified case, a basis of $\mathfrak{O} \cap \mathbb{Q}^{(\ell)}$ is given by the powers of a primitive ℓ-th root of unity. The result now follows on putting $L = 1 - N$ in (4.6). [More precisely, this shows that $\mathfrak{O} \subset \mathbb{Z}_p[M]$, so $\mathfrak{O} = \mathbb{Z}_p[M]$ and this has basis $1, \ldots, M^{\phi-1}$ since M is integral].

5 ACTION OF GALOIS

In this section we suppose that K/k is normal with galois group Γ and consider the action of Γ on the situation considered in the previous sections. We allow $|\ |$ to be arch. Instead of labelling the extensions of $|\ |$ to K with integers 1,2, ... it is convenient to label them with Fraktur capitals $\mathfrak{P}$, $\mathfrak{Q}$, ..., so we write $|\ |_{\mathfrak{P}}$ for the extension and $K_{\mathfrak{P}}$ for the corresponding completion.

Let $\mathfrak{P}$ be such an extension and $\sigma \in \Gamma$. We define $\sigma\mathfrak{P}$ by

$$|A|_{\sigma\mathfrak{P}} = |\sigma^{-1}A|_{\mathfrak{P}}. \tag{5.1}$$

This _is_ a valuation by "transport of structure". We have σ^{-1} (rather than σ) on the right hand side to ensure that

$$\sigma(\tau\mathfrak{P}) = (\sigma\tau)\mathfrak{P} \quad (\sigma,\tau \in \Gamma). \tag{5.2}$$

DEFINITION 5.1. The set of $\sigma \in \Gamma$ such that $\sigma\mathfrak{P} = \mathfrak{P}$ is the _splitting group_ (German: Zerlegungsgruppe) of $\mathfrak{P}$.

LEMMA 5.1. Let Λ _be the splitting group of_ $\mathfrak{P}$. _Then_ Λ _is the galois group of_ $K_{\mathfrak{P}}/\bar{k}$. _Every extension of_ $|\ |$ _to_ K _is_ $\sigma\mathfrak{P}$ _for some_ $\sigma \in \Gamma$.

Proof. The action of $\sigma \in \Lambda$ on K respects $|\ |_{\mathfrak{P}}$ and so extends uniquely to an action on $K_{\mathfrak{P}}/\bar{k}$. Hence $\Lambda \subset \mathrm{Gal}(K_{\mathfrak{P}}/\bar{k})$, and

$$\mathrm{card}(\Lambda) \leq [K_{\mathfrak{P}} : \bar{k}], \tag{5.3}$$

with equality only if Λ is the full galois group.

The valuation $\sigma\mathfrak{P}$ has splitting group $\sigma \Lambda \sigma^{-1}$ and so

$$\mathrm{card}(\Lambda) \leq [K_{\sigma\mathfrak{P}} : \bar{k}].$$

The number of distinct $\sigma\mathfrak{P}$ is $[\Gamma : \Lambda]$, and so

$$\begin{aligned} \mathrm{card}(\Gamma) &= [\Gamma : \Lambda]\ \mathrm{card}(\Lambda) \\ &\leq \sum_{\sigma\mathfrak{P}} [K_{\sigma\mathfrak{P}} : \bar{k}], \end{aligned} \tag{5.4}$$

where the summation is over distinct $\sigma\mathfrak{P}$. But

$$\mathrm{card}(\Gamma) = [K : k].$$

By (5.3) this is possible only if $\sigma\mathfrak{P}$ runs through all the extensions of $|\ |$ to K and if there is equality in (5.3). This concludes the proof.

Note After Lemma 5.1, we may refer to the inertia group Δ_o and the ramification groups Δ_j $(j \geq 1)$ of $\mathfrak{P}$ (Chapter 7, §8), as subgroups of Γ.

We shall also need some information about the behaviour of valuations at intermediate fields. As before, let K/k be normal with galois group Γ. Now let Θ be a (not necessarily normal) subgroup of Γ, with fixed field L. Let $\mathfrak{P}$ be a place of K and $P, \mathfrak{p}$ respectively the induced places in L, k.

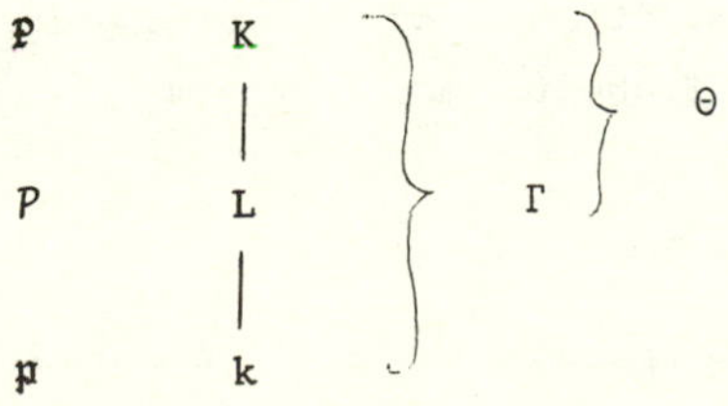

$$\begin{array}{ccc} \mathfrak{P} & K & \\ & | & \Theta \\ P & L & \Gamma \\ & | & \\ \mathfrak{p} & k & \end{array}$$

LEMMA 5.2. *Let* Λ *be the splitting group of* $\mathfrak{P}$ *over* k. *Then*

(i) *the splitting group of* $\mathfrak{P}$ *over* L *is* $\Theta \cap \Lambda$.

(ii) *suppose, further, that* Θ *is a normal subgroup of* Γ, *so* Γ/Θ *is the galois group of* L/k. *The splitting group of* P *over* k *consists of the cosets* $\sigma\Theta$, $\sigma \in \Lambda$.

Proof. Clear.

6 APPLICATION. QUADRATIC RECIPROCITY

Let $p > 2$ be prime and let $\mathbb{Q}^{(p)}$ be the field of the p-th roots of unity. The galois group of $\mathbb{Q}^{(p)}/\mathbb{Q}$ is cyclic of order $p - 1$, and so $\mathbb{Q}^{(p)}$ contains precisely one extension $K = \mathbb{Q}(\surd d)$ quadratic over $\mathbb{Q}$, where without loss of generality $d \in \mathbb{Z}$ is square free. If d were divisible by some prime $q \neq p$, then q would ramify in K and so in $\mathbb{Q}^{(p)}$, which we know not to be the case. Hence $d = \pm p$. If $d \equiv -1$ (4), then 2 would ramify in K, again a contradiction. We have proved

LEMMA 6.1. $\mathbb{Q}^{(p)}$ *contains precisely one quadratic extensions of* $\mathbb{Q}$, *namely*

$$\mathbb{Q}(\surd d), \qquad d = (-1)^{\frac{1}{2}(p-1)}\, p. \tag{6.1}$$

Note For explicit embeddings of $\mathbb{Q}(\surd d)$ in $\mathbb{Q}^{(p)}$ see Exercises 3 and 7 of Chapter 8.

Now let $q > 2$, $q \neq p$ be a rational prime. We shall apply Lemma 5.1 (ii) with

$$K = \mathbb{Q}^{(p)},\ L = \mathbb{Q}(\surd d),\ k = \mathbb{Q} \ \text{ and } \ \mathfrak{p} = q.$$

By Chapter 8, §2, the splitting group Λ of a place $\mathfrak{Q}$ of $\mathbb{Q}^{(p)}$ over q is generated by the Frobenius automorphism

$$\sigma : M \to M^q, \tag{6.2}$$

where M is a p-th root of unity. By Lemma 5.2 (ii), there are one or two places of $\mathbb{Q}(\sqrt{d})$ over q according as $\sigma \notin \Theta$ or $\sigma \in \Theta$, where Θ is the subgroup of index 2 of the galois group Γ of $\mathbb{Q}^{(p)}/\mathbb{Q}$. But Γ is canonically isomorphic to the multiplicative group mod p. Hence q splits in $\mathbb{Q}(\sqrt{d})$ precisely when

$$\left(\frac{q}{p}\right) = +1 \tag{6.3}$$

(in terms of the usual quadratic residuacity symbol).

On the other hand, by Lemma 2.1 the valuation q of $\mathbb{Q}$ splits in $\mathbb{Q}(\sqrt{d})$ precisely when $X^2 - d$ splits in $\mathbb{Q}_q[X]$, that is when

$$\left(\frac{d}{q}\right) = +1. \tag{6.4}$$

Hence

$$\left(\frac{q}{p}\right) = \left(\frac{d}{q}\right). \tag{6.5}$$

But this is just the law of <u>quadratic reciprocity</u>, since by (6.1)

$$\begin{aligned} \left(\frac{d}{q}\right) &= \left(\frac{-1}{q}\right)^{\frac{1}{2}(p-1)} \left(\frac{p}{q}\right) \\ &= (-1)^{\frac{1}{4}(p-1)(q-1)} \left(\frac{p}{q}\right). \end{aligned} \tag{6.6}$$

<u>Notes</u>

§6 Gauss gave 8 proofs of the law of quadratic reciprocity and Eisenstein gave 4. Some 25 proofs are reproduced in Baumgart (1885), and Bachmann (1921), pp47-48 gives a list of 56.

Gauss gave a law of quartic reciprocity for the field $\mathbb{Q}(i)$, $i^2 = -1$ and Eisenstein a law of cubic reciprocity for $\mathbb{Q}(\rho)$, $\rho^3 = 1$, see e.g. Bachmann (1872). Eisenstein then established a p-th power reciprocity law in $\mathbb{Q}(\xi)$, $\xi^p = 1$, see Eisenstein (1975) or the accounts in Hilbert (1897), Landau (1927). The proofs operate with Gauss sums. Now more general reciprocity laws follow from Class Field Theory, see Hasse (1926) or the Exercises in Cassels and Fröhlich (1967).

<u>Exercises</u>

1. (i) Let $K = \mathbb{Q}(\delta)$, where $\delta^2 = d \in \mathbb{Z}$ is square free.

Show that there are two extensions of $| \ |_2$ to K if $d \equiv 1 \ (8)$, but otherwise one.

(ii) Show that $\{1, \frac{1}{2}(1 + \delta)\}$ is an $\mathfrak{o}$-basis for the semilocal integers with respect to $| \ |_2$ if $d \equiv 1 \ (4)$; but that otherwise $\{1, \delta\}$ is a basis. Here $\mathfrak{o} = \mathbb{Z}_2 \cap \mathbb{Q}$.

(iii) Let $p \neq 2$. Show that $\{1, \delta\}$ is a basis for the semilocal integers with respect to $| \ |_p$.

2. (i) Let $K = \mathbb{Q}(\delta)$, where $\delta^3 = d \in \mathbb{Z}$ is cube-free. Show that if $d \not\equiv \pm 1 \ (9)$ there is one extension of $| \ |_3$ to K and that it is completely ramified. If $d \equiv \pm 1 \ (9)$ show that there are two extensions, one unramified and one completely ramified.

(ii) Show that a basis for the semi-local integers with respect to $| \ |_3$ is:

$\{1, \delta, (1 \pm \delta + \delta^2)/3$ if $d \equiv \pm 1 \ (9)$

$\{1, \delta, \delta^2/3$ if $d \equiv 0 \ (9)$

$\{1, \delta, \delta^2\}$ otherwise.

(iii) Find a basis for the semilocal integers with respect to $| \ |_p$ $(p \neq 3)$.

3. (i) Let $K = \mathbb{Q}(\delta)$, where $\delta^3 + 2\delta + 5 = 0$. Show that there are two extensions of $| \ |_5$ to K.

(ii) Determine a basis of the semilocal integers with respect to $| \ |_5$.

(iii) Similar to (i), (ii) but with $\delta^3 + 2\delta^2 + \delta + 5 = 0$.

4. Let $K = \mathbb{Q}(\rho)$, where ρ is a root of unity. For every prime p show that the semilocal integers with respect to $| \ |_p$ are $\mathfrak{o}[\rho]$, where $\mathfrak{o} = \mathbb{Z}_p \cap \mathbb{Q}$.

[Hint Consider first the cases (i) order of ρ is prime to p and (ii) order of ρ is a power of p].

5. Give an explicit map from the right hand side of (2.8) to the left.

6. Let K/k be normal and suppose that $\mathrm{Gal}(K/k)$ is simple (no normal subgroups) but not cyclic. Let $|\ |$ be a discrete valuation of k with finite residue class field. Show that there is always more than one extension of $|\ |$ to K.

[Hint Otherwise $\mathrm{Gal}(\bar{K}/\bar{k}) = \mathrm{Gal}(K/k)$, where $\bar{K}$, $\bar{k}$ are the completions. Now consider Theorem 8.1 of Chapter 7.]

7. In the notation of §4, suppose that $p \nmid m$. Let f be the smallest positive integer such that $p^f \equiv 1\ (m)$. Show that $\phi(m) = fg$ for some integer g and that there are precisely g extensions of $|\ |_p$ to $\mathbb{Q}^{(m)}$.

8. Suppose that K is the compositum of the fields $R \supset k$, $S \supset k$ and that the valuation $|\ |$ of k is completely ramified in R but unramified in S. Let r, s be the degrees of R/k, S/k. Show that the degree of K/k is rs. Let $B_1, \ldots, B_r$; $C_1, \ldots, C_s$ be bases of the semilocal integers of R, S respectively. Show that

$$B_i C_j \qquad (1 \leq i \leq r,\ 1 \leq j \leq s)$$

is a basis for the semilocal integers of K.

[Note Compare Lemma 4.2].

9. Let K be a finite extension of $\mathbb{Q}$ and suppose that the ring of semilocal integers for p is generated by a single element, say $\mathfrak{O} = \mathbb{Z}_p[E]$. Show that there can be at most p first-degree extensions of $|\ |_p$ to K.

When $K = \mathbb{Q}(7^{\frac{1}{2}}, 13^{\frac{1}{2}})$ and $p = 3$, show that $\mathfrak{O}$ is not generated by a single element.

[Note Contrast Chapter 7, Ex. 18. Primes, such as 3 in this example, which do not divide the discriminant of the field although they do divide the discriminant of every integer of it, are known as inessential divisors of the discriminant (ausserwesentliche Disckriminantenteiler). They were the topic of Hensel's doctoral dissertation in 1884, and their study may have been one impulse towards the development of p-adic theory].

9. bis. Let $K = \mathbb{Q}(C)$, where $C^3 - C^2 - 2C - 8 = 0$. Show that there are three extensions of the 2-adic valuation to K. Deduce that 2 is an inessential divisor of the discriminant for $K/\mathbb{Q}$.

[Note Due to Dedekind. Probably the first observed case of the phenomenon.]

10. (i) In the notation of §5 show that the splitting group of $\sigma\mathfrak{P}$ is $\sigma\Lambda\sigma^{-1}$, where Λ is the splitting group of $\mathfrak{P}$.

(ii) Suppose, further, that the galois group Γ of K/k is abelian. Show that the splitting group, inertia group, ramification groups of all the extensions $\mathfrak{P}$ of $|\ |$ to K are independent of the $\mathfrak{P}$ chosen.

10. bis. By considering the action of the galois group of $\mathbb{Q}^{(p)}/\mathbb{Q}$ (in the notation of §6) on the archimedean valuations, show that the quadratic subfield $\mathbb{Q}(\sqrt{d})$ of $\mathbb{Q}^{(p)}$ is real if $p \equiv 1\ (4)$ but complex if $p \equiv 3\ (4)$. Hence confirm Lemma 6.1.

10. ter. (i) In the notation of §5, show that $\mathbb{Q}^{(p)}$ contains precisely one subfield of degree $\frac{1}{2}(p - 1)$, namely $\mathbb{Q}(C)$, where $C = M + M^{-1}$ and $M^p = 1$.

(ii) For any prime q (including $q = p$) show that the ring of semilocal integers is $\mathbb{Q}_q[C]$.

(iii) Show that p ramifies completely in $\mathbb{Q}(C)$ but that no $q \neq p$ ramifies.

(iv) Show that $\mathrm{Norm}(2 - C) = p$.

(v) Show that the residue class degree f of an extension $\mathfrak{q}$ of $q \neq p$ to $\mathbb{Q}(C)$ is the least solution of $q^f \equiv \pm 1\ (p)$: in particular, q splits completely precisely when $q \equiv \pm 1\ (p)$.

(vi) Show that $\mathbb{Q}(C)$ is real at all archimedean places.

11. This exercise gives an explicit factorization of a prime

$$p = 3k + 1 > 0 \qquad (k \in \mathbb{Z})$$

in the ring $\mathbb{Z}[\rho]$, where ρ is a cube root of unity. We prove the statement:

(S). Let ℓ be the integer least in absolute value such that

$$\ell \equiv - \{(2k)!\}^3 \qquad (p).$$

Then there is an $m \in \mathbb{Z}$ such that

$$4p = \ell^2 + 27m^2.$$

Further,

$$\ell \equiv + 1 \ \ (3), \quad \text{and} \quad \ell \equiv m \ \ (2).$$

The Exercise will make use of Exercise 6 of Chapter 8 without always giving explicit reference.

There are two characters of order 3 on the multiplicative group of $\mathbb{Z}$ mod p taking values $1, \rho, \rho^2$. We denote one by χ and the other by $\bar{\chi} = \chi^2$. Let M be a primitive p-th root of 1 and put

$$T(\chi,M) = \sum_{a=1}^{p-1} \chi(a)\, M^a.$$

(i) Show that

$$T(\chi,M)\, T(\bar{\chi},M) = p.$$

(ii) Show that

$$\{T(\chi,M)\}^3 \in \mathbb{Z}[\rho],$$

and that

$$\{T(\chi,M)\}^3 \equiv -1 \qquad (3\sqrt{-3}).$$

[Hint $T(\chi,M) \equiv -1 \quad (1-\rho)$.]

There are two valuations on $\mathbb{Q}(\rho)$ extending the p-adic valuation and so embedding $\mathbb{Q}(\rho)$ in $\mathbb{Q}_p$. Denote by $\mathfrak{p}$ the valuation for which χ embeds as ω^k, where ω is the Teichmüller character, and denote the other valuation by $\bar{\mathfrak{p}}$. The unique extensions to $\mathbb{Q}(\rho,M)$ are denoted by $\mathfrak{P}$ and $\bar{\mathfrak{P}}$ respectively.

(iv) Show that

$$|T(\chi,M)|_{\mathfrak{P}} = p^{-2/3}, \quad |T(\chi,M)|_{\bar{\mathfrak{P}}} = p^{-1/3}.$$

[Hint Exercise 6 (iv) of Chapter 8].

(v) Deduce that

$$\{T(\chi,M)\}^3 = p\pi$$

where $\pi \in \mathbb{Z}(\rho)$ generates $\mathfrak{p}$ as a $\mathbb{Z}[\rho]$-module and $p\pi \equiv -1\ (3\sqrt{-3})$.

(vi) Show that

$$\left|\{T(\chi,M)\}^3 + \frac{\Pi^{p-1}}{(k!)^3}\right|_{\bar{\mathfrak{P}}} < p^{-1}.$$

[Hint See previous hint].

(vii) Deduce that

$$\pi \equiv \frac{1}{(k!)^3} \qquad (\bar{\pi}).$$

(viii) Put

$$2\pi = \ell + 3m\sqrt{-3},$$

where $\sqrt{-3} = \rho - \rho^2$, and complete the proof of Statement (S).

12. Let $p = 4k + 1$ be prime. Let ℓ, m be the numerically smallest solutions in $\mathbb{Z}$ of

$$2\{(2k)!\}\{k!\}^2 \ell \equiv 1 \quad (p) \qquad (*)$$

$$2m \equiv (2k!)\ \ell \quad (p) \qquad (\S)$$

By following the argument outlined below, show that

$$p = \ell^2 + 4m^2.$$

(i) The embeddings of $i = \sqrt{-1}$ in $\mathbb{Q}_p$ satisfy $i \equiv \pm(2k)! \bmod p$.

(ii) Let χ be a character of order 4 mod p taking values $\pm 1, \pm i$ and put

$$T(\chi,M) = \sum_{a=1}^{p-1} \chi(a)\, M^a.$$

Show that

$$\{T(\chi,M)\}^2 \equiv T(\chi^2,M) \equiv 1 \qquad (2)$$

and that

$$\{T(\chi,M)\}^2 = T(\chi^2,M)\ \alpha$$

where $\alpha \in \mathbb{Z}[i]$ has norm p.

(iii) Putting $\alpha = \ell + 2im$ show that $\pm \ell$ satisfies (*).

(iv) Deduce (§) from (i) above.

12. bis. Let $p = 2\ell + 1$ be prime, where ℓ is odd. Show that

$$\ell! \equiv \pm 1 \qquad (p)$$

and that both signs can occur.

[Note Cf. Mordell (1961).]

13. Let $p = 8k + 1$ be prime and let ℓ be the numerically smallest solution in $\mathbb{Z}$ of the congruence

$$2(k!)\{(4k)!\}\ell \equiv (5k)! \qquad (p).$$

Show that

$$p = \ell^2 + 2m^2$$

for some m.

14. (i) Let $p > 2$ be prime, let M be a primitive p-th root of unity, and let θ, χ be two characters on the multiplicative group mod p. Denote by $\theta\chi$ the character whose values are given by $(\theta\chi)(a) = \theta(a)\chi(a)$. On the hypothesis that none of $\theta, \chi, \theta\chi$ is the trivial character (= 1 identically), show that

$$T(\theta,M)\ T(\chi,M) = J(\theta,\chi)\ T(\theta\chi,M),$$

where

$$J(\theta,\chi) = \sum_{d \bmod p} \theta(d)\ \chi(1 - d) \qquad (£)$$

with the convention that $\theta(0) = \chi(0) = 0$.

[Hint See the hint for Exercise 6 (iii) of Chapter 8].

(ii) Let $p \equiv 1\ (3)$ and let χ be a character mod p of order 3 taking values $1, \rho, \rho^2$. Show that

$$J(\chi,\chi) \in \mathbb{Z}[\rho]$$

has norm p.

(iii) In (ii) above consider the embedding of $\mathbb{Q}(\rho)$ into $\mathbb{Q}_p$ for which $\chi = \omega^{2k}$ where ω^m is the Teichmüller character and $p = 3k + 1$. With this embedding, show that

$$J(\chi,\chi) \equiv \sum_{d \bmod p} (d - d^2)^{2k}$$

$$\equiv (-)^{k+1} \binom{2k}{k} \qquad (p)$$

on expanding by the binomial Theorem and interchanging the order of summation. Compare with the result in Exercise 11.

(iv) Consider similarly the results in Exercises 12, 13.

15. This exercise extends the notion of <u>different</u> to the semilocal situation, cf. Chapter 7, Exercise 20. Notation as in §3.

(i) Let

$$W = \{C \in K : S(AC) \in \mathfrak{o} \text{ for all } A \in \mathcal{O}\},$$

where $S = S_{K/k}$ is the trace. Show that W is a $\mathcal{O}$-module. For the extension $|\ |_j$ of $|\ |$ to K define $d(j)$ by

$$|\Pi_j|_j^{-d(j)} = \sup\{|C|_j : C \in W\}.$$

Show $d(j)$ is just the d of the local extension $K_j/\bar{k}$.

[Hint Lemma 3.1].

(ii) Show that

$$|\Pi_j|_j^{d(j)} = \sup_A |F_A'(A)|_j,$$

where A runs through $\mathfrak{O}$ and $F_A(X)$ is the characteristic polynomial of A over k.

[Note Contrast Exercise 9.

Hint Let B_j generate $\mathfrak{O}_j$ over $\bar{k}$, where without loss of generality $|B_j|_j = 1$. Consider A which are close to B_j with respect to $|\ |_j$ and to 0 with respect to the $|\ |_i$ $(i \neq j)$].

16. Let p_j $(1 \leq j \leq J)$ be distinct rational primes and let $B_j^2 = p_j$. Show that K (say) $= \mathbb{Q}(B_1,\ldots,B_J)$ has degree 2^J.

Let $C = \sum B_j$. Show that $K = \mathbb{Q}(C)$ and deduce that the minimum polynomial $F(T) \in \mathbb{Z}[T]$ of C has degree 2^J.

Show that F factorizes in $Z_p[T]$ into polynomials of degree ≤ 4 $(p \neq 2)$ and of degree ≤ 8 $(p = 2)$.

Note There are cribs to most of the exercises on cyclotomic fields in Bachmann (1872).

CHAPTER TEN: ALGEBRAIC NUMBER FIELDS

1 INTRODUCTION

In this book by an algebraic number field k we shall mean an extension of the rationals of finite degree. In this Chapter we shall investigate the valuations on an algebraic number field. The first objective is to establish the basic arithmetic properties of algebraic number fields and this, after the preparation in previous chapters, we can do in very short order. These properties were first found in the 19th century in a different way (ideal theory), and we explain the relation between the two approaches.

The structure theorems about algebraic number fields permit the techniques discussed earlier to be applied to a wide variety of diophantine problems and offers new approaches. This will be discussed here and in the next Chapter.

It is convenient to introduce notation and conventions which are rather different from those used earlier and which will now be explained. We refer to an equivalence class of valuations as a place and denote it by a (usually small) gothic letter: $\mathfrak{p}$, $\mathfrak{q}$, $\mathfrak{r}$. The corresponding completion of k is $k_{\mathfrak{p}}$, $k_{\mathfrak{q}}$, $k_{\mathfrak{r}}$. If $\mathfrak{p}$ is non-arch., the valuations induced in $\mathbb{Q} \subset k$ are equivalent to the p-adic valuation, where we write

$$\mathfrak{p} | p.$$

Then $k_{\mathfrak{p}}$ is a finite extention of $\mathbb{Q}_p$ and so is a $\mathfrak{p}$-adic field in the sense of Chapter 8 §1. We denote by $| \ |_{\mathfrak{p}}$ the renormalized valuation introduced there, so

$$|\pi_{\mathfrak{p}}|_{\mathfrak{p}} = q_{\mathfrak{p}}^{-1},$$

where $\pi_{\mathfrak{p}}$ is a prime element and $q_{\mathfrak{p}}$ is the cardinality of the residue class field. We shall also need to consider the arch. places $\mathfrak{r}$ on k. The relevant completion $k_{\mathfrak{r}}$ is either $\mathbb{R}$ ($\mathfrak{r}$ is a <u>real</u> place) or $\mathbb{C}$ (<u>complex place</u>). We denote by $| \; |_{\mathfrak{r}}$ the renormalized valuation, so $| \; |_{\mathfrak{r}}$ is the absolute value when $\mathfrak{r}$ is real but the square of the absolute value when $\mathfrak{r}$ is complex. An arch. place is an extension of the arch. place ∞ on $\mathbb{Q}$. We write this

$$\mathfrak{r}|\infty.$$

2 PRODUCT FORMULA

We showed in Chapter 2 that any finite number of valuations behave independently in a very strong sense. There is, however, even stronger independence involving *all* the valuations on an algebraic number field. Before proving it we need

LEMMA 2.1. *Let* $A \in k^*$, *where* k *is an algebraic number field. Then* $|A|_{\mathfrak{p}} = 1$ *for almost all* $\mathfrak{p}$.

Note By *almost all* we shall mean all except at most finitely many. As usual, k^* is the group of non-zero elements of k.

Proof. Let

$$\phi(X) = X^n + a_{n-1} X^{n-1} + \dots + a_o \in \mathbb{Q}[X] \tag{2.1}$$

be a minimum polynomial for A. Then

$$a_j \in \mathbb{Z}_p \qquad (0 \leq j \leq n - 1) \tag{2.2}$$

for almost all p. Now (2.2) implies $|A|_{\mathfrak{p}} \leq 1$ for all $\mathfrak{p}|p$. Hence $|A|_{\mathfrak{p}} \leq 1$ for almost all $\mathfrak{p}$. Similarly $|A^{-1}|_{\mathfrak{p}} \leq 1$ for almost all $\mathfrak{p}$.

THEOREM 2.1 (Product formula)

$$\prod_{\substack{\mathfrak{p} \\ \text{arch.} \\ \text{or non-arch.}}} |A|_{\mathfrak{p}} = 1 \tag{2.3}$$

for any $A \in k^*$.

Note Although the product in (2.3) is formally infinite, it is essentially finite by the previous lemma.

Proof. By Lemma 2.2 Corollary 3 we have

$$\prod_{\mathfrak{p}|w} |A|_{\mathfrak{p}} = |N_{k/\mathbb{Q}}(A)|_w \tag{2.4}$$

where w is a prime p or the symbol ∞. This reduces the theorem to the case $k = \mathbb{Q}$, which is easy.

For later purposes we note that the proof also gives the

COROLLARY 1.

$$\prod_{\mathfrak{p} \text{ non-arch.}} |A|_{\mathfrak{p}} = \pm 1/N_{k/\mathbb{Q}}(A) > 0 \tag{2.5}$$

$$\prod_{\mathfrak{p} \text{ arch.}} |A|_{\mathfrak{p}} = \pm N_{k/\mathbb{Q}}(A) > 0. \tag{2.6}$$

3 ALGEBRAIC INTEGERS

These are the analogue for algebraic number fields for the rational integers in $\mathbb{Q}$. They may be defined in a number of equivalent ways:

THEOREM 3.1 *Let* A *be algebraic over* $\mathbb{Q}$. *The following statements are equivalent:*

(i) *the minimum polynomial* $\phi(X) \in \mathbb{Q}[X]$ *for* A *(with top coefficient* 1) *is in* $\mathbb{Z}[X]$

(ii) *There is some* $\psi(X) \in \mathbb{Z}[X]$ *with top coefficient* 1 *such that* $\psi(A) = 0$.

(iii) $\mathbb{Z}[A]$ *is finitely generated as a* $\mathbb{Z}$*-module*.

(iv) $|A|_{\mathfrak{p}} \leq 1$ *for all non-arch. valuations* $\mathfrak{p}$ *of* $\mathbb{Q}(A)$.

DEFINITION 3.1. If one (and so all) of these statements is true, we say that A is an *algebraic integer*.

Proof. (i) → (ii). Trivial.

(ii) → (iii). If $A^m + \sum_{j<m} a_j A^j = 0$ $(a_j \in \mathbb{Z})$ then $\mathbb{Z}[A]$ is generated by $1, A, \ldots, A^{m-1}$.

(iii) → (iv). Let $B_1, \ldots, B_\ell$ be a set of generators of $\mathbb{Z}[A]$. Then

$$|C|_{\mathfrak{p}} \leq \max_j |B_j|_{\mathfrak{p}}$$

for all $C \in \mathbb{Z}[A]$. On putting $C = A^n$, $n \to \infty$, we have $|A|_{\mathfrak{p}} \leq 1$, as required.

(iv) → (i). Let K be a splitting field of $\phi(X)$, say

$$\phi(X) = \prod_j (X - A_j) = X^n + \sum_{j<n} a_j X^j. \quad (a_j \in \mathbb{Q}), \tag{3.1}$$

with $A_j \in K$ $(1 \leq j \leq n)$. Let p be a prime and let $\mathfrak{P}/p$ be an extension to K.

For $j = 1, \ldots, n$, the valuation $\mathfrak{P}$ induces a valuation $\mathfrak{p}_j$ on $\mathbb{Q}(A)$ via the embedding $A \mapsto A_j$ of $\mathbb{Q}(A)$ in K. Hence $|A_j|_{\mathfrak{P}} \leq 1$ $(1 \leq j \leq n)$. It follows that $|a_j|_{\mathfrak{P}} \leq 1$ $(0 \leq j \leq n-1)$, and so $|a_j|_p \leq 1$. Since p is any rational prime, we have $a_j \in \mathbb{Z}$, as required.

<u>COROLLARY 1</u>. <u>The algebraic integers in</u> $\mathbb{Q}$ <u>are just the rational integers</u>.

<u>Note</u> This is comforting. We shall feel free to use <u>integer</u> to mean "algebraic integer".

<u>COROLLARY 2</u>. <u>If</u> B <u>is algebraic over</u> $\mathbb{Q}$, <u>there is some</u> $\ell \in \mathbb{Z}$, $\ell \neq 0$ <u>such that</u> ℓB <u>is an algebraic integer</u>.

<u>Proof</u>. Easy, using (i) of Theorem 3.1.

<u>COROLLARY 3</u>. <u>The algebraic integers</u> A <u>in an algebraic number field</u> k <u>are characterized by</u> $|A|_{\mathfrak{p}} \leq 1$ <u>for all non-arch. valuations</u> $\mathfrak{p}$ <u>of</u> k.

<u>Proof</u>. For every valuation of $\mathbb{Q}(A)$ extends to k.

COROLLARY 4. The sets of integers $\mathfrak{o}$ in an algebraic number field k are a ring.

Proof. Previous corollary.

LEMMA 3.1. The integers $\mathfrak{o}$ of an algebraic number field k are a finitely generated $\mathbb{Z}$-module.

Proof. Let $n = [K : k]$. We shall require the discriminant

$$D(A_1,\ldots,A_n) = \det(S_{k/\mathbb{Q}}(A_iA_j))_{i,j} \tag{3.2}$$

of n elements $A_1,\ldots,A_n$ of k, which was introduced in a more general context in §6 of Chapter 7. It was shown there that $D(A_1,\ldots,A_n) \neq 0$ whenever $A_1,\ldots,A_n$ are linearly independent over $\mathbb{Q}$. If $A_1,\ldots,A_n$ are in the ring of integers of k, then $D(A_1,\ldots,A_n) \in \mathbb{Z}$ because it is (i) rational and (ii) an algebraic integer.

By Corollary 2 of the preceding theorem, there certainly are sets $A_1,\ldots,A_n$ of elements of $\mathfrak{o}$ which are linearly independent over $\mathbb{Q}$. If every $B \in \mathfrak{o}$ is a $\mathbb{Z}$-linear combination of $A_1,\ldots,A_n$ then we have a $\mathbb{Z}$-basis for $\mathfrak{o}$ and are done. Otherwise there is a $B \in \mathfrak{o}$ and a prime p such that $pB = a_1A_1 + \ldots + a_nA_n$ for some $a_j \in \mathbb{Z}$ but such that B itself is not of this shape. Clearly, not all the a_j can be divisible by p and without loss of generality $p \nmid a_1$. There are thus $b,c \in \mathbb{Z}$ such that

$$p(bB + cA_1) = A_1 + c_2A_2 + \ldots + c_nA_n$$

for some $c_2,\ldots,c_n \in \mathbb{Z}$.

Put $A_1^* = bB + cA_1$. Then

$$D(A_1^*,A_2,\ldots,A_n) = p^{-2}\, D(A_1,\ldots,A_n).$$

Since the discriminants are non-zero rational integers, this process can be repeated only a finite number of times. When it stops, we have a $\mathbb{Z}$-basis of $\mathfrak{o}$. This concludes the proof.

Let $\{A_1,\ldots,A_n\}$ and $\{B_1,\ldots,B_n\}$ be two bases of $\mathfrak{o}$. Then

$$B_i = \sum_j t_{ij} A_j$$

for $t_{ij} \in \mathbb{Z}$. Since the A_j can be integrally represented by the B_i, we have $\det(t_{ij}) = \pm 1$ and so

$$D(B_1,\ldots,B_n) = \{\det(t_{ij})\}^2 \; D(A_1,\ldots,A_n)$$

$$= D(A_1,\ldots,A_n).$$

We have thus proved

LEMMA 3.2. *The value of the discriminant* $D(A_1,\ldots,A_n)$ *is the same for all bases of* $\mathfrak{o}$.

DEFINITION 3.1. The common value D_k is the *discriminant* of k.

The discriminant is an important arithmetic invariant of k. We delay a further discussion of its properties until §11, where we shall meet a more general version. The discriminant defined above is then called the *absolute* discriminant since it refers to k as an extension of $\mathbb{Q}$, not of some intermediate field (and similarly with such expressions as "absolute norm".)

We now consider bases for the integers of certain types of field.

LEMMA 3.3. *Let* $k = \mathbb{Q}(B)$ *where* $B^2 = b \in \mathbb{Z}$ *is square free. Then a* $\mathbb{Z}$*-basis for the ring of integers is* $\{1, \frac{1}{2}(1 + B)\}$ *if* $b \equiv 1$ (4) *and* $\{1,B\}$ *otherwise.*

Proof. Suppose that

$$A = u + vB \quad (u,v \in \mathbb{Q})$$

is an integer. Its minimum polynomial is

$$X^2 - 2u\,X + (u^2 + v^2 b),$$

so $2u$, $u^2 - v^2 b \in \mathbb{Z}$. If $b \not\equiv 1$ (4) it really follows that $u,v \in \mathbb{Z}$

while if $b \equiv 1$ (4) we have $2u, 2v \in \mathbb{Z}$, $2u \equiv 2v$ (2).

The above case was so simple that we could use brute force. Usually it is more convenient to use

LEMMA 3.4. *Let $k/\mathbb{Q}$ have degree n and suppose that, for every prime p, the set $B_1,\dots,B_n$ is a $\mathbb{Z}_p$-basis for the semilocal integers of k with respect to $|\ |_p$. Then $B_1,\dots,B_n$ is a $\mathbb{Z}$-basis for the integers of k.*

Proof. Clear, e.g. by Theorem 3.1, Corollary 3. cf. Lemma 11.1.

As applications we have

LEMMA 3.5. *Let $k = \mathbb{Q}(M)$ where M is a root of unity. Then a basis for the integers is $\{1, M,\dots,M^{n-1}\}$, where n is the degree.*

Proof. Follows from the preceding lemma and Lemma 4.2 of Chapter 9.

LEMMA 3.6. *Let $k = \mathbb{Q}(B)$ where $B^3 = b$ and $b \in \mathbb{Z}$ is cube free, say $b = ef^2$ where e,f are square free. Put $C = f^{-1}B^2$. Then $\{1,B,C\}$ is a $\mathbb{Z}$-basis for the integers of k unless $b \equiv \pm 1$ (9); in which case a basis is $\{1,B, (1 + eB + sC)/3\}$ with $s^2 = 1$, $s \equiv f$ (3).*

Proof. We again apply Lemma 3.4. For $p \nmid 3b$ the sets specified are $\mathbb{Z}_p$-bases of the semilocal integers by Lemma 3.3 of Chapter 9. If $p|b$, then p is completely ramified, and it is easily checked that we have a $\mathbb{Z}_p$-basis. Finally, for $p = 3$ it can again be verified that we have a $\mathbb{Z}_3$-basis (cf. Exercise 2 of Chapter 9).

We conclude this section by noting that there is an alternative approach to Lemma 3.1 using the Product Theorem 2.1, which is, however, in some sense less constructive than the argument using the discriminant. If $A \in \mathfrak{o}$, $A \neq 0$, we have $|A|_{\mathfrak{p}} \leq 1$ for all non-arch. $\mathfrak{p}$, and so $\prod_j |A|_j \geq 1$ by Theorem 2.1, where for simplicity, we have denoted the arch. valuations by $|\ |_j$. Hence there is a neighbourhood of 0 in

$$\mathbb{R} \otimes_{\mathbb{Q}} k = \bigoplus_j k_j \tag{3.3}$$

which contains no element of $\mathfrak{o}$ except 0. Here k_j is the completion of k with respect to $|\ |_j$ and we have invoked Theorem 1.1 of Chapter 9. But (3.3) is isomorphic to $\mathbb{R}^n$ as an $\mathbb{R}$-module. The existence of a $\mathbb{Z}$-basis of $\mathfrak{o}$ now follows from Appendix C, Theorem 2.

4 STRONG APPROXIMATION THEOREM

The following theorem generalizes the "Chinese Remainder Theorem" to algebraic number fields.

THEOREM 4.1. (Strong approximation) *Let P be a finite set of non-arch. places of the algebraic number field k. Let $\varepsilon > 0$ and $A_{\mathfrak{p}} \in k_{\mathfrak{p}}$ $(\mathfrak{p} \in P)$ be given. Then there is an $A \in k$ such that*

$$|A - A_{\mathfrak{p}}|_{\mathfrak{p}} < \varepsilon \quad (\mathfrak{p} \in P) \tag{4.1}$$

$$|A|_{\mathfrak{p}} \leq 1 \quad (\mathfrak{p} \text{ non-arch. } \notin P). \tag{4.2}$$

Note If $A_{\mathfrak{p}} \in \mathfrak{o}_{\mathfrak{p}}$ $(\mathfrak{p} \in P)$, we can replace (4.2) by $A \in \mathfrak{o}$ (Definition 3.1).

Proof. Let P_o be the set of rational primes p such that $\mathfrak{p}|p$ for some $\mathfrak{p} \in P$. We may suppose without loss of generality that P consists of all the $\mathfrak{p}$ which extend the $p \in P_o$ (putting, say, $A_{\mathfrak{p}} = 0$ for the $\mathfrak{p}$ which were not in the original P).

By the "Weak Approximation Theorem" (Theorem 3.1 of Chapter 2) there is a $B \in k$ such that

$$|B - A_{\mathfrak{p}}| < \varepsilon \quad (\mathfrak{p} \in P).$$

By Lemma 2.1 the set S of non-arch. places $\mathfrak{p} \notin P$ for which $|B|_{\mathfrak{p}} > 1$ is finite. Let S_o be the set of rational primes p for which there is a $\mathfrak{p} \in S$ with $\mathfrak{p}|p$. Then P_o and S_o are disjoint by the first paragraph of the proof. Let $\eta > 0$ be given. By the Chinese Remainder Theorem there is an $\ell \in \mathbb{Z}$ such that

$$|\ell - 1|_p < \eta \qquad (p \in P_o)$$

$$|\ell|_p < \eta \qquad (p \in S_o).$$

Then $A = \ell B$ satisfies the conclusions of the Theorem provided that η is chosen sufficiently small.

5 DIVISORS. RELATION TO IDEAL THEORY

For a given algebraic number field k a divisor $\underline{d} = \{d_{\mathfrak{p}}\}$ is a map from the non-arch. places $\mathfrak{p}$ of k to $\mathbb{Z}$ such that $d_{\mathfrak{p}} = 0$ for almost all $\mathfrak{p}$. Addition and subtraction of divisors is defined elementwise:

$$\{d_{\mathfrak{p}}\} \pm \{e_{\mathfrak{p}}\} = \{d_{\mathfrak{p}} \pm e_{\mathfrak{p}}\}. \tag{5.1}$$

For reasons which will be apparent in a moment, we denote the divisor all of whose elements are 0 by [1]. There is a partial order on divisors. We write

$$\{d_{\mathfrak{p}}\} \geq \{e_{\mathfrak{p}}\} \tag{5.2}$$

if $d_{\mathfrak{p}} \geq e_{\mathfrak{p}}$ for all $\mathfrak{p}$. If $\underline{d} \geq [1]$ we say that $\underline{d}$ is effective.

We shall use the order notation which was introduced briefly in §7 of Chapter 4. For $A \in k_{\mathfrak{p}}$ we define

$$\operatorname{ord}_{\mathfrak{p}} A = \begin{cases} m & \text{if } A \neq 0 \text{ and } |A|_{\mathfrak{p}} = |\pi_{\mathfrak{p}}|^m \\ \infty & \text{if } A = 0. \end{cases} \tag{5.3}$$

For $A \in k^*$ we have $\operatorname{ord}_{\mathfrak{p}} A = 0$ for almost all m by Lemma 2.1, and so

$$[A] = \{\operatorname{ord}_{\mathfrak{p}} A\}. \tag{5.4}$$

is a divisor. We call it the principal divisor belonging to A. In particular, all the elements of [1] are 0, justifying the notation introduced above. Clearly

$$[A^{-1}] = -[A] \tag{5.4.bis}$$

and $$[AB] = [A] + [B] \tag{5.4.ter}$$

so we have a homomorphism of k^* into the group of divisors. Further,

$$A \in \mathfrak{o} \iff A = 0 \quad \text{or} \quad [A] \geq [1] \tag{5.5}$$

by Definition 3.1.

We note in passing that there is a more picturesque "multiplicative" notation for divisors:

$$\Pi\, \mathfrak{p}^{d_{\mathfrak{p}}} \tag{5.5.bis}$$

instead of $\{d_{\mathfrak{p}}\}$. This stresses the analogy between the principal divisor $[A]$ and the expression of a rational number up to sign as the product of prime powers, and is sometimes more convenient in practice.

The arithmetic theory of algebraic numbers was originally developed in terms of the ideals of $\mathfrak{o}$ and is usually still discussed in that language. We shall now establish the correspondence between the language of ideals and that of divisors.

Let $\underline{e}$ be an effective divisor and define

$$\mathfrak{a}(\underline{e}) = \{A : A = 0 \quad \text{or} \quad [A] \geq \underline{e}\}. \tag{5.6}$$

Clearly $\mathfrak{a}(\underline{e})$ is an $\mathfrak{o}$-ideal by (5.4) and (5.5). Further, $\mathfrak{a}(\underline{e})$ contains elements of $\mathfrak{o}$ other than 0.

Now let $\mathfrak{b}$ be an ideal of $\mathfrak{o}$ other than the ideal consisting of 0 alone (which we shall agree in future to exclude from consideration). Define

$$\operatorname{ord}_{\mathfrak{p}} \mathfrak{b} = \inf_{\substack{B \in \mathfrak{b} \\ \neq 0}} \operatorname{ord}_{\mathfrak{p}} B. \tag{5.7}$$

Clearly

$$[\mathfrak{b}] = \{\operatorname{ord}_{\mathfrak{p}} \mathfrak{b}\} \tag{5.8}$$

is an effective divisor.

<u>THEOREM 5.1</u> <u>There is a</u> 1 - 1 <u>correspondence between effective divisors and ideals of</u> $\mathfrak{n}$ <u>given by</u> (5.6) <u>and</u> (5.8). <u>More precisely:</u>

$$[\mathfrak{a}(\underline{e})] = \underline{e} \tag{5.9}$$

<u>and</u>

$$\mathfrak{a}([\mathfrak{h}]) = \mathfrak{h}. \tag{5.10}$$

<u>Proof.</u> (i) Let $\underline{e}$ be an effective divisor. Clearly

$$[\mathfrak{a}(\underline{e})] \geq \underline{e}. \tag{5.11}$$

Let B be any non-zero element of $\mathfrak{a}(\underline{e})$ and let P be the set of $\mathfrak{p}$ for which $\mathrm{ord}_{\mathfrak{p}} B > 0$, so P is finite by Lemma 2.1. By the strong Approximation Theorem 4.1 there is a $C \in \mathfrak{n}$ such that

$$\mathrm{ord}_{\mathfrak{p}} C = e_{\mathfrak{p}} \quad (\text{all } \mathfrak{p} \in P).$$

Then $C \in \mathfrak{a}(e)$, and so

$$\mathrm{ord}_{\mathfrak{p}}\, \mathfrak{a}(\underline{e}) \leq \max\{\mathrm{ord}_{\mathfrak{p}} B, \mathrm{ord}_{\mathfrak{p}} C\}$$

$$= e_{\mathfrak{p}}.$$

This with (5.11) gives (5.9).

(ii) Let $\mathfrak{h}$ be any ideal of $\mathfrak{n}$. Clearly

$$\mathfrak{a}([\mathfrak{h}]) \supset \mathfrak{h}. \tag{5.12}$$

Let B be any non-zero element of $\mathfrak{h}$ and denote by P the set of $\mathfrak{p}$ for which $\mathrm{ord}_{\mathfrak{p}} B > 0$, so P is finite. By (5.7) for every $\mathfrak{p} \in P$ there is a $C_{\mathfrak{p}} \in \mathfrak{h}$ such that

$$\mathrm{ord}_{\mathfrak{p}}\, C_{\mathfrak{p}} = \mathrm{ord}_{\mathfrak{p}}\, \mathfrak{h}. \tag{5.13}$$

Let A be any non-zero element of $\mathfrak{a}([\mathfrak{h}])$. We have to show that $A \in \mathfrak{h}$. By (5.13) we have

$$\mathrm{ord}_{\mathfrak{p}}(A/C_{\mathfrak{p}}) \geq 0$$

and so by the Strong Approximation Theorem there are $L_{\mathfrak{p}} \in \mathfrak{o}$ such that

$$\mathrm{ord}_{\mathfrak{p}}\{L_{\mathfrak{p}} - (A/C_{\mathfrak{p}})\} \geq \mathrm{ord}_{\mathfrak{p}} B$$

$$\mathrm{ord}_{\mathfrak{q}} L_{\mathfrak{p}} \geq \mathrm{ord}_{\mathfrak{q}} B \quad (\mathfrak{q} \in P, \quad \mathfrak{q} \neq \mathfrak{p}).$$

Then

$$\mathrm{ord}_{\mathfrak{p}}\{A - \sum_{\mathfrak{p} \in P} L_{\mathfrak{p}} C_{\mathfrak{p}}\} \geq \mathrm{ord}_{\mathfrak{p}} B \qquad (5.14)$$

for all $\mathfrak{p} \in P$. Indeed (5.14) holds for all $\mathfrak{p}$, since $\mathrm{ord}_{\mathfrak{p}} B = 0$ for $\mathfrak{p} \notin P$ by the definition of P.

Put

$$A - \sum_{\mathfrak{p}} L_{\mathfrak{p}} C_{\mathfrak{p}} = BM.$$

Then $M \in \mathfrak{o}$ by (5.14) and Definition 3.1. Since $B \in \mathfrak{h}$ and $C_{\mathfrak{p}} \in \mathfrak{h}$, we have $A \in \mathfrak{h}$, as required.

COROLLARY 1. *Every ideal $\mathfrak{h}$ of $\mathfrak{o}$ is generated as an $\mathfrak{o}$-module by at most 2 elements, one of which may be any given element $B \neq 0$ in $\mathfrak{h}$.*

Proof. Let C be as in part (i) of the proof with $\mathfrak{h} = \mathfrak{a}(\underline{e})$. Then in (ii) of the proof we can take $C_{\mathfrak{p}} = C$ for all $\mathfrak{p} \in P$.

Note After Theorem 5.1 there is no longer any point in using a different type of notation for divisors and ideals. In what follows, we shall often bow to tradition and use small Fraktur letters for divisors (even when they are not effective). We will not be consistent, but (appropriately enough for a volume with this title) our conventions will be locally constant. In the same spirit, we shall sometimes use *prime divisor* (or even *prime ideal*) as a synonym for: place.

We say that the field k has a *euclidean algorithm* if for every $A \in k$ there is some $B \in \mathfrak{o}$ such that

$$|N_{k/\mathbb{Q}}(A - B)| < 1. \qquad (5.11 \text{ bis}).$$

COROLLARY 2. If k has a euclidean algorithm, then every divisor is principal.

Proof. It is enough to consider effective divisors and so, by the Theorem, $\mathfrak{o}$-ideals.

Let $\mathfrak{a}$ be an $\mathfrak{o}$-ideal. Since the norm takes integral values on $\mathfrak{o}$, there is at least one $C \in \mathfrak{a}$ such that

$$|N_{k/\mathbb{Q}}(C)|_\infty = \inf_{\substack{D \in \mathfrak{a} \\ D \neq 0}} |N_{k/\mathbb{Q}}(D)|_\infty \tag{5.11 ter}$$

Let $E \in \mathfrak{a}$. By (5.11 bis) with $A = E/C$ there is some $B \in \mathfrak{o}$ such that

$$|N_{k/\mathbb{Q}}(E - BC)|_\infty < |N_{k/\mathbb{Q}}(C)|_\infty.$$

But $E - BC \in \mathfrak{a}$, so $E - BC = 0$ by (5.11 ter). Hence $\mathfrak{a}$ is the principal ideal generated by C.

An example of a field with a euclidean algorithm is $\mathbb{Q}(\sqrt{-2})$. Here $1, \sqrt{-2}$ is a basis for $\mathfrak{o}$. If

$$A = x + y\sqrt{-2} \quad (x, y \in \mathbb{Q})$$

we take

$$B = u + v\sqrt{-2} \quad (u, v \in \mathbb{Z})$$

with

$$|u - x|_\infty \leq \tfrac{1}{2}, \quad |v - y|_\infty \leq \tfrac{1}{2}.$$

Then

$$|N(A - B)| = (x - u)^2 + 2(y - v)^2$$

$$\leq \tfrac{3}{4} < 1,$$

as required.

We now require

DEFINITION 5.1. The norm $\mathfrak{N}(\underline{d})$ of a divisor $\underline{d} = \{d(\mathfrak{p})\}$ is given by

$$\mathfrak{N}(\underline{d}) = \prod_{\mathfrak{p}} q_{\mathfrak{p}}^{d(\mathfrak{p})}, \tag{5.15}$$

where $q_{\mathfrak{p}}$ is the cardinality of the residue class field.

An immediate consequence is

$$\mathfrak{N}(\underline{d} + \underline{e}) = \mathfrak{N}(\underline{d})\, \mathfrak{N}(\underline{e}). \tag{5.16}$$

LEMMA 5.1. Let $A \in k^*$. Then

$$\mathfrak{N}([A]) = \pm N_{k/\mathbb{Q}}(A) > 0. \tag{5.17}$$

Proof. Follows at once from (2.4) and (5.4).

LEMMA 5.2. Let $\underline{d}$ be an effective divisor and $\mathfrak{a} = \mathfrak{a}(\underline{d})$ in the notation (5.6). Then $\mathfrak{N}(\underline{d})$ is the number of residue classes in $\mathfrak{o}$ mod $\mathfrak{a}$.

Proof. Let $\mathfrak{p}$ be a place and $d = d(\mathfrak{p}) > 0$. As in the proof of Lemma 1.4 of Chapter 4, the cardinality of $\mathfrak{o}_{\mathfrak{p}}$ modulo $\pi_{\mathfrak{p}}^{d}\, \mathfrak{o}_{\mathfrak{p}}$ is $q_{\mathfrak{p}}^{d(\mathfrak{p})}$, where $\mathfrak{o}_{\mathfrak{p}}$ is the ring of $\mathfrak{p}$-integers. Let P be the set of $\mathfrak{p}$ for which $d(\mathfrak{p}) > 0$. By the strong approximation theorem 4.1, $\mathfrak{o}$ runs independently through all the residue classes for the $\mathfrak{p} \in P$. Hence the total number of possibilities is (5.15), as required.

COROLLARY 1. $\mathfrak{a}(\underline{d})$ is a free $\mathbb{Z}$-module on $n = [k : \mathbb{Q}]$ generators $B_1, \ldots, B_n$. If $A_1, \ldots, A_n$ is a basis of $\mathfrak{o}$ then

$$B_i = \sum t_{ij} A_j \tag{5.18}$$

where $t_{ij} \in \mathbb{Z}$ and

$$\det(t_{ij}) = \pm \mathfrak{N}(\underline{d}). \tag{5.19}$$

In particular,

$$\det S_{K/k}(B_i B_j) = \{\mathfrak{N}(\underline{d})\}^2 D_k \tag{5.20}$$

where D_k *is the discriminant of* k.

Proof. Follows from Lemma 1 of Appendix C with $m = \mathfrak{N}(\underline{d})$, since $\mathfrak{a}(\underline{d})$ is of index $\mathfrak{N}(\underline{d})$ in $\mathfrak{o}$.

COROLLARY 2. Suppose that $\underline{d}$ *is not necessarily effective, and denote by* $\mathfrak{a}(\underline{d})$ *the set of* $A \in k$ *with* $(A) \geq \underline{d}$ *together with* 0. *Then the conclusions of Corollary* 1 *continue to hold except that now the* $t_{ij} \in \mathbb{Q}$ *are not necessarily integral*.

Proof. There is a positive rational integer m such that $\underline{d}^* = [m] + \underline{d} \geq [1]$. If $B_1^*, \ldots, B_n^*$ is a $\mathbb{Z}$-basis for $\mathfrak{a}(\underline{d}^*)$ then $B_j = m^{-1} B_j^*$ $(1 \leq j \leq n)$ is a $\mathbb{Z}$-basis for $\mathfrak{a}(\underline{d})$.

LEMMA 5.3. Let $c > 0$ *be given. There are only finitely many effective divisors* $\underline{d}$ *with* $\mathfrak{N}(\underline{d}) \leq c$.

Proof. Since $q_{\mathfrak{p}}$ is divisible by the residue class characteristic, there are only finitely many $\mathfrak{p}$ with $q_{\mathfrak{p}} \leq c$. The Lemma is now immediate by (5.15).

6 EXISTENCE THEOREMS

To obtain the structure theorems for units and for the divisor class group in the next two sections we shall need the existence of elements of the algebraic number field k with prescribed behaviour with respect to both the arch. and to non-arch. places. An appropriate tool is Minkowski's convex body Theorem, which is proved in Appendix C.

As usual, $n = [k : \mathbb{Q}]$. We suppose that there are r real and s complex places, so $r + 2s = n$. The notation r,s is standard in this context. We denote the arch. places by $\mathfrak{r}_j$, where $\mathfrak{r}_j$ is real for $1 \leq j \leq r$ and complex for $r + 1 \leq j \leq r + s$. To simplify notation, we denote the (renormalized) valuation at $\mathfrak{r}_j$ by $| \; |_j$ and the completion by k_j.

The following theorem can be regarded as a counterpart to the Product Theorem 2.1.

THEOREM 6.1 There is a constant $c > 0$ *depending only on the field* k

with the following property:

Let $\underline{d}$ be a divisor (not necessarily effective) and let $t_j > 0$ $(1 \leq j \leq r + s)$ satisfy

$$\Pi\, t_j \geq c\, N(\underline{d}). \tag{6.1}$$

Then there is an $A \in k^*$ such that

$$[A] \geq \underline{d} \tag{6.2}$$

and

$$|A|_j \leq t_j \quad (1 \leq j \leq r + s). \tag{6.3}$$

A permissible value of c is

$$c = (2/\pi)^s\, |D_k|_\infty^{\frac{1}{2}}. \tag{6.4}$$

Here $\pi = 3.14159\ldots$, D_k is the discriminant (Definition 3.1) and $|\ |_\infty$ is the absolute value.

To emphasize the connection with the product formula, we give a reformulation. Let $\underline{d} = \{d(\mathfrak{p})\}$, where $\mathfrak{p}$ runs through the non-arch. valuations and put

$$t_{\mathfrak{p}} = |\pi_{\mathfrak{p}}|^{d(\mathfrak{p})}. \tag{6.5}$$

Then

$$\prod_{\mathfrak{p}\text{ non-arch.}} t_{\mathfrak{p}} = \{N(\underline{d})\}^{-1} \tag{6.6}$$

by (5.15). When $\mathfrak{p} = \mathfrak{r}_j$ is an arch. valuation we put $t_{\mathfrak{p}} = t_j$. Clearly the above enunciation is equivalent to

THEOREM 6.1 (second formulation). Let $t_{\mathfrak{p}}$ be given in the value group of $\mathfrak{p}$ for all places $\mathfrak{p}$, arch. and non-arch. Suppose that $t_{\mathfrak{p}} = 1$ for almost all $\mathfrak{p}$ and that

$$\prod_{\substack{\mathfrak{p}\text{ arch. and}\\ \text{non-arch.}}} t_{\mathfrak{p}} \geq c. \tag{6.7}$$

<u>Then there is an</u> $A \in k^*$ <u>such that</u>

$$|A|_{\mathfrak{p}} \leq t_{\mathfrak{p}} \quad \text{(all } \mathfrak{p}\text{, arch. and non-arch.)} \tag{6.8}$$

<u>Proof</u>. We prove the Theorem in the first formulation. Corresponding to the r_j we have embeddings

$$\sigma_j : k \hookrightarrow \mathbb{R} \quad (1 \leq j \leq r) \tag{6.9$_1$}$$

$$: k \hookrightarrow \mathbb{C} \quad (r+1 \leq j \leq r+s). \tag{6.9$_2$}$$

Let

$$\sigma_{j+s} = \bar{\sigma}_j \quad (r+1 \leq j \leq r+s) \tag{6.10}$$

where, in this proof, a bar denotes the complex conjugate. We have thus $n = r + 2s$ embeddings of k into $\mathbb{C}$.

Let $B_1, \ldots, B_n$ be the $\mathbb{Z}$-basis of the A satisfying (6.2) which is given by Lemma 5.2 Corollary 2, and put

$$\Delta = \det(\sigma_j B_\ell). \tag{6.11}$$

Then

$$\Delta^2 = \{N(\underline{d})\}^2 D_k \tag{6.12}$$

by (5.20).

Let the real linear forms $L_j(X_1, \ldots, X_n)$ in the indeterminates $X_1, \ldots, X_n$ be given by

$$\sum_\ell (\sigma_j B_\ell) X_\ell = L_j(X_1, \ldots, X_n) \quad (1 \leq j \leq r) \tag{6.13$_1$}$$

$$\sum_\ell (\sigma_j B_\ell) X_\ell = L_j(X_1, \ldots, X_n) + iL_{j+s}(X_1, \ldots, X_n)$$

$$(r+1 \leq j \leq r+s). \tag{6.13$_2$}$$

Then

$$\det(L_1, \ldots, L_n) = \pm(2i)^{-s}\, \Delta \; (\in \mathbb{R}) . \tag{6.14}$$

To prove the theorem we have to show that there are $b_1, \ldots, b_n$ not all 0 such that

$$\left.\begin{aligned} &|L_j(b_1, \ldots, b_n)|_\infty \leq t_j \quad (1 \leq j \leq r) \\ &\{L_j(b_1, \ldots, b_n)\}^2 + \{L_{j+s}(b_1, \ldots, b_n)\}^2 \leq t_j \\ &\qquad\qquad\qquad (r+1 \leq j \leq r+s) \end{aligned}\right\} \tag{6.15}$$

We can now apply Minkowski's Convex Body Theorem. The set C of $(y_1, \ldots, y_n) \in \mathbb{R}^n$ with

$$\left.\begin{aligned} &|y_j|_\infty \leq t_j \quad (1 \leq j \leq r) \\ &y_j^2 + y_{j+s}^2 \leq t_j \quad (r+1 \leq j \leq r+s) \end{aligned}\right\} \tag{6.16}$$

is clearly symmetric and convex. It has volume

$$V(C) = 2^r\, \pi^s\, \Pi t_j . \tag{6.17}$$

There is thus certainly a solution of (6.15) provided that

$$V(C) \geq 2^n\; |\det(L_1, \ldots, L_n)| . \tag{6.18}$$

By (6.12), (6.14) and (6.17), this is condition (6.1) of the Theorem and concludes the proof.

If we choose the t_j in Theorem 6.1 so that there is equality in (6.1) it follows that there is an A with $[A] \geq \underline{d}$ and such that $|N_{k/\mathbb{Q}}(A)|_\infty \leq \Pi t_j \leq c\mathbb{N}(\underline{d})$. For most theoretical applications the value of c is irrelevant but in practical numerical applications the value is quite important. In this context (but not in the Theorem itself) it turns out that we can do much better than (6.4). The determination of the best possible constant (considered as a function of r and s) is a difficult problem in the Geometry of Numbers which has been solved only for a few values of r and s. There is, however, an estimate due to Minkowski which is very useful:

THEOREM 6.2 Let $\underline{d}$ be a divisor. There is an $A \in k^*$ with $[A] \geq \underline{d}$ such that

$$|N_{k/\mathbb{Q}}(A)|_\infty \leq c_o \, \mathbb{N}(\underline{d}), \tag{6.19}$$

where

$$c_o = (4/\pi)^s \, (n!/n^n) \, |D_k|_\infty^{\frac{1}{2}} . \tag{6.20}$$

Proof. This is as for the preceding Theorem except that instead of (6.16) we consider C_o given by

$$\sum_{j=1}^{r} |y_j| + \sum_{j=r+1}^{r+s} \sqrt{(y_j^2 + y_{j+s}^2)} \leq t \tag{6.21}$$

for an appropriate t, where we have written $|\ |$ for $|\ |_\infty$ as no other valuations will occur. This is symmetric and convex and has volume

$$V(C_o) = (2\pi)^s \, t^n/n! \tag{6.22}$$

The reader will have no difficulty in verifying (6.22) on introducing polar co-ordinates

$$y_j = R_j \cos\theta_j, \quad y_{j+s} = R_j \sin\theta_j \quad (r + 1 \leq j \leq r + s) \tag{6.23}$$

Put $R_j = |y_j|$ $(j \leq r)$ and then make the change of co-ordinates

$$\left.\begin{aligned} R_1 + \ldots + R_{r+s} &= z_o \\ R_1 + \ldots + R_{r+s-1} &= z_o z_1 \\ R_1 + \ldots + R_{r+s-2} &= z_o z_1 z_2 \\ &\ldots \\ R_1 &= z_o z_1 z_2 \ldots z_{r+s-1}. \end{aligned}\right\} \tag{6.24}$$

(Cf. the section 12.5 on "Dirichlet's integrals" in E.T. Whittaker and G.N. Watson, Modern Analysis).

We choose t so that

$$V(C_o) = 2^n \, |\det(L_1, \ldots, L_n)|$$

in the notation of the previous proof. There are then $b_1, \ldots, b_n \in \mathbb{Z}$, not all 0 such that $y_j = L_j(b_1, \ldots, b_n)$ satisfy (6.21).

Applying the inequality of the arithmetic and geometric means to (6.21), we obtain

$$\prod_{j=1}^{r} |y_j| \prod_{j=r+1}^{s} (y_j^2 + y_{j+s}^2) \leq 2^{2s}(t/n)^n \tag{6.25}$$

This is just (6.19), (6.20).

<u>COROLLARY 1.</u>

$$|D_k|_\infty \geq (\pi/4)^{2s} \, (n^n/n!)^2. \tag{6.26}$$

<u>Proof</u>. Put $\underline{d} = [1]$, so $A \in \mathfrak{o}$ and $|N_{K/k}(A)|_\infty \geq 1$. Hence $c_o \geq 1$ which gives (6.26) by (6.20).

7 <u>FINITENESS OF THE CLASS NUMBER</u>

As already noted when they were defined in §5, the divisors $\underline{d}$ of a number field k form an abelian group, which we shall denote by $\underline{D}$. The principal divisors $[A]$, $A \in k^*$ are a subgroup $\underline{P}$. The quotient group $\underline{D}/\underline{P}$ is the <u>divisor class group</u> (or <u>ideal class group</u>).

<u>THEOREM 7.1</u> <u>The divisor class group $\underline{D}/\underline{P}$ is finite</u>.

<u>Proof</u>. Let $\underline{d}$ be any divisor and let A be given by Theorem (6.2). Then

$$[A] = \underline{d} + \underline{e} \tag{7.1}$$

where $\underline{e} \geq [1]$; i.e., $\underline{e}$ is effective. Further

$$|N_{k/\mathbb{Q}}(A)|_\infty = \mathbb{N}([A]) = \mathbb{N}(\underline{d}) \, \mathbb{N}(\underline{e}) \tag{7.2}$$

by (5.16) and Lemma 5.1. Hence $\mathfrak{N}(\underline{e}) \leq c_o$ by (6.19). There are only a finite number of possibilities for $\underline{e}$ by Lemma 5.3. This proves the Theorem and also

COROLLARY 1. *Every divisor class contains an effective divisor* $\underline{e}$ *with* $\mathfrak{N}(\underline{e}) \leq c_o$, *where* c_o *is given by* (6.20).

We conclude with a concrete example to illustrate the results so far. Let

$$k = \mathbb{Q}(C), \quad C^3 = 6. \tag{7.3}$$

Consider first the places of k. The places $2, 3$ of $\mathbb{Q}$ ramify completely. There is thus just one place $\mathfrak{p}_2|2$ and just one $\mathfrak{p}_3|3$ and they are of the first degree (the *degree* of a place is the degree of the corresponding residue class field extension). Adopting the multiplicative notation (5.5 bis) for divisors, we have

$$[2] = \mathfrak{p}_2^3, \quad [3] = \mathfrak{p}_3^3. \tag{7.4}$$

If $p \equiv 2 \ (3)$, $p \neq 2$, there is just one cube root of 6 in $\mathbb{Q}_p$. Hence there are two extensions of p to k, one $\mathfrak{p}_p$ of degree 1 and one $\mathfrak{q}_p$ of degree 2, where

$$[p] = \mathfrak{p}_p \, \mathfrak{q}_p. \tag{7.5}$$

If $p \equiv 1 \ (3)$, there are two possibilities. The first is that there are no cube roots of 6 in $\mathbb{Q}_p$: in this case there is just one extension $[p]$ of p to k and it is of degree 3. Otherwise, there are three cube roots in $\mathbb{Q}_p$, so three first-degree extensions $\mathfrak{p}_{pj}$ $(j = 1, 2, 3)$ and

$$[p] = \mathfrak{p}_{p1} \, \mathfrak{p}_{p2} \, \mathfrak{p}_{p3}. \tag{7.6}$$

The next thing is to find a basis for the ring $\mathfrak{o}$ of integers. Clearly C is an integer. Its conjugates over $\mathbb{Q}$ are CW, CW^2 where $W^3 = 1$, and so

$$D(1, C, C^2) = \left\{ \det \begin{pmatrix} 1 & C & C^2 \\ 1 & CW & C^2W^2 \\ 1 & CW^2 & C^2W \end{pmatrix} \right\}^2$$

$$= -\ 2^2.3^5 \tag{7.7}$$

The only prime squares dividing this are 2^2, 3^2. If $(1, C, C^2)$ is not a basis, then for either $p = 2$ or $p = 3$ there are $a, b, c \in \mathbb{Z}$, not all divisible by p such that $A = p^{-1}(a + bC + cC^2)$ is an integer. Here a, b, c need to be considered only modulo p. One has only to check at the $\mathfrak{p}|p$ and in this case one verifies readily that no such a,b,c exist. Hence $(1, C, C^2)$ is a basis and

$$D_k = -\ 2^2.3^5. \tag{7.8}$$

In the notation of (6.20) we have $s = 1$, $n = 3$, and so

$$c_o = 8.82126 \ldots \ .$$

By the Corollary to Theorem 3.1, in order to find the class group we can confine attention to places of norm $\leq c_o$. By (7.5) we need consider only first-degree places and so only $\mathfrak{p}_2, \mathfrak{p}_3, \mathfrak{p}_5, \mathfrak{p}_{7,1}, \mathfrak{p}_{7,2}, \mathfrak{p}_{7,3}$ come into consideration.

At this stage, preferably with the help of a calculator, one compiles a table of integers of small norm and of the corresponding divisors. In our case we obtain, for example

$$N(C) = 6 \qquad [C] = \mathfrak{p}_2\,\mathfrak{p}_3,$$

$$N(1 - C) = -\ 5 \qquad [1 - C] = \mathfrak{p}_5,$$

$$N(2 - C) = 2 \qquad [2 - C] = \mathfrak{p}_2,$$

$$N(3 - C) = 21 \qquad [3 - C] = \mathfrak{p}_3\,\mathfrak{p}_{7,1},$$

$$N(1 + C) = 7 \qquad [1 + C] = \mathfrak{p}_{7,2}.$$

Here $N = N_{k/\mathbb{Q}}$. It follows that all the places are principal divisors, and that the class number is 1.

Anticipating the next section, we note that since $[2] = \mathfrak{p}_2^3$ and $[2 - C] = \mathfrak{p}_2$, we have

$$(2 - C)^3 = 2(1 - 6C + 3C^2)$$

where $[1 - 6C + 3C^2] = [1]$. We have therefore found a unit of $\mathbb{Q}(C)$.

If the class number is actually greater than 1, then its determination requires a technique to prove that a given divisor is NOT principal. For thus we require some knowledge of the unit group, and so we defer the discussion until the end of §8.

8 THE UNIT GROUP

LEMMA 8.1. *Let $E \in k$. The following statements are equivalent:*

(i) E *and* E^{-1} *are both in* $\mathfrak{o}$.

(ii) $|E|_{\mathfrak{p}} = 1$ *for all non-arch. places* $\mathfrak{p}$.

(iii) *The principal divisor* $[E] = [1]$.

(iv) $E \in \mathfrak{o}$ *and* $N_{k/\mathbb{Q}}E = \pm 1$.

Proof. It is trivial that (i), (ii), (iii) are equivalent and imply (iv).

Suppose that E satisfies (iv). Then the principal divisor $[E]$ has $\mathfrak{N}(E) = 1$ by Lemma 5.1. Let $[E] = \prod_{\mathfrak{p}} \mathfrak{p}^{d(\mathfrak{p})}$, where $d(\mathfrak{p}) \geq 0$ since $E \in \mathfrak{o}$. Then $d(\mathfrak{p}) = 0$ for all $\mathfrak{p}$ by Definition 5.1. Hence E satisfies (iii).

DEFINITION 8.1. If E satisfies one (and so all) of the conditions of Lemma 8.1, we say that E is a *unit*.

The units form a group under multiplication which we denote by $U = U_k$.

Note The condition $E \in \mathfrak{o}$ in (iv) of Lemma 8.1. is essential. For example, $(3 + 4i)/5 \in \mathbb{Q}(i)$ $(i^2 = -1)$ has norm 1 but is not a unit.

THEOREM 8.1. *The unit group U is the direct product of a finite cyclic group consisting of roots of unity and a free group on $r + s - 1$ generators, where r, s are respectively the number of real and of complex places. In other words, there are $H_1, \ldots, H_{r+s-1} \in U$ such that every*

$E \in U$ is uniquely of the shape

$$E = R H_1^{m(1)} \ldots H_{r+s-1}^{m(r+s-1)} , \qquad (8.1)$$

where $m(1), \ldots, m(r+s-1) \in \mathbb{Z}$ and R is a root of unity from a finite set.

Note When $r + s = 1$, the statement is to be interpreted as that U is finite and consists of roots of unity. Apart from $r = 1$, $s = 0$, which happens only for $k = \mathbb{Q}$, this occurs for the "complex quadratic" fields with $s = 1$, $r = 0$.

We require two lemmas. As in §6 the arch. places are r_j $(1 \leq j \leq r+s)$, the corresponding valuations and completions being $| \ |_j$ and k_j.

LEMMA 8.2. Let j be fixed, $1 \leq j \leq r+s$. There is a unit $E = E_j$ such that

$$|E|_j > 1, \quad |E|_i < 1 \quad (\text{all } i \neq j). \qquad (8.2)$$

Proof. We construct a sequence of $A_m \in \mathfrak{o}$ $(m = 0, 1, 2 \ldots)$ such that

$$|A_{m+1}|_i < |A_m|_i \quad (i \neq j) \qquad (8.3)$$

and

$$|N_{k/\mathbb{Q}} A_m|_\infty \leq c \quad (m = 0, 1, 2 \ldots \quad , \qquad (8.4)$$

where c is given by (6.4). Take $A_o = 1$. If $A_o, \ldots, A_m$ have already been constructed, put

$$t_i = \tfrac{1}{2} |A_m|_i \quad (i \neq j) \qquad (8.5)$$

and define t_j by $\prod_{\ell=1}^{r+s} t_\ell = c$. By Theorem 6.1 with $\underline{d} = [1]$, there is an $A_{m+1} \neq 0$ in $\mathfrak{o}$ with

$$|A_{m+1}|_i \leq \tfrac{1}{2} |A_m|_i \quad (i \neq j) \qquad (8.6)$$

and

$$|N_{k/\mathbb{Q}} A_{m+1}|_\infty = \prod_{\ell} |A_{m+1}|_\ell \leq \prod_{\ell} t_\ell = c. \qquad (8.7)$$

This concludes the induction step.

By (8.4) and Lemma 5.3 there are only finitely many possibilities for the principal ideal $[A_m]$. Hence there are m, M with $m < M$ such that $[A_m] = [A_M]$. Put

$$E = A_M/A_m.$$

Then $[E] = [1]$, that is, E is a unit. Further, $|E|_i < 1$ for $i \neq j$ by (8.3). Finally, $|E|_j > 1$ since $\prod_{\ell} |E|_\ell = |N_{k/\mathbb{Q}} E|_\infty = 1$.

The next lemma has independent interest.

<u>LEMMA 8.3.</u> (Minkowski) <u>Let</u> $m > 0$ <u>be an integer and let</u> $t_{ij} \in \mathbb{R}$ $(1 \leq i, j \leq m)$. <u>Suppose that</u> $t_{jj} > 0$ $(1 \leq j \leq m)$, <u>that</u> $t_{ij} < 0$ $(i \neq j)$ <u>and that</u>

$$\sum_j t_{ij} > 0.$$

<u>Then</u>

$$\det(t_{ij}) \neq 0.$$

<u>Proof</u>. If $\det(t_{ij}) = 0$, there would be $a_1, \ldots, a_m \in \mathbb{R}$, not all 0 such that

$$\sum_j t_{ij} a_j = 0 \quad (1 \leq i \leq m).$$

Without loss of generality, $a_1 = \max|a_j|_\infty > 0$. Then

$$\sum t_{1j} a_j \geq a_1 \sum_j t_{1j} > 0.$$

Contradiction!

We now revert to the proof of Theorem 8.1 and consider the map $U \to \mathbb{R}^{r+s}$ given by

$$\Theta: E \mapsto (\log|E|_j)_{j=1, \dots, r+s} \tag{8.8}$$

Denoting co-ordinates in $\mathbb{R}^{r+s}$ by $(x_1, \dots, x_{r+s})$, we see that the image of Θ is in the hyperplane

$$T: \sum_j x_j = 0. \tag{8.9}$$

<u>LEMMA 8.4</u>. <u>The image of</u> U <u>under</u> Θ <u>contains</u> $r + s - 1$ <u>linearly independent points</u>.

<u>Proof</u>. Let $E_1, \dots, E_{r+s-1}$ be the units given by Lemma 8.2, so the image of E_i is

$$(t_{21}, \dots, t_{i,r+s}), \quad t_{ij} = \log|E_i|_j.$$

Then

$$t_{ii} > 0, \quad t_{ij} < 0 \quad (i \neq j)$$

and

$$\sum_{j=1}^{r+s-1} t_{ij} = - t_{i,r+s} > 0.$$

Hence, by Lemma 8.3 with $m = r + s - 1$, the images $\Theta(E_i)$ $(1 \leq i \leq r + s - 1)$ are linearly independent.

<u>LEMMA 8.5</u>. <u>The kernel of</u> Θ <u>is finite and the image of</u> U <u>under</u> Θ <u>is discrete</u>.

<u>Proof</u>. We have to show that the inverse image $\Theta^{-1} S$ of any bounded $S \subset \mathbb{R}^{r+s}$ is finite. Clearly $\Theta^{-1} S$ is contained in a bounded subset $S_o \subset \prod_j k_j$. But (c.f. end of §3 or proof of Theorem 6.1) S_o contains only finitely many elements of $\mathfrak{O} \supset U$.

We can now complete the proof of the Theorem. Since the kernel of Θ is finite, the elements of the kernel are of finite order, i.e., they are roots of unity. Any discrete subgroup of an $\mathbb{R}^m$ is free on $\leq m$ generators by Theorem 2 of Appendix C. Since $\Theta(U) \subset T$ contains $r + s - 1$ linearly independent elements by Lemma 8.3 and T

has dimension $r + s - 1$, it follows that $\Theta(U)$ is free on $r + s - 1$ generators. This concludes the proof of Theorem 8.1.

As a consequence of Lemma 8.5 we have

COROLLARY 1. *A necessary and sufficient condition that* $A \in k$ *be a root of unity is that* $|A|_{\mathfrak{p}} = 1$ *for all* $\mathfrak{p}$, *arch. and non-arch.*

The proof of Lemma 8.2 is constructive in the logical sense, but at best tedious in practice. Usually one can find the requisite number of independent units by trial and error, manipulating integers of small norm in the way exemplified by the example at the end of §7. We now discuss how one can then find a set of generators of U and also how we can decide whether a given divisor $\underline{d}$ is principal.

Once one has $r + s - 1$ independent units $E_1, \ldots, E_{r+s-1}$, the full group of units can be found by a finite search. Let E be any unit, so in the notation (8.8) we have

$$\Theta(E) = \sum_{\ell=1}^{r+s-1} t_\ell \, \Theta(E_\ell)$$

for some $t_\ell \in \mathbb{R}$. On replacing E by

$$E \prod_\ell E_\ell^{m(\ell)}$$

for suitable $m(\ell) \in \mathbb{Z}$ we may suppose that

$$-\frac{1}{2} \leq t_\ell \leq \frac{1}{2} .$$

Then we have bounds for $|E|_j$ $(1 \leq j \leq r + s)$ and so E belongs to a finite set.

By much the same technique, one can decide whether or not a given divisor $\underline{d}$ is principal. Rather than formulate general theorems, which would require the setting up of a cumbersome notational apparatus with little intellectual content, we discuss some special cases.

LEMMA 8.6. *Let* k *be a "real quadratic" field, that is,* $k = \mathbb{Q}(C)$ *with* $C^2 \in \mathbb{Q}$, $C^2 > 0$. *Let* σ_1, σ_2 *be the two embeddings of* k *into* $\mathbb{R}$ *and let* E *be a unit of* k *with*

$$1 < \sigma_1 E = e \text{ (say)}. \tag{8.10}$$

<u>Let</u> $A \in \mathfrak{a}$. <u>Then there is an</u> $m \in \mathbb{Z}$ <u>such that</u> $B = E^m A$ <u>satisfies</u>

$$|\sigma_i B| \leq |e\, N(A)|^{\frac{1}{2}}, \quad (i = 1,2), \tag{8.11}$$

<u>where</u> $|\ |$ <u>is the ordinary absolute value.</u>

<u>Proof</u>. We can find m such that

$$|N(A)|^{\frac{1}{2}}\, e^{-\frac{1}{2}} \leq \pm\, \sigma_1(E^m A) \leq |N(A)|^{\frac{1}{2}}\, e^{\frac{1}{2}}. \tag{8.12}$$

Now

$$\begin{aligned} \sigma_1(E^m A).\sigma_2(E^m A) &= N(E^m A) \\ &= \{N(E)\}^m\, N(A) \\ &= \pm\, N(A), \end{aligned}$$

and so the left hand inequality in (8.12) gives

$$|\sigma_2(E^m A)| \leq |N(A)|^{\frac{1}{2}}\, e^{\frac{1}{2}}.$$

This concludes the proof.

As an application we show that the prime divisor (= place) of norm 2 in $\mathbb{Q}(\sqrt{82})$ is not principal. By inspection, $E = 9 + \sqrt{82}$ is a unit. It is fundamental (i.e., ± 1 and E generate the whole group of units): in fact if E were not fundamental we should have $E = F^q$ for some unit F and some integer $q \geq 2$ and arguing as below with $\sigma_1 F$, $\sigma_2 F$ we should get a contradiction. However, for our present purpose we do not need to know that E is fundamental. We take for σ_1 the map for which $\sigma_1\sqrt{82} > 0$, so $\sigma_2\sqrt{82} < 0$. By abuse of notation we shall use $\sqrt{82}$ to denote both the element of $\mathbb{Q}(\sqrt{82})$ and the positive real number.

Suppose, if possible, that there is an integer A of $\mathbb{Q}(\sqrt{82})$ with $N(A) = \pm 2$. By the last lemma, we have without loss of generality

$$|\sigma_i A| \leq |N(A)e|^{\frac{1}{2}}$$

$$= |2(9 + \sqrt{82})|^{\frac{1}{2}} \quad (i = 1,2).$$

Now $A = u + v\sqrt{82}$ for some $u,v \in \mathbb{Z}$, so

$$|u \pm v\sqrt{82}| \leq |2(9 + \sqrt{82})|^{\frac{1}{2}},$$

and hence

$$|v| \leq |2(9 + \sqrt{82})|^{\frac{1}{2}}/\sqrt{82}$$

$$= 0.66 \ldots < 1 .$$

Hence $v = 0$, a contradiction.

We can express what has just been proved in another way: there are no $u,v \in \mathbb{Z}$ such that

$$u^2 - 82v^2 = \pm 2.$$

The reader will readily verify that there are solutions (for each choice of sign) in $\mathbb{Z}_2$ and $\mathbb{Z}_{41}$ and indeed in $\mathbb{Z}_p$ for every p. We thus have a diophantine equation which has an integral solution everywhere locally but not globally. (For $p \neq 2, 41$ cf. Appendix D).

In the example just concluded, we showed that a divisor is not principal by using the real (and complex) embeddings. One can also use the non-archimedean embeddings. The technique is tentative but, in practice, is usually more rapid. We illustrate it by finding the class number of $k = \mathbb{Q}(C)$ with $C^3 = 22$. Here the discriminant is

$$D_k = - 3^3.22^2$$

and so by Theorem 7.1 we need look only at the effective divisors $\mathfrak{c}$ with

$$\mathfrak{N}(\mathfrak{c}) \leq 32.34463\ldots \quad .$$

We are thus concerned only with the prime divisors (places) $\mathfrak{p}$ with $\mathfrak{N}(\mathfrak{p}) \leq 31$.

The rational primes $p = 2, 3, 11$ ramify completely, say $[2] = \mathfrak{p}_2^3$, $[3] = \mathfrak{p}_3^3$, $[11] = \mathfrak{p}_{11}^3$. The remaining $p \equiv 2$ (3) split into a first degree prime, which we denote by $\mathfrak{p}_p$, and a second degree prime. A prime $p \equiv 1$ (3) either splits into three first degree primes or is <u>inert</u> (i.e., has only one, unramified, extension). In the relevant range, 13, 19 and 31 are inert and 7 splits, say $[7] = \mathfrak{p}_7\mathfrak{p}_7'\mathfrak{p}_7''$.

We first look at the principal divisors given by elements of k with small norm;

$$[2 + C] = \mathfrak{p}_2\mathfrak{p}_3\mathfrak{p}_5; \quad [1 + C] = \mathfrak{p}_{23}; \quad [C] = \mathfrak{p}_2\mathfrak{p}_{11};$$

$$[1 - C] = \mathfrak{p}_3\mathfrak{p}_7; \quad [2 - C] = \mathfrak{p}_2\mathfrak{p}_7'; \quad [3 - C] = \mathfrak{p}_5;$$

$$[4 - C] = \mathfrak{p}_2\mathfrak{p}_3\mathfrak{p}_7''; \quad [6 + C] = \mathfrak{p}_2\mathfrak{p}_7\mathfrak{p}_{17}; \quad [13 - C] = \mathfrak{p}_3\mathfrak{p}_5^2\mathfrak{p}_{29}.$$

It is readily verified that the class group is generated by the class of $\mathfrak{p}_3$, and so is of order 1 or 3.

We shall require a unit and look for an element of k whose principal divisor is [1]. Using the above table, we have

$$(1 - C)(2 - C)(3 - C) = -7(2 + 2C - C^2).$$

Now

$$(2 + 2C - C^2)^2 = 6(-14 + 5C)$$

and

$$(2 + 2C - C^2)(-14 + 5C) = -6(23 + 3C - 4C^2),$$

$$= -6E \text{ (say)},$$

where

$$N(E) = +1.$$

Our next objective is to show that E is not a cube. For this, we consider the map of the ring $\mathfrak{o}$ of integers of k into the finite field $\mathbb{F}_{43}$ of 43 elements given by the first degree prime divisor $\mathfrak{p}_{43}$ for which $C \to 19 \bmod 43$. Then $E \to 12 \bmod 43$ and 12 is not a cubic residue mod 43.

Next consider the map into $\mathbb{F}_7$ given by $\mathfrak{p}_7$, so $C \to 1 \bmod 7$. Then $E \to 1 \bmod 7$. Hence none of 3, 3E, $3E^2$ is a cube.

Now suppose, if possible that $\mathfrak{p}_3$ is principal, say $[B] = \mathfrak{p}_3$. Then $B^3 = 3H$ for some unit H. Now the only roots of unity in k are ±1, since there is a real embedding, and so the group of units is the product of {±1} by an infinite cyclic group (Theorem 8.1). We now have a contradiction in the facts in the last two paragraphs.

Hence $\mathfrak{p}_3$ is not principal and so the class number of $\mathbb{Q}(C)$ is 3.

We have determined the class number without completely determining the group of units. If we do wish to do this as well, the local considerations can greatly facilitate the calculation. In the first place, the map into $\mathbb{F}_{43}$ used above shows that E is not a square, since 12 is not a quadratic residue mod 43. [We could alternatively have used $\mathfrak{p}_7'$ or $\mathfrak{p}_7''$]. Further, $-E$ is not a square, since $N(-E) = -1$. It follows that if E is not a fundamental unit we should have

$$E = G^g \quad (g \geq 5)$$

for some unit G. Now consider the complex embedding

$$\sigma_2 : C \to 22^{1/3}(-1 + \sqrt{-3})/2.$$

Then

$$|\sigma_2 E| = 48.77469\ldots$$

where $|\ |$ is the absolute value. Hence

$$|\sigma_2 G| \leq 2.1759 = \lambda \quad \text{(say)}.$$

Clearly,

$$|\sigma_1 G| < 1,$$

where σ_1 is the real embedding. Now let

$$G = u + vC + wC^2 \quad (u,v,w \in \mathbb{Z}) .$$

Then

$$3.22^{1/3} v = \sigma_1 G + \Omega^2 \sigma_2 G + \Omega \bar{\sigma}_2 G,$$

where $\Omega = (-1 + \sqrt{-3})/2$. Hence

$$|v| \leq (1 + 2\lambda)/3.22^{1/3} = 0.636... < 1$$

and so $v = 0$. Similarly $w = 0$ and we have a contradiction. Hence E is a fundamental unit, as required.

9 APPLICATION TO DIOPHANTINE EQUATIONS. RATIONAL SOLUTIONS

We saw (Chapter 4, §3 bis) that consideration of factorizations can make the solution of a diophantine equation depend on that of a more tractable one. Now that we understand factorization in algebraic number fields, our armoury is increased. We give two instructive examples.

LEMMA 9.1. *The equation*

$$3X^3 + 4Y^3 + 5Z^3 = 0 \tag{9.1}$$

has non-trivial rational solutions everywhere locally but not globally.

Proof. It is readily verified that there are solutions in $\mathbb{R} = \mathbb{Q}_\infty$, $\mathbb{Q}_2$, $\mathbb{Q}_3$, $\mathbb{Q}_5$. For the remaining $\mathbb{Q}_p$ the theory of equations over finite fields gives a solution mod p which can be lifted to $\mathbb{Q}_p$ by Hensel's Lemma (cf. Appendix D).

Now suppose, if possible, that there is a solution (x,y,z) in $\mathbb{Q}$. Then without loss of generality $x,y,z \in \mathbb{Z}$ are coprime in pairs. It follows that $3x^3$, $4y^2$, $5z^3$ are coprime in pairs.

We write our equation as

$$(2y)^3 + 6x^3 = -10z^3 \tag{9.2}$$

and consider factorization in the field $\mathbb{Q}(C)$ with $C^3 = 6$ considered at the end of §7. We have

$$(2y + xC)(4y^2 - 2xyC + x^2C^2) = - 10z^3. \tag{9.3}$$

Any prime divisor which divides the two bracketed expressions on the left hand side must divide both $3x^2C^2$ and $12y^2$, and so must be $\mathfrak{p}_2$ or $\mathfrak{p}_3$. It cannot be $\mathfrak{p}_3$ since that divides C but does not divide $2y$. On considering the principal divisors corresponding to the factors in (9.3) we see that

$$[2y + xC] = \mathfrak{p}_2\mathfrak{p}_5\mathfrak{a}^3 \tag{9.4}$$

for some divisor $\mathfrak{a}$, where $\mathfrak{p}_2$ and $\mathfrak{p}_5$ are the first-degree prime divisors of 2,5.

It was shown in §7 that $\mathbb{Q}(C)$ has class number 1 and that

$$E = (2 - C)^3/2 = 1 - 6C + 3C^2 \tag{9.5}$$

is a unit. In fact E is a fundamental unit, but all we need is that E is not a cube, and this is immediate since 2 is not a cube in $\mathbb{Q}(C)$.

From (9.4) and the facts about $\mathbb{Q}(C)$ just rehearsed, we have

$$2y + xC = (2 - C)(1 - C)\ E^j\ H^3$$

for $j = 0$, 1 or 2 and some $H \in \mathbb{Q}(C)$. On using (9.5), this implies that

$$2y + xC = 2^j(2 - C)(1 - C)\ G^3 \tag{9.6}$$

where

$$\begin{aligned} G &= (2 - C)^j\ H \\ &= u + vC + wC^2 \end{aligned}$$

for some $u,v,w \in \mathbb{Q}$ [indeed, in $\mathbb{Z}$]. On equating the coefficient of C^2 on both sides of (9.6), we have

$$0 = u^3 + 6v^3 + 36w^3 + 18uvw$$

$$- 3(3u^2v + 18uw^3 + 18v^2w)$$

$$+ 2(3uv^2 + 3u^2v + 18vw^2).$$

It is readily verified that this has no nontrivial solution in $\mathbb{Q}_3$. We have thus reached a contradiction on the assumption that there is a rational solution.

The enunciation of the next lemma is similar to that of the preceding one, but the crunch in the proof is different.

LEMMA 9.2. *The equation*

$$X^3 + 22Y^3 + 3Z^3 = 0$$

has nontrivial rational solutions everywhere locally but not globally.

Proof. (Sketch). Suppose that there is a solution $x,y,z \in \mathbb{Z}$ coprime in pairs. We work in the field $\mathbb{Q}(C)$ with $C^3 = 22$ discussed at the end of §8. As in the previous proof, we have the equation

$$[x + yC] = \mathfrak{p}_3\mathfrak{a}^3$$

in divisors, for some $\mathfrak{a}$. This is impossible, because we proved that the class-number is 3 and that $\mathfrak{p}_3$ is not principal.

10 APPLICATION TO DIOPHANTINE EQUATION. INTEGRAL SOLUTIONS

The structure theorems we have proved for the integers of algebraic number fields can often be used to attach equations requiring solutions in rational integers. We have already met one example of this in Lemma 6.1, Corollary of Chapter 4, which finds the solutions of $x^2 + 7 = 2^m$. Our first example goes back to Fermat.

LEMMA 10.1. *The only solutions in integers of*

$$y^2 + 2 = x^3 \qquad (10.1)$$

<u>are</u> $x = 3$, $y = \pm 5$.

<u>Proof</u>. Clearly y is odd. We work in the field $\mathbb{Q}(\sqrt{-2})$, which has a euclidean algorithm and so class number 1 (Theorem 5.1, Corollary 2). In the factorization

$$(y + \sqrt{-2})(y - \sqrt{-2}) = x^3,$$

the two factors on the left hand side are coprime, and so their divisors are cubes. The only units in $\mathbb{Q}(\sqrt{-2})$ are $\pm 1 = (\pm 1)^3$, and so we have

$$y + \sqrt{-2} = (a + b\sqrt{-2})^3$$

for some $a,b \in \mathbb{Z}$.

On equating coefficients of $\sqrt{-2}$, we have

$$1 = 3a^2b - 2b^3$$

$$= b(3a^2 - 2b^2).$$

Hence $b = 3a^2 - 2b^2 = \pm 1$ and the rest follows.

We give the following theorem of Skolem because of the elegance of the proof although (as we explain in the notes at the end of the Chapter), stronger results are known.

<u>THEOREM 10.1</u> <u>For given</u> $d \in \mathbb{Z}$ <u>there is at most one pair</u> $x,y \in \mathbb{Z}$ <u>with</u> $y \neq 0$ <u>such that</u>

$$x^3 + dy^3 = 1. \qquad (10.2)$$

<u>Note</u> Thus the only solution of $x^3 + 7y^3 = 1$ with $y \neq 0$ is $x = 2$, $y = -1$.

<u>Proof</u>. If d is a rational cube, say $d = c^3$, the left hand side of (10.2) is divisible by $x + cy$, so $x + cy = \pm 1$ and the result follows.

We work in the field $\mathbb{Q}(C)$ with $C^3 = d$. The group of units is the product of $\{\pm 1\}$ and an infinite cyclic group. Suppose, if possible, that there are two integral solutions (x_1,y_1), (x_2,y_2) of (10.2) with $g_1 \neq 0$, $y_2 \neq 0$. Then $x_1 + y_2C$ and $x_2 + y_2C$ are units of norm + 1. Since Norm (-1) = -1, there are thus $n(1), n(2) \in \mathbb{Z}$ such that

$$(x_1 + y_1C)^{n(1)} = (x_2 + y_2C)^{n(2)}; \quad \gcd(n(1), n(2)) = 1. \qquad (10.3)$$

Without loss of generality, $3 \nmid n(2)$ and so

$$N \text{ (say)} = n(1)/n(2) \in \mathbb{Z}_3. \qquad (10.4)$$

For ease of notation we put

$$x = x_2, \quad y = y_2. \qquad (10.5)$$

By (10.2) we have

$$(x + yC)^3 = 1 + 3xy\,G, \qquad (10.6)$$

where

$$G = xC + yC^2. \qquad (10.7)$$

We work in $\mathbb{Q}_3[C]$ which (Chapter 9) is the direct sum of the completions of $\mathbb{Q}(C)$ at the extensions of the 3-adic valuation. From (10.3) we have

$$x_2 + y_2C = (x + yC)^N, \qquad (10.8)$$

where the right hand side is defined for $N \in \mathbb{Z}_3$ like the exponential function in Chapter 4, §5. We consider three cases.

(i) $N \equiv 0$ (3), say $N = 3M$. Then

$$x_2 + y_2C = (1 + 3xy\,G)^M$$
$$= 1 + m.3xy\,G + \dots \binom{M}{j} (3xy)^j\, G^j + \dots,$$

where the right hand side converges in the 3-adic topology. Equating coefficients of C^2 we have

$$0 = 3xy^2M + \sum_{j\geq 2} \binom{M}{j} (3xy)^j a_j,$$

where $a_j \in \mathbb{Z}$ depends on x,y. If $M = 0$, we have $x_2 + y_2C = 1$, which is excluded. Otherwise, we may divide by $3xy^2M$ and obtain

$$0 = 1 + \sum_{j\geq 2} \binom{M-1}{j-1} \frac{3^{j-1}}{j} x^{j-1} y^{j-2} a_j.$$

Here all the terms in the sum are divisible by 3. Contradiction.

(ii) $N \equiv 1$ (3), say $N = 3M + 1$. Here

$$x_2 + y_2C = (x + yC)(1 + 3xy\ G)^M.$$

On expanding and equating the coefficients of C^2 we have

$$0 = 6x^2y^2M + \sum_{j\geq 2} \binom{M}{j} (3xy)^j a_j$$

with $a_j \in \mathbb{Z}$. If $M = 0$, we have $x_2 + y_2C = x_1 + yC$, which is excluded. Otherwise, on dividing by $3x^2y^2M$, we have

$$0 = 2 + \sum_{j\geq 2} \binom{M-1}{j-1} \frac{3^{j-1}}{j} (xy)^{j-2} a_j.$$

Again all the terms in the sum are divisible by 3.

(iii) $N \equiv 2$ (3), say $N = 3M + 2$. Now

$$x_2 + y_2C = (x^2 + 2xyC + y^2C^2)(1 + 3xyG)^M.$$

Expanding and equating coefficients of C^2 we have by (10.7) that

$$0 = y^2 + 9x^3y^2M + \sum_{j\geq 2} \binom{M}{j} (3xy)^j a_j$$

with $a_j \in \mathbb{Z}$. On dividing by $y^2 \neq 0$, we have

$$0 = 1 + 9x^3M + \sum_{j\geq 2} \binom{M}{j} (3x)^j y^{j-2} a_j;$$

again a contradiction.

This concludes the proof.

With a knowledge of the units in the relevant field one can sometimes show that there are no solutions of (10.2) with $y \neq 0$.

<u>LEMMA 10.2</u> <u>There is no integral solution of</u>

$$x^3 - 11y^3 = 1 \tag{10.9}$$

<u>with</u> $y \neq 0$.

<u>Proof</u>. We work in $\mathbb{Q}(C)$ with $C^3 = 11$. It can be shown that

$$E = 1 + 4C - 2C^2 \tag{10.10}$$

is a fundamental unit, where $\mathrm{Norm}(E) = +1$. By (10.9) we have

$$x - yC = E^n \tag{10.11}$$

for some $n \in \mathbb{Z}$.

In the spirit of Chapter 5, we look for a p-adic field into which there are three distinct embeddings of $\mathbb{Q}(C)$. In $\mathbb{Q}_{19}$ there are three solutions of $c^3 = 11$, namely

$$c_1 \equiv -3 + 5.19 \quad (19^2)$$

$$c_2 \equiv -2 + 8.19 \quad (19^2)$$

$$c_3 \equiv 5 + 6.19 \quad (19^2).$$

The corresponding values of (10.10) are

$$e_1 \equiv 9 + 2.19 \quad (19^2)$$

$$e_2 \equiv 4 + 0.19 \quad (19^2)$$

$$e_3 \equiv 9 + 6.19 \quad (19^2).$$

By (10.11) we have

$$x - yc_j = e_j{}^n \quad (j = 1,2,3),$$

and so

$$c_1 e_1^n + c_2 e_2^n + c_3 e_3^n = 0.$$

Since $e_1 e_2 e_3 = \text{Norm}(E) = 1$, we can write this as

$$c_1 + c_2 (e_2^2 e_3)^n + c_3 (e_2 e_3^2)^n = 0. \qquad (10.12)$$

Now

$$e_2^2 e_3 \equiv 11 \quad (19), \qquad e_2 e_3^2 \equiv 1 \quad (19);$$

so we must have $n = 3m$, $m \in \mathbb{Z}$.

Finally,

$$(e_2^2 e_3)^3 \equiv 1 + 7.19 \quad (19^2)$$

$$(e_2 e_3^2)^3 \equiv 1 + 11.19 \quad (19^2).$$

On putting $n = 3m$ in (10.12) and expanding in powers of m, the left hand side is

$$a_1 m + a_2 m^2 + \dots$$

where $|a_1|_{19} = 19^{-1}$, $|a_j|_{19} < 19^{-1}$ $(j \geq 2)$. Hence the only solution is $m = 0$, by Strassmann's theorem.

10 bis. APPLICATION TO DIOPHANTINE EQUATIONS. INTEGRAL SOLUTIONS (Contd).

In the previous section, the rank of the unit group of the algebraic number field under consideration was always one. We now consider cases where the rank is higher, taking rank two as typical. For this we require a generalization of Hensel's Lemma:

LEMMA 10 bis.1. (Skolem). *Let* k *be a field complete with respect to a non-archimedean valuation* $|\ |$, *with ring of valuation integers* $\mathfrak{o}$. *Let* $F(X,Y)$, $G(X,Y)$ *be functions given by power series with coefficients in* $\mathfrak{o}$ *and convergent for all values of* X,Y *in* $\mathfrak{o}$. *Let* $a,b \in \mathfrak{o}$ *with* $|F(a,b)| < 1$, $|G(a,b)| < 1$ *and suppose that*

$$|F_1(a,b)\ G_2(a,b) - F_2(a,b)\ G_1(a,b)| = 1, \qquad (10\ \text{bis}.\ 1)$$

where $F_1(X,Y) = \partial F/\partial X$ *etc. Then there is precisely one pair* $x,y \in \mathfrak{o}$ *such that*

$$F(x,y) = 0, \quad |x - a| < 1, \quad |y - b| < 1. \qquad (10\ \text{bis}.\ 2)$$

Proof. This is completely analogous to that of Lemma 3.1 and its Corollary 1 of Chapter 4, and so is left to the reader. There is an obvious generalization to n variables.

One type of problem to which this Lemma may be applicable is to equations

$$f(a,b) = c \qquad (10\ \text{bis}.\ 3)$$

where $f(X,Y) \in \mathbb{Z}[X,Y]$ is a given quartic form and where $c \in \mathbb{Z}$ is given. This leads to equations of the type

$$a - bC = DE \qquad (10\ \text{bis}.\ 4)$$

where C is given by $f(C,1) = 0$, $D \in \mathbb{Q}(C)$ belongs to a finite set and E is a unit. If C is neither totally real nor totally complex, the rank of the unit group is two, so there are units E_1, E_2 of $\mathbb{Q}(C)$ such that

$$E = \pm E_1^u E_2^v \quad (u,v \in \mathbb{Z}) \tag{10 bis. 5}$$

The fact that there are no terms in C^2, C^3 on the left hand side of (10 bis. 4) gives two conditions on u,v. In appropriate cases this can be reduced to consideration of the situation considered in Lemma 10 bis. 1.

We shall, however, discuss a rather different problem:

THEOREM 10 bis. 1. *Let* $C^5 = 2$. *The only solutions* $u,v,w \in \mathbb{Z}$ *of*

$$\begin{aligned} &\mathrm{Norm}(u + vC + wC^2) \\ &= u^5 + 2v^5 + 4w^5 - 10uv^3w + 10u^2vw^2 \\ &= 1 \end{aligned} \tag{10 bis. 6}$$

are

$$(u,v,w) = (1,0,0),\ (-1,1,0),\ (1,-2,1). \tag{10 bis. 7}$$

COROLLARY. *The only solutions* $u,v \in \mathbb{Z}$ *of* $u^5 + 2v^5 = 1$ *are* $(u,v) = (1,0)$ *and* $(-1,1)$.

Proof. (Sketch). The group of units of $\mathbb{Q}(C)$ is generated by ± 1, E, H, where

$$E = -1 + C, \quad H = 1 + C + C^3 \tag{10 bis. 8}$$

have norm +1. The general unit D of norm +1 is thus

$$D = E^r H^s \quad (r,s \in \mathbb{Z}). \tag{10 bis. 9}$$

We have to show that D is of the shape $u + vC + wC^2$ only when $s = 0$, $r = 0$, 1 or 2.

We work in a splitting field of $X^5 - 2$. There are 5 embeddings of $\mathbb{Q}(C)$ given, say, by $C \to C_j$ $(1 \leq j \leq 5)$. We denote the corresponding images of D, E, H etc. by D_j, E_j, H_j etc. Clearly a necessary and sufficient condition that D have the required shape is that

$$\sum C_j D_j = \sum C_j^2 D_j = 0. \qquad (10\ \text{bis}.\ 10)$$

We choose the first p-adic field in which $X^5 - 2$ splits, namely Q_{151}. The following table gives to modulus 151^2 the C_j, E_j, H_j and some other quantities we shall need (cf. Appendix F). We put $p = 151$.

C_j	E_j	H_j	E_j^{150}	$59H_j^{30}$
25 + 114p	24 + 114P	98 + 0p	1 + 94p	1 + 31p
49 + 85p	48 + 85p	70 + 59p	1 + 24p	1 + 25p
90 + 98p	89 + 98p	63 + 73p	1 + 147p	1 + 63p
116 + 42p	115 + 42p	126 + 139p	1 + 46p	1 + 143p
22 + 112p	21 + 112p	101 + 28p	1 + 142p	1 + 33p

We first check (10 bis. 10) as a congruence modulo 151, where r,s in (10 bis. 9) run over $0 \leq r,s \leq 149$. It is preferable to use a computer. It turns out that the solutions are $r = 0,1,2$ and $s \equiv 0 \pmod{30}$. Inspection shows that $H_j^{30} \equiv 64$ for all j. We therefore put

$$r = R + 150x, \quad s = 30y \quad (R = 0,1,2;\ x,y \in \mathbb{Z}).$$

Since $59.64 \equiv 1\ (151)$ we consider

$$59^y D = E^R\ (E^{150})^x\ (59H^{30})^y$$

which, for fixed R, can be expanded as a power series in x,y. For $R = 0$ the criterion (10 bis. 10) becomes

$$0 = ax + by + \text{higher degree},$$

$$0 = cx + dy + \text{higher degree},$$

where all the remaining coefficients are divisible by 151^2 and

$$a \equiv 69, \quad b \equiv 58, \quad c \equiv 108, \quad d \equiv 122 \quad (151),$$

so

$$ad - bc \equiv 40 \not\equiv 0 \quad (151).$$

Hence $x \equiv y \equiv 0$ (151) and Lemma 10 bis. 1 shows that the only solution is $x = y = 0$.

Similarly, for $R = 1$ we have

$$a \equiv 150, \quad b \equiv 68, \quad c \equiv 69, \quad d \equiv 26,$$

$$ad - bc \equiv 114 \not\equiv 0;$$

and for $R = 2$:

$$a \equiv 140, \quad b \equiv 89, \quad c \equiv 14, \quad d \equiv 48,$$

$$ad - bc \equiv 38 \not\equiv 0.$$

Hence again, in both cases the only solution is $x = y = 0$.

11 THE DISCRIMINANT

Let k be an algebraic number field. We recall that the discriminant D_k is given by

$$D_k = \det\{S_{k/Q}(A_i A_j)\}_{i,j}$$

where $A_1, \ldots, A_n$ is any basis for $\mathfrak{o}$. (Lemma 3.2 and Definition 3.1). In this section we show that D_k is closely related to the semilocal discriminants introduced in §3 of Chapter 9. It will follow, for example, that a prime p divides D_k if and only if there is some $\mathfrak{p}|p$ which is ramified.

We must first translate the notation of §3 of Chapter 9 into conformity with our present notation. We have the following dictionary, where we introduce some new notation

Chapter 9	Here
K	k
k	$\mathbb{Q}$
$\lvert\ \rvert$	$\lvert\ \rvert_p$
$\lvert\ \rvert_j$	$\lvert\ \rvert_{\mathfrak{p}}$ for some $\mathfrak{p}\mid p$
U	$U(p)$: the group of $a \in \mathbb{Q}$ with $\lvert a\rvert_p = 1$
$\bar{U}$	U_p : the group of $a \in \mathbb{Q}_p$ with $\lvert a\rvert = 1$.
$\bar{k}$	$\mathbb{Q}_p$
$\mathfrak{o}$	$\mathbb{Z}(p)$: the set of $a \in \mathbb{Q}$ with $\lvert a\rvert_p \leq 1$.
$\bar{\mathfrak{o}}$	$\mathbb{Z}_p$
$\mathfrak{O}$	$\mathfrak{o}(p)$: the set of $A \in k$ with $\lvert A\rvert_{\mathfrak{p}} \leq 1$ for all $\mathfrak{p}\mid p$
$D_{K/k} \in \mathfrak{o}/U^2$	$D_k(p) \in \mathbb{Z}(p)/\{U(p)\}^2$

<u>THEOREM 11.1</u> (i) <u>$D_k(p)$ is the image of D_k under the map $\mathbb{Z} \to \mathbb{Z}(p)/\{U(p)\}^2$</u>

(ii) <u>the sign of D_k is $(-1)^s$, where s is the number of complex places.</u>

Before proving Theorem 11.1, we note that it shows that the knowledge of the semilocal invariants $D_k(p)$ and s gives the value of D_k; and indeed with considerable overkill. In fact D_k is uniquely determined already by the $\lvert D_k\rvert_p = \lvert D_k(p)\rvert_p$ and s.

Part (i) of the Theorem is an immediate consequence of

LEMMA 11.1. Let $\{A_1,\ldots,A_n\}$ be a $\mathbb{Z}$-basis of $\mathfrak{a}$. Then it is a $\mathbb{Z}(p)$-basis of $\mathfrak{a}(p)$.

Proof. (cf. Lemma 3.4). If not, there would be some

$$B = b_1A_1 + \ldots + b_nA_n \in \mathfrak{a}(p) \tag{11.1}$$

where $b_1,\ldots,b_n \in \mathbb{Q}$ but not every b_j is in $\mathbb{Z}(p)$.

By the Chinese Remainder Theorem (or by the Strong Approximation Theorem 4.1 for $\mathbb{Q}$) there are $c_j \in \mathbb{Q}$ such that

$$|c_j - b_j|_p \leq 1; \quad |c_j|_q \leq 1 \quad \text{(all primes } q \neq p).$$

Put

$$C = \sum c_jA_j.$$

Then

$$C - B = \sum(c_j - b_j)A_j \in \mathfrak{a}(p) \tag{11.2}$$

and

$$C = \sum c_jA_j \in \mathfrak{a}(q) \quad \text{all} \quad q \neq p. \tag{11.3}$$

Then $C \in \mathfrak{a}(p)$ by (9.1) and (9.2), so

$$C \in \bigcap_{\substack{\text{all } q \\ \text{inc } p}} \mathfrak{a}(q) = \mathfrak{a}. \tag{11.4}$$

This contradicts the hypothesis that $\{A_1,\ldots,A_n\}$ is a basis of $\mathfrak{a}$, since not every c_j is in $\mathbb{Z}$. The contradiction proves the lemma.

Part (i) of Theorem 11.1 follows on comparing Definition 3.1 of Chapter 9 with Definition 3.1 of this Chapter.

Part (ii) of Theorem 11.1 is essentially a rather simple analogue of Part (i) together with Corollary 1 below. In fact $k \otimes_{\mathbb{Q}} \mathbb{R}$ is the direct sum of r copies of $\mathbb{R}$ and s copies of $\mathbb{C}$. Then

Part (ii) of the theorem now follows from a simple analogue of the proof of Lemma 3.2 of Chapter 9.

COROLLARY 1. *The image of* D_k *under* $\mathbb{Z} \to \mathbb{Z}_p/U_p^2$ *is*

$$\prod_{\mathfrak{p}|p} D_{k_\mathfrak{p}/\mathbb{Q}_p} . \tag{11.5}$$

Proof. For, by Lemma 3.2 of Chapter 9, this is the image of $D_k(p)$ under $\mathbb{Z}(p)/\{U(p)\}^2 \to \mathbb{Z}_p/U_p^2$.

COROLLARY 2. $p|D_k$ *if and only if some* $\mathfrak{p}|p$ *ramifies.*

Proof. Follows from Theorem 6.1 of Chapter 7 and the above Corollary.

COROLLARY 3. *If* $k \neq \mathbb{Q}$ *there is always a ramified prime.*

Proof. $D_k \neq \pm 1$ by Theorem 6.2, Corollary 1, and the previous Corollary applies.

COROLLARY 4. $D_k \equiv 0$ or 1 (4).

Proof. Follows from Lemma 6.4 of Chapter 7 and Corollary 1 above.

We conclude with a few words about the notion of a discriminant for an extension K/k where K,k are algebraic number fields. In general the ("global") integers $\mathfrak{O}$ of K are not a free $\mathfrak{o}$-module, and so Definition 3.1 does not generalize. However for every place $\mathfrak{p}$ of k there is a semilocal discriminant $D_{k/k}(\mathfrak{p}) \in \mathfrak{o}(\mathfrak{p})/\{U(\mathfrak{p})\}^2$ in a notation analogous to that used above. Classically, the discriminant of K/k is defined to be the k-divisor

$$\mathcal{D}_{K/k} = \{\mathrm{ord}_\mathfrak{p}\, D_{K/k}(\mathfrak{p})\}. \tag{11.6}$$

Clearly the analogue of Corollary 2 above holds. We have also

LEMMA 11.2. (Hecke). *The class of the divisor* (11.6) *is the square of a divisor class.*

<u>Proof</u>. Let $B_1,\dots,B_n$ be any (field-) basis of K/k. For any non-arch. place $\mathfrak{p}$ of k let $A_1(\mathfrak{p}),\dots,A_n(\mathfrak{p})$ be a basis of the semilocal integers $\mathfrak{O}(\mathfrak{p})$ over $\mathfrak{o}(\mathfrak{p})$. Then

$$A_i(\mathfrak{p}) = \sum_j t_{ij}(\mathfrak{p})\, B_j$$

for some $t_{ij}(\mathfrak{p}) \in k_{\mathfrak{p}}$. Hence

$$D(A_1(\mathfrak{p}),\dots,A_n(\mathfrak{p})) = T(\mathfrak{p})^2\, D(B_1,\dots,B_n)$$

where $T(\mathfrak{p}) = \det(t_{ij}(\mathfrak{p}))$ and the notation is analogous to (3.2). It follows that

$$\mathrm{ord}_{\mathfrak{p}}\, D_{K/k}(\mathfrak{p}) = 2\, \mathrm{ord}_{\mathfrak{p}}\, T(\mathfrak{p}) + \mathrm{ord}_{\mathfrak{p}}\, D(B_1,\dots,B_n)$$

This concludes the proof.

The definition (11.6) does not encapsulate all the information from the semilocal discriminants. Fröhlich (1960) has shown how this can be remedied by using "ideles".

12 <u>THE KRONECKER-WEBER THEOREM</u>

We recall that a field extension K/k is <u>abelian</u> if it is normal with abelian galois group: it is <u>cyclotomic</u> if $K = k(u)$ for some root of unity u. Clearly an extension which is contained in a cyclotomic extension is abelian. The theorem of Kronecker-Weber states that over the rational field the converse is true: every abelian extension of $\mathbb{Q}$ is contained in a cyclotomic one. We have already proved the corresponding result for local fields (Chapter 8, §4), and so done the hard work. Following Shafarevich (1951), we show now that the global Kronecker-Weber follows straightforwardly from the local one and from a "local to global" theorem of independent interest.

<u>THEOREM 12.1</u> ("Arithmetic monodromy"). <u>Let</u> $k/\mathbb{Q}$ <u>be a normal extension, with galois group</u> Γ. <u>Then</u> Γ <u>is generated (as a group) by the inertia subgroups of the places</u> $\mathfrak{p}$ <u>of</u> k.

<u>Note</u> Here we use the language of Chapter 7, §8 and Chapter 9, §5. The inertia group is trivial except for the $\mathfrak{p}$ which ramify; and these are

finite in number by Theorem 11.1, Corollary 2.

Proof. Let Λ be the subgroup generated by the inertia subgroups. Clearly Λ is normal in Γ. Let K be the fixed field of Λ. Clearly every place of K is unramified over $\mathbb{Q}$ and so $K = \mathbb{Q}$ by Theorem 11.1, Corollary 3. Hence $\Gamma = \Lambda$, as required.

THEOREM 12.2 (Kronecker-Weber). Let $k/\mathbb{Q}$ be an abelian extension. Then k is contained in a cyclotomic field.

Proof. Let $\mathfrak{p}$ be a place of k and $\mathfrak{p}|p$. By the local "Kronecker-Weber" theorem (Theorem 4.1 of Chapter 8), the local completion $k_{\mathfrak{p}}$ is contained in a cyclotomic extension of $\mathbb{Q}_p$, say

$$k_{\mathfrak{p}} \subset \mathbb{Q}_p(u_p, v_p), \tag{12.1}$$

where u_p is a root of unity whose order is a power $p^{s(p)}$ of p (say) and v_p is a root of unity of order prime to p. Since for fixed p the $k_{\mathfrak{p}}$ are permuted transitively by the action of the galois group Γ of $k/\mathbb{Q}$ (Chapter 9, §4), the order of u_p depends only on p and not on $\mathfrak{p}$.

Let L be the cyclotomic extension of $\mathbb{Q}$ generated by the $p^{s(p)}$-th roots of unity for all the ramified primes p, and let $K = kL$ be the compositum. Then $K/\mathbb{Q}$ is an abelian extension. For any prime $\mathfrak{P}$ of K over p we have now

$$K_{\mathfrak{P}} \subset \mathbb{Q}_p(u_p, w_p)$$

where w_p is a root of unity of order prime to p. It is clearly enough to prove the theorem with K instead of k. Hence without loss of generality we may suppose that

$$L \subset k. \tag{12.2}$$

For the degrees we have now

$$[k : \mathbb{Q}] \geqslant [L : \mathbb{Q}] = \prod_p \phi(p^{s(p)}), \tag{12.3}$$

where ϕ is Euler's totient function.

Since the galois group Γ of $k/\mathbb{Q}$ is abelian, the inertia group of a prime divisor $\mathfrak{p}$ of k depends only on the underlying rational prime (cf. Chapter 9, Exercise 10), and so may be denoted by $\Delta_o(p)$. By (12.1) the order satisfies

$$\mathrm{card}(\Delta_o(p)) \leq \phi(p^{s(p)}).$$

By Theorem 12.1, the group Γ is generated by the $\Delta_o(p)$, so

$$\mathrm{card}(\Gamma) \leq \prod_p \mathrm{card}(\Delta_o(p)) \leq \prod_p \phi(p^{s(p)}).$$

But $\mathrm{card}(\Gamma) = [k : \mathbb{Q}]$, and hence there is equality everywhere. Equality in (12.3) implies that $k = L$: and this concludes the proof.

13 STATISTICS OF PRIME DECOMPOSITION

Let K/k be an extension of algebraic number fields. The decomposition of an individual prime divisor $\mathfrak{p}$ of k in the extension is described in Chapter 9. Here we are concerned with "how often" (in some sense) the different kinds of decomposition occur as $\mathfrak{p}$ runs through the prime divisors of k. We give merely a brief statement of the facts, which are most easily proved by (comparatively elementary) analytic number theory. For proofs we refer, for example, to the article by Heilbronn in the Brighton Book (Cassels and Fröhlich (1967)) and, at a more advanced level, the article by Lagarias and Odlysko in Fröhlich (1977).

At most finitely prime divisors $\mathfrak{p}$ of k ramify in K/k, since they must divide the discriminant of the extension. In what follows we ignore these finitely many divisors.

We denote by $\mathbb{N}(\mathfrak{p})$ the (absolute) norm of the prime divisor $\mathfrak{p}$ (cf. §5). Denote by $\pi_k(x)$ the number of prime divisors $\mathfrak{p}$ of the field k for which $\mathbb{N}(\mathfrak{p}) \leq x$. The generalization to algebraic number fields of the prime number theorem asserts that $\pi_k(x)$ is asymptotically equal to $x/\log x$, which we write

$$\pi_k(x) \sim x/\log x. \tag{13.1}$$

(The classical prime number theorem is just (13.1) for $k = \mathbb{Q}$). Now

$N(\mathfrak{p}) = p^f$ where p is the underlying rational prime and f is the degree of the residue field extension. The $\mathfrak{p}$ with $f \geq 2$ clearly give a contribution of the order of $x^{\frac{1}{2}}$ to $\pi_k(x)$. Hence in considering the statistics of the behaviour of $\mathfrak{p}$ we can restrict attention to those of (absolute) first degree.

Now let K/k be an extension. Then (13.1) holds also for K. It follows immediately that infinitely many $\mathfrak{p}$ must have a first-degree prime divisor $\mathfrak{P}$ in K. The simplest case is when $[K : k] = 2$. Then (ignoring ramification) there are just the two cases

(i) $\mathfrak{p}$ splits into two divisors $\mathfrak{P}_1, \mathfrak{P}_2$ (say)

(ii) $\mathfrak{p}$ is inert.

The primes $\mathfrak{p}$ inert for K/k give a negligeable contribution to $\pi_K(x)$ for the reason given at the end of the preceding paragraph. Hence on comparing (13.1) for k and K we see that approximately <u>half</u> of the $\mathfrak{p}$ are of type (i) and, consequently, about half are of type (ii). More generally, if K/k is a normal extension of degree $[K : k] = n$, then any $\mathfrak{p}$ which has one first degree prime divisor $\mathfrak{P}$ in K must split completely in K. Hence a fraction n^{-1} of the $\mathfrak{p}$ will split completely and the remainder will either be inert or will decompose in ways that do not involve first-degree divisors. In particular, if $[K : k] = 3$ and the extension is normal, then one third of the $\mathfrak{p}$ will split completely, and the remainder will be inert.

To obtain further statistical information one must in general use (13.1) with further fields than k, K. Let $\mathfrak{K}$ be the least normal extension of k which contains K (we do not exclude $\mathfrak{K} = K$). If $\mathfrak{p}$ has one first degree extension to $\mathfrak{K}$ then (by an argument we have already used) it splits completely in $\mathfrak{K}$; and it does this precisely when it splits completely in K. The proportion of $\mathfrak{p}$ which do this is $1/m$, where $m = [\mathfrak{K} : k]$. This accounts for a proportion n/m of the first degree prime divisors of K. Hence at most a proportion $(1 - n/m)$ of the $\mathfrak{p}$ in K do not split completely in K but do have at least one first degree extension, where $n = [K : k]$. It follows that a proportion $(n - 1)/m > 0$ of $\mathfrak{p}$ do not have any first degree extension to K. To summarize:

<u>THEOREM 13.1</u> (i) <u>A positive proportion of prime divisors</u> $\mathfrak{p}$ <u>in</u> k <u>split completely into</u> $n = [K : k]$ <u>prime divisors of</u> K.

(ii) <u>A positive proportion of the</u> $\mathfrak{p}$ <u>do not have any first degree extension in</u> K.

<u>Note 1</u> When $[K : k] = 3$ but K/k is not abelian the above argument gives all the statistical information. Here $m = 6$ and 1/6 of the $\mathfrak{p}$ split completely, 1/2 of them have precisely two extensions (one of degree 1 and one of degree 2) and 1/3 are inert. For example, if $k = \mathbb{Q}$, $K = \mathbb{Q}(D)$ with $D^3 = d$, then a rational prime $p \equiv 2(3)$ has precisely one first degree extension to K while a $p \equiv 1\ (3)$ either splits completely or is inert.

<u>Note 2</u> It can happen that no $\mathfrak{p}$ is inert in K/k, which excludes a possible strengthening of (ii) of the Theorem. This is the case, for example, when K/k is normal of degree 6 and the galois group is non-cyclic (e.g., when K is the normal closure of a cubic extension of k). (Cf. also Chapter 6, Exercise 3).

<u>Note 3</u> There is an easy elementary proof of the weaker form of (i) which merely asserts that there are infinitely many $\mathfrak{p}$ which split completely, Cf. Lemma 2.2 of Chapter 5. There does not, however, seem to be any very simple proof of the corresponding weakening of (ii).

The case when K/k is of degree 3 is especially simple. In general, one has to use (13.1) not merely for k, K and the normal closure $\mathcal{K}$ but also for other fields between k and $\mathcal{K}$. This was worked out by Frobenius. A more precise result was given by Chebotarev, which gives information about the distribution of the Frobenius automorphisms when K/k is normal. We shall not go into this but refer the reader to the references cited at the beginning of this section.

One can reformulate Theorem 13.1 in terms of polynomials:

<u>COROLLARY.</u> <u>Let</u> $F(X) \in k[X]$ <u>be irreducible of degree</u> $n > 1$. <u>Then</u>

(i) <u>for a positive proportion of</u> $\mathfrak{p}$ <u>there are</u> n <u>roots of</u> F <u>in</u> $k_{\mathfrak{p}}$

(ii) <u>for a positive proportion of</u> $\mathfrak{p}$ <u>there are no roots of</u> F <u>in</u> $k_{\mathfrak{p}}$.

<u>Note</u> The condition that F be irreducible is not necessary for (i) [take K to be the splitting field of F] but it is necessary for (ii),

cf. Lemma 3 bis. 2 of Chapter 4.

We conclude with a methodological note. We have used (13.1) for ease of exposition. All that has been proved above can be derived from

$$\lim_{s\to 1+} (s-1) \sum \mathbb{N}(\mathfrak{p})^{-s} = 1. \tag{13.2}$$

This is much more readily proved than (13.1), and the proofs using (13.2) essentially the same as those using (13.1).

Notes

§§7,8 We have adopted the traditional way of proving the finiteness of the class number and the structure theorem for the units. There is an alternative approach using adeles (valuation vectors) and ideles which has claims to be considered more in the spirit of this book. For an account, see Chapter 2 of the Brighton Book (Cassels and Fröhlich (1967)).

§§10, 10 bis. There is a vast literature dealing with equations

$$F(x,y) = c \qquad (\pounds)$$

where $F(X,Y) \in \mathbb{Z}[X,Y]$ is homogeneous of degree $n \geq 3$, $c \in \mathbb{Z}$ is given and x,y are required to be in $\mathbb{Z}$. The only interesting case is when F is irreducible. The use of the structure theorems for the units of $\mathbb{Q}(C)$, where $F(C,1) = 0$ together with local considerations is usually referred to as "Skolem's method". It was applied by Skolem himself and, particularly, by W. Ljunggren to many problems. Some of their papers are listed in the references at the end of the book, for others, see the Zentralblatt and/or Mathematical Reviews. A recent paper worked out in detail is N. Tzanakis (1984).

It was proved by Chabauty, using these methods, that the equation (£) has only finitely many solutions provided that the relevant number field is not totally real. This is primarily of methodological interest, since Thue had already proved finiteness using diophantine approximation without this restriction. However the local method is often better suited to finding all the solutions in practice. In some cases Skolem and Ljunggren could also use local methods in the totally real case by introducing auxiliary equations (e.g. Ljunggren (1942)).

Until recently, all methods of treating (£) were ineffective, that is, they could not be guaranteed in advance to determine all the solutions: though, of course, they might do so in special cases. Recently Baker has given an effective approach to this and similar problems, see his book. In appropriate cases, the method can actually be used to determine all solutions, at least if one has a sufficiently powerful computer (Baker and Davenport (1969), Ellison et al. (1972)).

As noted in the text, Theorem 10.1 is not the best known. The proof (due to Skolem) was given as a simple and elegant application of local techniques. It was shown, however, by B. Delaunay (= Delone) in 1915 that if $x^3 + dy^3 = 1$ where $x,y,d \in \mathbb{Z}$, and $y \neq 0$, then $x + Cy$ is the fundamental unit of the ring $\mathbb{Z}[C]$, where $C^3 = d$, for which $0 < x + Cy < 1$ in the real embedding. This gives an effective procedure for determining all solutions. His proof is fairly complicated but elementary (with a p-adic flavour). The original publication was in an obscure Russian journal, but it is reproduced as Delaunay (1928). There are versions in Leveque (1956) and Mordell (1969). See also Nagell (1925).

Nagell (1928) gave bounds for the number of integral solutions of $F(X,Y) = 1$, where F is a cubic form with one real root. The case when there are three real roots has been considered by a number of authors, the latest being Evertse (1983).

The problem considered in Theorem 10 bis. 1 was treated by Skolem (1934) in $\mathbb{Q}_5$, but the bound obtained is larger than the number of solutions.

Finally, we observe that there are treatments of some of the subject matter of this note in Skolem (1938), Leveque (1956) and Mordell (1969).

§12 The Kronecker-Weber theorem gives a very explicit description of the abelian extensions of $\mathbb{Q}$. Class field theory originated in an attempt to give an equally explicit description of the abelian extensions of any algebraic number field k. In contrast to Theorem 11.1, Corollary 3, it is possible for there to exist extensions K/k which are unramified everywhere. This can happen only when the class number of k is > 1, and then there is a one-one correspondence between the unramified abelian extensions and the divisor classes. (Hence the term "class fields"). Later, a more general description of abelian extensions K/k developed, which took account of possible ramification. Later still, the theory developed to describe aspects of more general normal extensions K/k, but now it uses the language of homological algebra. It had also been discovered that there is a related "class field theory" of local fields (cf. notes to Chapter 8, §4) and that the global class field theory of a global field k and the local class field theories of its localizations $k_{\mathfrak{p}}$ mesh intimately together. There is a good account by Tate in the Brighton Book (Cassels and Fröhlich (1967)). Other good references are Chevalley (1953-4) (highly eccentric but extremely suggestive), Artin and Tate (1968) and Neukirch (1969) (an "axiomatic" approach).

Exercises

1. Let $k = \mathbb{Q}(C)$, where $C^2 = -7$. Show that a basis for the integers is $1, B$ where $B = (1 + C)/2$, and that $N(a + bB) = a^2 + ab + 2b^2$. Deduce that k has a euclidean algorithm.

Determine the unit group and divisor class group.

2. Let $k = \mathbb{Q}(C)$, where $C^2 = -5$. Show that 3 splits in k but that 3 is not a norm. Show, further, that 9 is the norm of an integer of k other than ± 3. Deduce that the class number is even.

Determine the unit group and class group.

3. Let $k = \mathbb{Q}(C)$, where $C^2 = -23$. Determine a base for the integers and show that 2 is not the norm of an integer.

Show that the divisor $[(3 + C)/2]$ is a cube but that $(3 + C)/2$ itself is not a cube. What does this imply for the class group?

Determine the unit group and the class group completely.

4. Let $k = \mathbb{Q}(C)$, $C^2 = 10$. Show that 3 splits but that the equations

$$N(x + yC) = \pm 3$$

have no solutions as congruences modulo 5.

Find the unit group and class group.

5. Let $k = \mathbb{Q}(C)$, $C^2 = 79$. Show that $E = 80 + 9C$ is a unit. By considering it at the $\mathfrak{p}|7$, show that E is not a cube.

Let $F = 17 - 2C$. Show that the divisor of F is a cube. Show that none of F, EF, E^2F is a cube by considering them at the $\mathfrak{p}|7$ and at the $\mathfrak{p}|13$.

Deduce that the class number is divisible by 3. Determine the unit group and the class group.

5 bis. Let $K = k(A_1,A_2)$ where k is an algebraic number field and $A_1^2 \in k$, $A_2^2 \in k$. Put $A_3 = A_1A_2$ and suppose that $A_j \notin k$ $(j = 1,2,3)$.

(i) For any $C \in K$ show that

$$C^2 \, N_{K/k}(C) = \prod_j N_j(C),$$

where N_j is the norm for $K/k(A_j)$.

(ii) Let V be the subgroup of K^* generated by U_1, U_2, U_3, where U_j is the group of units of $k(A_j)$. Let E be any unit of K. Show that $E^2 \in V$.

(iii) Determine the group of units of $\mathbb{Q}(A_1,A_2)$, where $A_1^2 = 2$, $A_2^2 = 3$.

[Hint To show that an element of V is not a square, consider it in $K_{\mathfrak{P}}$ for a suitable place $\mathfrak{P}$].

6. Show that $1 - 6C + 3C^2$ is a fundamental unit of $\mathbb{Q}(C)$, where $C^3 = 6$.

7. Let $k = \mathbb{Q}(C)$, where $C^3 = 11$. Show that $E = 1 + 4C - 2C^2$ is a unit. By considering the completion of k at the first-degree prime divisor of 5, show that E, $-E$ are not squares.

Let $B = 9 - 4C$. Show that its divisor is a square. By considering appropriate completions of k, show that none of B, $-B$, EB, $-EB$ is a square. Deduce that the class number is even.

Determine completely the unit group and the divisor class group.

8. Let M be a primitive p^n-th root of unity for some prime p and some $n \geq 1$. Show that $(M^j - 1)/(M - 1)$ is a unit for all integers j prime to p.

9. Let M be a primitive m-th root of unity, where m is not a prime power. Show that $M - 1$ is a unit.

[Hint Let $m = p\ell$, $C = M^\ell$. Show that $M - 1$ divides $C - 1$. Do this for two distinct primes p dividing m.]

9 bis. (i) Let $\mathbb{Q}^{(p)}$ be the p-th cyclotomic field, where p is an odd prime, and let $\mathfrak{a}$ be an integral divisor of $\mathbb{Q}^{(p)}$ prime to p. Show that

$$\mathbb{N}(\mathfrak{a}) \equiv 1 \quad (p).$$

(ii) Let K be the subfield of $\mathbb{Q}^{(p)}$ of index 2. If $\mathfrak{b}$ is an integral divisor of K prime to p show that

$$\mathbb{N}(\mathfrak{b}) \equiv \pm 1 \quad (p)$$

[Hint Chapter 9, Exercise 10 ter].

10. Let t be a square-free positive integer, $t \equiv 1$ or $2 \bmod 4$. Suppose that $\mathbb{Q}(\sqrt{-t})$ has class number prime to 3. Show that

$$y^2 + t = x^3$$

has no solutions in integers unless $t = 3a^2 \pm 1$ for some integer a. Determine all the solutions in this exceptional case.

11. Show that there are no solutions of $x^2 + 1 = 2y^n$ with $x, y \in \mathbb{Z}$, $x \neq \pm 1$ and n odd > 1 by completing the following argument:

(i) x, y are odd

(ii) $y = u^2 + v^2$ for some $u, v \in \mathbb{Z}$.

[Hint Consider factorization in $\mathbb{Z}[i]$, $i^2 = -1$]

(iii) $1 + ix = i^k(1 + i)(u + iv)^n$ (*)

for some $u,v \in \mathbb{Z}$ and some $k = 0,1,2,3$.

(iv) $\pm u \pm v = 1$.

[Hint Equate real parts and show that right hand side is divisible by $u \pm v$ for some sign. Recall n is odd.]

(v) Hence

$$1 \pm ix = i^\ell(1 + i)\{1 + 2(1 + i)w\}^n$$

for some $w \in \mathbb{Z}$ and some ℓ.

(vi) By equating real parts and working in $\mathbb{Q}_2$, deduce a contradiction.

12. Suppose that the discriminants of the extensions $L/\mathbb{Q}$, $K/\mathbb{Q}$ are coprime. Let $\{A_1,\ldots,A_n\}$, $\{B_1,\ldots,B_m\}$ be bases of integers. Show that the compositum KL has degree mn and that a basis for integers is

$$A_iB_j \quad (1 \leq i \leq n, \ 1 \leq j \leq m).$$

13. Let $F(X) \in k[X]$ have degree ≤ 4, where k is an algebraic number field. If F has a zero in all except finitely many local completions $k_\mathfrak{p}$, show that it has a zero in k.

[Fujwara (1973). Cf. Chapter 4, Exercise 4. Hint: If $F(X) = F_1(X)F_2(X)$, where $F_j(X) \in k[X]$ is quadratic, consider the proportion of primes of k which split in $k(\theta_1)$, $k(\theta_2)$ and $k(\theta_1,\theta_2)$ where $F_j(\theta_j) = 0$. Cf. proof of Theorem 13.1.]

14. Let $K = \mathbb{Q}(\sqrt{-3},\sqrt{13})$.

(i) Show that 2 is unramified in $K/\mathbb{Q}$.

(ii) Find a basis for the integers of K.

(iii) Hence find the discriminant $D_{K/\mathbb{Q}}$ and confirm that it is odd (Lemma 3.2, Corollary of Chapter 9).

(iv) Determine the unit group of K.

(v) Show that $[3] = \mathfrak{P}_3^2(\mathfrak{P}_3')^2$, where the prime divisors $\mathfrak{P}_3$, $\mathfrak{P}_3'$ are not principal.

(vi) Show that the class number of K is 2.

[Hint Exercise 5 bis].

15. Let K/k be an extension of number fields. Denote by λ the map of the divisor group of k into that of K by

$$\lambda(\mathfrak{p}) = \prod_{\mathfrak{P}} \mathfrak{P}^{e(\mathfrak{P})}$$

where $\mathfrak{P}$ runs through the places of K extending the place $\mathfrak{p}$ of k and where $e(\mathfrak{P})$ is the ramification of $K_{\mathfrak{P}}/k_{\mathfrak{p}}$.

(i) Show that the absolute norms of $\mathfrak{a}$ and $\lambda(\mathfrak{a})$ are the same.

(ii) Let $a \in k$ and let $\mathfrak{a}$, $\mathfrak{A}$ be the principal divisors of a considered as an element of k, K respectively. Show that $\mathfrak{A} = \lambda(\mathfrak{a})$.

[Note All older, and many modern writers, identify $\mathfrak{a}$ and $\lambda(\mathfrak{a})$, just as they identify $b \in k$ with b considered as an element of K. Some pernickety algebraists do, however, distinguish between $\mathfrak{a}$ and $\lambda(\mathfrak{a})$. They call λ the <u>conorm</u> map.]

16. This exercise deals with the global <u>different</u>, Cf. Chapter 7, Exercise 20 and Chapter 9, Exercise 15. Let K/k be an extension of algebraic number fields.

(i) Let

$$W = \{C \in K : S(AC) \in \mathfrak{o} \text{ for all } A \in \mathfrak{O}\},$$

where $S = S_{K/k}$ is the trace. For a place $\mathfrak{P}$ of K define the integer $d(\mathfrak{P}) \geq 0$ by

$$- d(\mathfrak{P}) = \inf_{C \in W} \operatorname{ord}_{\mathfrak{P}} C.$$

Show that $d(\mathfrak{P})$ is the same as the corresponding d of the local situation (Chapter 7, Exercise 20) for $K_{\mathfrak{P}}/k_{\mathfrak{p}}$, where $\mathfrak{P}$ extends the place $\mathfrak{p}$ of k.

(ii) Show that $d(\mathfrak{P}) > 0$ precisely when $\mathfrak{P}$ is ramified over k. Deduce that $d(\mathfrak{P}) = 0$ for almost all $\mathfrak{P}$, so that the global different

$$\mathfrak{D}_{K/k} = \prod_{\mathfrak{P}} \mathfrak{D}_{\mathfrak{P}}$$

is well-defined as a divisor, where $\mathfrak{D}_{\mathfrak{P}} = \mathfrak{P}^{d(\mathfrak{P})}$ is the local different.

(iii) Show that $\mathfrak{D}_{K/k}$ is the divisor belonging to the $\mathfrak{O}$-ideal generated by the $F_A'(A)$, where A runs through $\mathfrak{O}$ and $F_A(X)$ is the characteristic polynomial of A for the extension K/k.

(iv) Show that

$$\operatorname{Norm}_{K/k} \mathfrak{D}_{K/k} = \mathcal{D}_{K/k},$$

where $\mathcal{D}_{K/k}$ is the discriminant (given by (11.6)).

(v) If $k \subset L \subset K$, show that

$$\mathfrak{D}_{K/k} = \mathfrak{D}_{K/L} \mathfrak{D}_{L/k},$$

where the L-divisor $\mathfrak{D}_{L/k}$ is considered as a K-divisor as in the preceding exercise.

17. Fill in the steps of the suggested proof of the following statement:

(S). Let k be an algebraic number field and m a positive integer. Let $a \in k^*$ and suppose that $a \in (k_{\mathfrak{p}}^*)^m$ for almost all places $\mathfrak{p}$. Then one of the two following is true

(I) $a \in (k^*)^m$

(II) $m = 2^t m'$ where $2 \nmid m'$. Let K be the field generated over k by the 2^t-th roots of unity. The galois group of K/k is not cyclic (so $t \geq 3$). Further, $a \in (k^*)^{\frac{1}{2}m}$.

[Note Compare Chapter 4, Exercise 23. Statement (S) is closely related to the Grunwald-Wang Theorem. As the Grunwald Theorem, it lasted for some fifteen years, and was quite widely used until Wang (1950) pointed out that it was not true in the generality claimed. See also Chapter 9 of Artin and Tate (1968).]

(i) If m,n are coprime and $a \in (k^*)^m$, $a \in (k^*)^n$ then $a \in (k^*)^{mn}$. Hence it is enough to consider the case when $m = \ell^s$ is a power of a prime ℓ, which we now suppose. Let L be the field obtained by adjoining the m-th roots of 1 to k. In what follows $a \in (k_{\mathfrak{p}}^*)^m$ for almost all places $\mathfrak{p}$.

(ii) Suppose $k = L$. Then $a \in (k^*)^m$.

[Hint Theorem 13.1 (ii). Question. Where have you used $k = L$?].

(iii) In any case, $a = \alpha^m$, $\alpha \in L$.

(iv) If $[L : k]$ is prime to ℓ, then $a \in (k^*)^m$.

[Hint Take norms in (iii)].

(v) If the galois group of L/k is cyclic, of order a power of ℓ, then $a \in (k^*)^m$.

[Hint Let $X^m - a = \Pi\, g_j(X)$ be the decomposition into irreducibles in $k[X]$. Each $g_j(X)$ splits completely in L and so its galois group is a quotient group of $gal(L/k)$. Let g_J have minimal degree. For almost all $\mathfrak{p}$ some g_j has a root in $k_{\mathfrak{p}}$, so g_j and hence g_J splits completely in $k_{\mathfrak{p}}[X]$. Now see hint for (ii)]

(vi) If $gal(L/k)$ is cyclic, then $a \in (k^*)^m$.

[Hint Take an appropriate intermediate field between k, L and combine last two arguments].

(vii) Hence (I) of statement (S) holds if $\ell \neq 2$ or if $\ell = 2$, L/k cyclic.

(viii) If $\ell = 2$, L/k is not cyclic, then $L/k(i)$ is cyclic, $i^2 = -1$. Hence $a \in (k^*)^{\frac{1}{2}m}$.

[Hint cf (iv).

Note It can be shown that there is an $a_o \in k$ depending only on k and m such that either $a \in (k^*)^m$ of $a \in a_o(k^*)^m$].

CHAPTER ELEVEN: DIOPHANTINE EQUATIONS

1 INTRODUCTION

By a Diophantine equation we shall mean an equation where we require the solution to lie in a given field (e.g. $\mathbb{Q}$) or ring (e.g. $\mathbb{Z}$). In this chapter we shall be concerned only with the first of these two cases, although often the equations concerned will be homogeneous, so the distinction between rational and integral solutions disappears.

As already noted in Chapter 4, §3 bis, a necessary condition that an equation have a solution in an algebraic number field k is that it have a solution at all the local completions k (including those at the archimedean places). ("solutions everywhere locally"). We have also given (Chapter 4, §3 bis; Chapter 10, §9) examples where this condition is not sufficient. There are however some general situations where the existence of solutions everywhere locally implies the existence of a global solution: when this is the case, there is said to be a *Hasse principle*.

Perhaps the two most important examples of Hasse principles are

THEOREM A. (Hasse). *Let*

$$F(X_1,\ldots,X_n) \in k[X_1,\ldots,X_n] \tag{1.1}$$

be a quadratic form, where k *is an algebraic number field. Suppose that the equation* $F = 0$ *has a nontrivial solution everywhere locally. Then it has a nontrivial solution in* k.

Note The trivial solution of $F = 0$ is, of course, that in which all the variables are 0.

<u>THEOREM B</u>. (Hasse). <u>Let</u> K/k <u>be a cyclic extension of number fields and let</u> $b \in k$. <u>If</u> b <u>is a norm everywhere locally then it is a norm globally</u>.

By a <u>cyclic</u> extension, we mean a normal extension with cyclic galois group. Being a norm everywhere locally is to be interpreted in the semilocal sense of Chapter 9, §2. Equivalently,

$$N_{K/k}(x_1A_1 + \dots + x_nA_n) \tag{1.2}$$

for a basis $A_1,\dots,A_n$ of K/k and for $x_1,\dots,x_n \in k$ is given by a polynomial $G(x_1,\dots,x_n) \in k[x_1,\dots,x_n]$ (say). The hypothesis of Theorem B is that $G = b$ has a solution everywhere locally and the conclusion is that there is a global solution.

Theorem A and Theorem B have an important case in common. When $n = 3$, the equation $F = 0$ is readily reduced by a linear transformation of the variables defined over k [i.e., with coefficients in k] to the shape

$$a_1X_1^2 + a_2X_2^2 + a_3X_3^2 = 0, \tag{1.3}$$

and so to

$$X^2 - aY^2 = b, \tag{1.4}$$

where

$$a = -a_2/a_1, \quad b = -a_3/a_1. \tag{1.5}$$

But now a solution of (1.4) in k expresses b as a norm for the extension $k(\sqrt{a})/k$.

For those with a background in algebraic geometry, we note that every curve of genus 0 defined over k is birationally equivalent over k to a curve (1.3). Hence the Hasse principle for (1.3) implies that for curves of genus 0. We revert to this point at the end of §2.

We shall not prove Theorems A,B. Indeed even a proof of the Hasse principle for (1.3) and general field k is beyond the modest scope of this book. There is, however, a rather simple proof of the Hasse principle for (1.3) over $\mathbb{Q}$; and this we give in §2.

In situations where the Hasse principle does not hold, there can nevertheless be a very interesting interaction between the local and the global situation. There is a particularly nice theory for curves of genus 1. It would take us too far even to explain the general theory. In §§3, 4, however, we shall discuss the particularly simple case of the curves

$$aX^3 + bY^3 + cZ^3 = 0. \tag{1.6}$$

This is associated with its <u>jacobian</u>

$$X^3 + Y^3 + dZ^3 = 0, \tag{1.7}$$

where $d = abc$. The existence or non-existence of points on (1.6) gives us information about the set of rational points on (1.7). In some cases, this will permit us to conclude, for example, that the only rational points on (1.7) are those with $Z = 0$.

2 HASSE PRINCIPLE FOR TERNARY QUADRATICS

THEOREM 2.1 Let

$$F(X_1,X_2,X_3) \in \mathbb{Q}[X_1,X_2,X_3] \tag{2.1}$$

be a quadratic form, and suppose that in every $\mathbb{Q}_p$ *there is a nontrivial solution of* $F = 0$. *Then there is a nontrivial solution in* $\mathbb{Q}$.

Proof. After a linear transformation of the variables, we may suppose without loss of generality that

$$F(X_1,X_2,X_3) = a_1X_1^2 + a_2X_2^2 + a_3X_3^2. \tag{2.2}$$

If any a_j is zero, there is certainly a solution of $F = 0$ in $\mathbb{Q}$. Hence we suppose that they are all non-zero. On replacing F by bF and the X_j by c_jX_j for suitable $b, c_j \in \mathbb{Q}$, we may suppose without loss of generality that

$$a_j \in \mathbb{Z}\,; \quad a_1a_2a_3 \text{ square free.} \tag{2.3}$$

We must now express the hypothesis that $F = 0$ has everywhere locally a nontrivial solution in a suitable form for application.

LEMMA 2.1. *Suppose that* (2.2), (2.3) *hold and that* $F = 0$ *has a nontrivial solution in* $\mathbb{Q}_p$.

(i) *If* $p \neq 2$ *and* $p \mid a_3$, *then there is an* $r \in \mathbb{Z}$ *such that*

$$r^2a_1 + a_2 \equiv 0 \quad (p). \tag{2.4}$$

(ii) *If* $p = 2$ *and* $2 \nmid a_1a_2a_3$, *then, after permuting the indices* 1,2,3 *if necessary, we have*

$$a_1 + a_2 \equiv 0 \ (4). \tag{2.5}$$

(iii) *If* $p = 2$ *and* $2 \mid a_3$ *then there is an* $s = 0$ *or* 1 *such that*

$$a_1 + a_2 + a_3s^2 \equiv 0 \ (8). \tag{2.6}$$

Proof. Left to the reader.

We now define a set H of triples $(x_1,x_2,x_3) \in \mathbb{Z}^3$ by imposing congruence conditions corresponding to the primes p dividing $2a_1a_2a_3$.

(i) If $p \neq 2$ and (i) of the Lemma holds, then

$$x_1 \equiv rx_2 \quad (p). \tag{2.7}$$

(and similarly if p divides a_1 or a_2)

(ii) If $p = 2$ and (ii) holds, then

$$x_3 \equiv 0 \quad (2).$$

(iii) If $p = 2$ and (iii) holds, then

$$x_1 \equiv x_2 \ (4) \qquad x_3 \equiv sx_2 \ (2).$$

We denote by $|\ |$ the absolute value.

LEMMA 2.2. H is a subgroup of $\mathbb{Z}^3$ of index (at most) $4\ |a_1a_2a_3|$.

Proof. Clear.

LEMMA 2.3. If $(x_1,x_2,x_3) \in H$, then

$$F(x_1,x_2,x_3) \equiv 0 \quad (4a_1a_2a_3). \tag{2.8}$$

Proof. Clear.

We now apply Minkowski's Convex Body Theorem, in the variant given as Lemma 1, Corollary of Appendix C, to the body $C \subset \mathbb{R}^3$ given by

$$C: \ |a_1|\ x_1^2 + |a_2|\ x_2^2 + |a_3|x_3^2 < 4\ |a_1a_2a_3|. \tag{2.9}$$

Clearly C is convex and symmetric. It has volume

$$V(C) = (\pi/3).2^3.|4a_1a_2a_3|$$

$$> 2^3.4|a_1a_2a_3|. \tag{2.10}$$

Minkowski's Theorem asserts the existence of a $\underline{c} \neq \underline{0}$ in $H \cap C$. But then $F(\underline{c}) = 0$ by (2.8) and (2.9). This concludes the proof of Theorem 2.1.

Note 1 The proof uses only the solubility of $F = 0$ in $\mathbb{Q}_p$ for $p|2a_1a_2a_3$. For other p, however, solubility is guaranteed, since there is a nontrivial solution of $F \equiv 0\ (p)$ by Warning's theorem (Appendix D).

Note 2 Theorem 2.1 does not require the existence of a solution in $\mathbb{Q}_\infty = \mathbb{R}$, and so goes beyond the Hasse principle. In particular, the hypotheses of Theorem 2.1 imply that F is indefinite or singular. This is closely related to the law of quadratic reciprocity. For example, let $p > 0$, $q > 0$ be distinct odd primes and suppose that $q \equiv -1 \ (4)$ and that $-p$ is a quadratic residue of q. Then

$$X_1^2 + pX_2^2 + qX_3^2 = 0$$

has a nontrivial solution in $\mathbb{Q}_2$ and in $\mathbb{Q}_q$. There is clearly no solution in $\mathbb{Q}$ and so, by Theorem 2.1, there is no solution in $\mathbb{Q}_p$: that is, $-q$ is a quadratic non-residue of p. For a fuller discussion cf. Cassels (1978).

It was noted in §1 that Theorem 2.1 implies the Hasse principle for curves of genus 0 defined over $\mathbb{Q}$. More generally, we have

LEMMA 2.1. *Let* C *be a curve of genus* 0 *defined over a field* k. *Then* C *is birationally equivalent over* k *to a conic*.

Proof. (Sketch). The differential of any non-constant element of the function-field $k(C)$ provides a divisor D of degree -2 defined over k. By the Riemann-Roch theorem, there are three k-linearly independent elements s_1, s_2, s_3 of $k(C)$ which are $\succ D$. Hence $s_1^2, s_2^2, s_3^2, s_2s_3, s_3s_1, s_1s_2$ are all $\succ 2D$, where $2D$ has degree -4. By Riemann-Roch, they must be k-linearly dependent, that is

$$F(s_1, s_2, s_3) = 0,$$

where $F(X_1, X_2, X_3) \in k[X_1, X_2, X_3]$ is some quadratic form. Thus we have a birational equivalence with $F(\underline{X}) = 0$.

From Theorem 2.1 and Lemma 2.1 we have at once the

COROLLARY. *Let* C *be a curve of genus* 0 *defined over* $\mathbb{Q}$ *and suppose that for every prime* p *there is a place on* C *defined over* $\mathbb{Q}_p$. *Then there is a place defined over* $\mathbb{Q}$.

Note It is important that the Corollary refers to places, which is a birational concept, and not to points. It is quite possible for a multiple point to be defined over a field k (in the sense that all the co-ordinates are in that field) without the places underlying it being so defined.

To illustrate this, consider the curve

$$G(x,y,z) = (3x^2 - yz)^2 + 5(y^2 - zx)^2 - 2(z^2 - 3xy)^2 = 0 \quad (2.11)$$

which is equivalent by the Cremona transformation

$$\xi:\eta:\zeta = 3x^2 - yz : y^2 - zx : z^2 - 3xy; \quad (2.12_1)$$

$$x:y:z = \xi^2 - \eta\zeta : 3\eta^2 - \zeta\xi : \zeta^2 - 3\xi\eta \quad (2.12_2)$$

to

$$\xi^2 + 5\eta^2 - 2\zeta^2 = 0. \quad (2.13)$$

The quartic curve (2.11) has three double points, namely $(x,y,z) = (1,\delta,\delta^2)$, where $\delta^3 = 3$. Since $X^3 - 3$ has a root in $\mathbb{Q}_2$ and in $\mathbb{Q}_5$ there are thus nontrivial solutions of $G(x,y,z) = 0$ in those fields. For $p \neq 2,5$ (including $p = \infty$) there are infinitely many solutions in $\mathbb{Q}_p$ of (2.13) and so of (2.11). On the other hand, there is no nontrivial solution of $G(x,y,z)$ in $\mathbb{Q}$ since the singular points do not have rational coefficients and there are no rational points on (2.13).

There are forms of fixed degree in an arbitrarily large number of variables for which the Hasse principle fails. Examples are

$$(x^2 - 2y^2 + \sum_j z_j^2)^2 + (x^2 - 2y^2 - \sum_j t_j^2)^2 = 0 \quad (2.14)$$

in the $2n + 2$ variables x,y,z_j,t_j and, less trivially,

$$3(\sum_j x_j^2)^3 + 4(\sum_j y_j^2)^3 - 5(\sum_j z_j^2)^3 = 0 \quad (2.15)$$

in the $3n$ variables x_j,y_j,z_j. The proof that this is so is left to the reader (cf. Chapter 10, Lemma 9.1).

3 CURVES OF GENUS 1. GENERALITIES

There is a very rich theory. We shall illustrate a few basic ideas by considering non-singular cubic plane curves.

We start with some general theory. Let k be any field and let

$$F(X_1,X_2,X_3) \in k[X_1,X_2,X_3]. \tag{3.1}$$

be a cubic form. Then

$$C: F(X_1,X_2,X_3) = 0 \tag{3.2}$$

is the equation of a cubic plane curve. We suppose that it has no singularities (double points or cusps). A point $\underline{a} = (a_1,a_2,a_3)$ in the plane is said to be defined over k if the homogeneous co-ordinates can be chosen so that $a_j \in k$.

Let $\underline{a},\underline{b}$ be two points on C, both defined over k. Then the line joining them meets C in a third point $\underline{c}$, which is also defined over C since it is given by a cubic equation of which two roots already lie in k. It is possible to take $\underline{b} = \underline{a}$, when the line is the tangent at $\underline{a}$. The points $\underline{a},\underline{b},\underline{c}$ can indeed all be the same: thus is the case when $\underline{a}$ is a point of inflexion (we shall meet a specific case later on). In general, however, one may expect to obtain new points on C, defined over k from some already known. This is called the chord and tangent process. It goes back to Diophantos, Fermat and Newton.

LEMMA 3.1. *If there is a point on C defined over a quadratic extension of k, then there is a point defined over k.*

Proof. Let $\underline{A}$ be the point and $\underline{A}'$ its conjugate over k. The line joining $\underline{A},\underline{A}'$ cuts C in a third point, which must be defined over k.

We also quote a useful formula of Desboves.

LEMMA 3.2. *Suppose that*

$$F(X_1,X_2,X_3) = a_1X_1^3 + a_2X_2^3 + a_3X_3^3 + bX_1X_2X_3 \tag{3.3}$$

(i) Let $\underline{u} = (u_1, u_2, u_3)$, $\underline{v} = (v_1, v_2, v_3)$ be distinct points on C. The third intersection with C of the line joining them is $\underline{w}$, where

$$w_1 = u_1^2 v_2 v_3 - v_1^2 u_2 u_3, \quad w_2 = u_2^2 v_3 v_1 - v_2^2 u_3 u_1,$$

$$w_3 = u_3^2 v_1 v_2 - v_3^2 u_1 u_2 \tag{3.4}$$

(ii) The third point of intersection $\underline{x}$ of the tangent at $\underline{u}$ is given by

$$x_1 = u_1(a_2 u_2^3 - a_3 u_3^3), \quad x_2 = u_2(a_3 u_3^3 - a_1 u_1^3),$$

$$x_3 = u_3(a_1 u_1^3 - a_2 u_2^3). \tag{3.5}$$

Proof. Left to the reader. [Hint: express the required vector $\underline{w}$ as $\lambda\underline{u} + \mu\underline{v}$ for scalar λ, μ and substitute in F].

We now revert to a general F.

It is a fundamental fact that the chord and tangent process gives the set $\mathfrak{G}$ of points on C defined over k an abelian group structure (provided, of course, that $\mathfrak{G}$ is not empty). Let $\underline{0}$ be any element of $\mathfrak{G}$: we fix it once for all and make $\mathfrak{G}$ into an abelian group with $\underline{0}$ as the neutral element (or zero). For our group law,

$$\underline{a} + \underline{b} + \underline{c}$$

will be the same for any collinear triple $\underline{a}, \underline{b}, \underline{c}$ of points of $\mathfrak{G}$. Hence to find $\underline{a} + \underline{b}$ we first find the third point of intersection $\underline{d}$ of the line

$$\underline{\ell} : \ell_1 X_1 + \ell_2 X_2 + \ell_3 X_3 = 0 \tag{3.6}$$

through $\underline{a}$ and $\underline{b}$ with the curve C and then take the third intersection $\underline{e}$ of the line

$$\underline{t} : t_1 X_1 + t_2 X_2 + t_3 X_3 = 0 \tag{3.7}$$

through $\underline{0}$ and $\underline{d}$. Then, by definition, $\underline{e} = \underline{a} + \underline{b}$.

LEMMA 3.3. *Under the above law of composition,* $\mathfrak{G}$ *is an abelian group.*

Proof. The only nontrivial thing to verify is the associative law:

$$(\underline{a} + \underline{b}) + \underline{c} = \underline{a} + (\underline{b} + \underline{c}). \tag{3.8}$$

We give two proofs: the first is more basic but involves a little algebraic geometry, the second is more special.

First proof. Let $\ell(\underline{X})$, $t(\underline{X})$ be the linear forms occurring in (3.6), (3.7) and let $\underline{a},\underline{b},\underline{d},\underline{e}$ have the meanings they have there. Then

$$f(\underline{X}) = \ell(\underline{X})/t(\underline{X})$$

is homogeneous of degree 0 in $\underline{X}$, and so defines a function on the curve C. It has zeros at $\underline{a},\underline{b}$ and poles at $\underline{0},\underline{e}$ (the point $\underline{d}$, being a zero of both linear forms, is neither a pole nor a zero of $f(\underline{X})$). Hence the pair $\{\underline{a},\underline{b}\}$ is linearly equivalent to the pair $\{\underline{0},\underline{e}\}$. This serves as an alternative characterization of $\underline{e}$, and (3.8) now follows from elementary facts about linear equivalence.

Second proof. Consider the diagram:

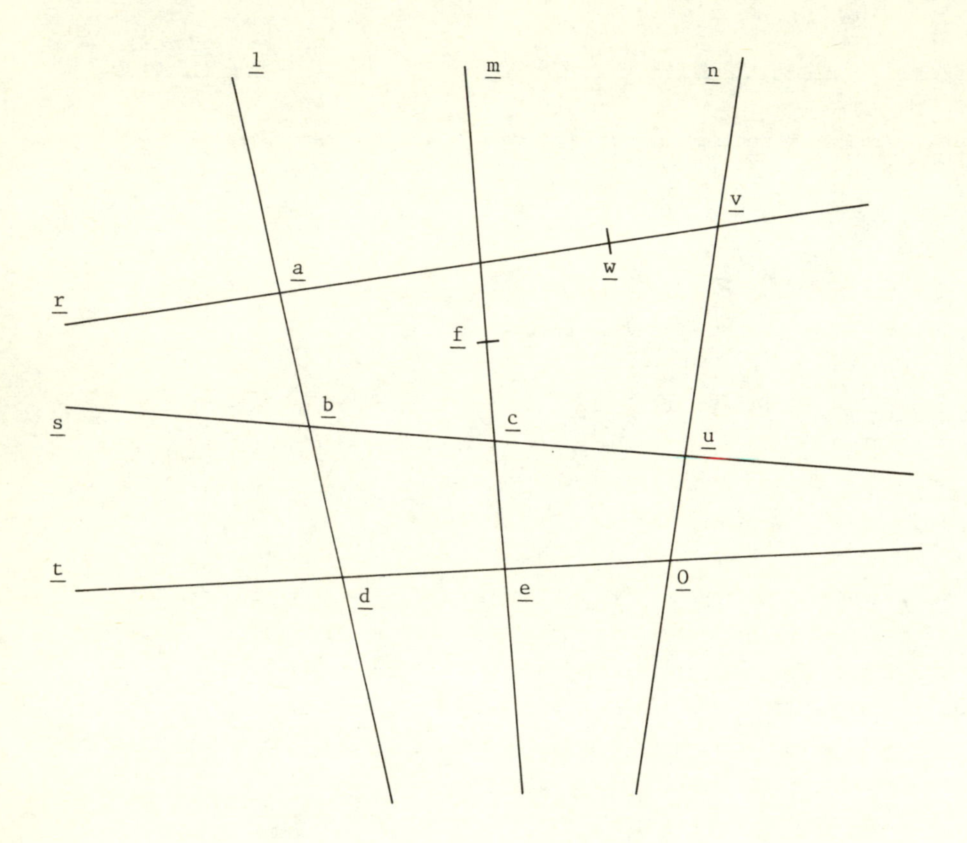

There are six lines $\underline{\ell},\underline{m},\underline{n},\underline{r},\underline{s},\underline{t}$ and ten points $\underline{0},\underline{a},\underline{b},\underline{c},\underline{d},\underline{e},\underline{f},\underline{u},\underline{v},\underline{w}$. The points all lie on the given non-singular cubic C, which has not been drawn in, to avoid confusion. To prove (3.8) we have to show that the points $\underline{w},\underline{f}$ are the same (and so that they are not as shown, but both coincide with the intersection of the lines $\underline{m}$ and $\underline{r}$).

We can regard the triplet of lines $\underline{\ell},\underline{m},\underline{n}$ as a singular cubic

$$G(\underline{X}) = \ell(\underline{X})m(\underline{X})n(\underline{X}) = 0$$

and similarly for $\underline{r},\underline{s},\underline{t}$:

$$H(\underline{X}) = r(\underline{X})s(\underline{X})t(\underline{X}) = 0.$$

To show that $\underline{f} = \underline{w}$ we need now invoke

LEMMA 3.4. *Suppose that the three cubic curves* $F(\underline{X}) = 0$, $G(\underline{X}) = 0$, $H(\underline{X}) = 0$ *have 8 points in common. Then they have 9 points in common.*

Proof. The general cubic form has 10 coefficients. The condition that it vanish at a given point is a linear condition on the coefficients. Hence the set of cubic forms vanishing at 8 given points is a vector space of dimension 2. In particular, the forms $F(\underline{X})$, $G(\underline{X})$, $H(\underline{X})$ are linearly dependent over k. Two cubic curves $F = 0$, $G = 0$ have 9 points in common; and all 9 must lie on $H = 0$.

We can now enunciate

THEOREM C. ("Mordell's finite basis Theorem".) *Let* k *be an algebraic number field. Then the group* $\mathfrak{G}$ *defined above is finitely generated.*

Unfortunately a proof would take us too far from the principal aims of this book, but we sketch a proof of a special case at the end of the next section.

4 CURVES OF GENUS 1. A SPECIAL CASE

We consider the determination of the rational points on

$$C = C(d): \quad X_1^3 + X_2^3 + dX_3^3 = 0, \tag{4.1}$$

where

$$d \in \mathbb{Z}; \quad d > 0, \quad \text{cube free.} \tag{4.2}$$

It turns out to be very advantageous to work in the "Eisenstein" field

$$k = \mathbb{Q}(\rho); \quad \rho^2 + \rho + 1 = 0. \tag{4.3}$$

The ring of integers is

$$\mathfrak{o} = \mathbb{Z}[\rho]. \tag{4.4}$$

The class number is 1 (there is a Euclidean algorithm). The units are

$\pm 1,\ \pm\rho,\ \pm\rho^2$.

We denote by $\mathfrak{G}$ the group of points on (4.1) defined over k, and take

$$\underline{0} = (1, -1, 0) \tag{4.5}$$

as the zero of the group. Then

$$-(a_1, a_2, a_3) = (a_2, a_1, a_3) \tag{4.6}$$

in the sense of the group law. Two further elements of $\mathfrak{G}$ are

$$(\rho, -\rho^2, 0), \quad (\rho^2, -\rho, 0). \tag{4.7}$$

They are of order 3. We note that

$$(\rho, -\rho^2, 0) + (a_1, a_2, a_3) = (\rho a_1, \rho^2 a_2, a_3). \tag{4.8}$$

For $\underline{a} \in \mathfrak{G}$ other than the points (4.7) we write

$$m(\underline{a}) = (\rho a_1 + \rho^2 a_2)/(\rho^2 a_1 + \rho a_2). \tag{4.9}$$

LEMMA 4.1. _Suppose that_ $\underline{a},\underline{b},\underline{c} \in \mathfrak{G}$ _are distinct from (4.7) and that_

$$\underline{a} + \underline{b} + \underline{c} = \underline{0}. \tag{4.10}$$

Then

$$m(\underline{a})m(\underline{h})m(\underline{c}) \in (k^*)^3 \tag{4.11}$$

Note k^* is the multiplicative group of non-zero elements of k, so (4.11) means that the left hand side is a non-zero cube.

Proof. We introduce new co-ordinates

$$U_1 = \rho X_1 + \rho^2 X_2, \quad U_2 = \rho^2 X_1 + \rho X_2, \tag{4.12}$$

so C has the equation

$$U_1U_2(U_1 + U_2) - dX_3^3 = 0. \tag{4.13}$$

By definition, (4.10) means that $\underline{a},\underline{b},\underline{c}$ lie on a line, say

$$X_3 = r_1U_1 + r_2U_2 \qquad (r_1,r_2 \in k). \tag{4.14}$$

The intersection of (4.13), (4.14) is given by

$$U_1U_2(U_1 + U_2) - d(r_1U_1 + r_2U_2)^3 = 0 \tag{4.15}$$

The product of the values of U_1/U_2 at the three intersections is thus $-(r_2/r_1)^3$. This implies (4.11).

In view of (4.11), we write

$$\tilde{m}(\underline{a}) = m(\underline{a})k^*, \tag{4.16}$$

the class of $k^*/(k^*)^3$ to which $m(\underline{a})$ belongs. We note that

$$\frac{\rho^2a_1 + \rho a_2}{a_1 + a_2} \in d^{-1}\tilde{m}(\underline{a}); \qquad \frac{a_1 + a_2}{\rho a_1 + \rho^2a_2} \in d\tilde{m}(\underline{a}), \tag{4.17}$$

since, for example,

$$\frac{\rho^2a_1 + \rho a_2}{a_1 + a_2} \Bigg/ \frac{\rho a_1 + \rho^2a_2}{\rho^2a_1 + \rho a_2} = \frac{(\rho^2a_1 + \rho a_2)^3}{a_1^3 + a_2^3} = -\frac{(\rho^2a_1 + \rho a_2)^3}{da_3^3}.$$

This motivates the definitions

$$\tilde{m}((\rho,-\rho^2,0)) = d^{-1}(k^*)^3; \; \tilde{m}((\rho^2,-\rho,0)) = d(k^*)^3 \tag{4.18}$$

for the points (4.7).

<u>LEMMA 4.2.</u> <u>The map</u>

$$\underline{a} \to \tilde{m}(\underline{a}) \tag{4.19}$$

<u>is a group homomorphism from</u> $\mathfrak{G}$ <u>to</u> $k^*/(k^*)^3$.

<u>Proof</u>. Follows from Lemma 4.1 except when one of $\underline{a},\underline{b},\underline{c}$ in (4.10) is a point (4.7). In the exceptional cases it is left to the reader to verify the Lemma, either by using (4.8) or by introducing co-ordinates (U_1,U_3,X_3), (U_2,U_3,X_3) instead of (U_1,U_2,X_3) in the proof of Lemma 4.1, where $U_3 = X_1 + X_2$.

Suppose now that

$$\tilde{m}(\underline{a}) = m(k^*)^3 \tag{4.20}$$

for some $m \in k^*$. By (4.9), (4.16), (4.17) there are $\alpha_1,\alpha_2,\alpha_3 \in k$ such that

$$\rho a_1 + \rho^2 a_2 : \rho^2 a_1 + \rho a_2 : a_1 + a_2$$
$$= m^{-1}\alpha_1^3 : m\alpha_2^3 : d\alpha_3^3, \tag{4.21}$$

and so

$$m^{-1}\alpha_1^3 + m\alpha_2^3 + d\alpha_3^3 = 0. \tag{4.22}$$

This leads us to introduce the curve

$$\mathcal{D} = \mathcal{D}(m) : m^{-1} Y_1^3 + mY_2^3 + dY_3^3 = 0. \tag{4.23}$$

If m,m' are in the same coset of $k^*/(k^*)^3$, then clearly $\mathcal{D}(m),\mathcal{D}(m')$ can be taken into one another by multiplying Y_1,Y_2 by constant elements of k, and so for our present purposes the curves may be identified.

The curve $\mathcal{D}(m)$ is not, in general, birationally equivalent to $\mathcal{C}$ over k. They are equivalent, however, over the field $\bar{\mathbb{Q}}$ of algebraic numbers by taking

$$X_1 = (m^{1/3})^{-1}Y_1, \quad X_2 = m^{1/3}Y_2, \quad X_3 = Y_3, \tag{4.24}$$

where $m^{1/3}$ is any cube root of m. There is also a map

$$\phi: \mathcal{D}(m) \to \mathcal{C} \tag{4.25}$$

defined over k given by

$$X_1 : X_2 : X_3 = \tag{4.26}$$

$$\rho^2 m^{-1} Y_1^3 + \rho m Y_2^3 + dY_3^3 : \rho m^{-1} Y_1^3 + \rho^2 m Y_2^3 + dY_3^3 : -3Y_1 Y_2 Y_3.$$

We have thus proved:

LEMMA 4.2. *Let* $\underline{a}$ *be a point on* $\mathcal{C}$ *defined over* k *and suppose that* $\tilde{m}(\underline{a}) = m(k^*)^3$ *for some* $m \in k^*$. *Then there is a point* $\underline{\alpha}$ *on* $\mathcal{D}(m)$ *defined over* k *such that*

$$\phi(\underline{\alpha}) = \underline{a}. \tag{4.27}$$

We are thus led to consider for given d and m whether there are points on $\mathcal{D}(m)$ defined over k. As at the moment of writing (3.43 p.m. on 4 March, 1985), there is no algorithm which can be guaranteed to decide this. We are thus driven to consider the weaker statement that there is a point on $\mathcal{D}(m)$ everywhere locally.

LEMMA 4.3. *The set of* m *such that there is a point on* $\mathcal{D}(m)$ *everywhere locally is a group under multiplication.*

DEFINITION. The corresponding subgroup of $k^*/(k^*)^3$ is called the *Selmer group* of $\mathcal{C}$.

Lemma 4.3 is an immediate consequence of

LEMMA 4.4. *Let* $K \supset k$ *be a field. Then the set of* $m \in k$ *such that there is a point on* $\mathcal{D}(m)$ *defined over* K *is a group.*

Proof. If m has the property then clearly so does m^{-1}. Hence it is enough to show that if $\ell mn = 1$ and ℓ, m have the property, then so does n. It would not be difficult to construct a proof by modifying the proof of Lemma 4.1 to work over K. However Lemma 3.2 gives the following more explicit proof:

Let $\underline{a}, \underline{b}$ be points on $\mathcal{D}(\ell), \mathcal{D}(m)$ respectively defined over k, so

$$\underline{u} = (\ell^{-1/3}a_1, \ell^{1/3}a_2, a_3), \quad \underline{v} = (m^{-1/3}b_1, m^{1/3}b_2, b_3)$$

are on $\mathcal{C}$ (defined over $K(\ell^{1/3}, m^{1/3})$). Defining $\underline{w}$ by (3.4) we have

$$w_1 = (\ell m)^{1/3}c_1, \quad w_2 = (\ell m)^{-1/3}c_2, \quad w_3 = c_3,$$

where c_1, c_2, c_3 are in K and $\underline{c}$ lies on $\mathcal{D}(\ell^{-1}m^{-1}) = \mathcal{D}(n)$.

LEMMA 4.5. *Let* $\mathrm{ord}_{\mathfrak{p}}$ *be the order function for some non-arch. place* $\mathfrak{p}$ *of* k. *If there is a point on* $\mathcal{D}(m)$ *defined over* $k_{\mathfrak{p}}$, *then*

$$\mathrm{ord}_{\mathfrak{p}} m \equiv 0 \quad \text{or} \quad \equiv \pm\, \mathrm{ord}_{\mathfrak{p}} d \qquad (3).$$

Proof. For if $\underline{a} = (a_1, a_2, a_3)$ is the point, then two at least of

$$|m^{-1}a_1^3|_{\mathfrak{p}}, \quad |ma_2^3|_{\mathfrak{p}}, \quad |da_3^3|_{\mathfrak{p}}$$

must be equal.

What we have done so far has used only that k is an algebraic number field with $\rho, d \in k$. We now require the hypotheses (4.2), (4.3), (4.4) (though they could still be weakened).

COROLLARY 1. *Suppose that* $m(k^*)^3$ *is in the Selmer group. Then there is a factorization*

$$d = rst \tag{4.28}$$

with

$$r,s,t \in \mathfrak{o}; \quad \text{g.c.d.}(r,s) = \text{g.c.d.}(s,t) = \text{g.c.d.}(t,r) = 1 \tag{4.29}$$

such that

$$r/s \in m(k^*)^3 \tag{4.30}$$

Proof. We have a factorization

$$d = \varepsilon \prod \pi_j^{\ell(j)}$$

where π_j are prime elements for the valuations $\mathfrak{p} = \mathfrak{p}_j$ for which d is not a unit and where ε is a (global) unit. Let

$$R = \Pi\, \pi_j^{\ell(j)} \quad (\mathrm{ord}_j m \equiv \mathrm{ord}_j d \quad (3))$$

$$S = \Pi\, \pi_j^{\ell(j)} \quad (\mathrm{ord}_j m \equiv -\,\mathrm{ord}_j d \quad (3))$$

where, for simplicity, we have denoted the order at $\mathfrak{p}_j$ by ord_j. Then by the lemma

$$\mathrm{ord}_{\mathfrak{p}}(mR^{-1}S) \equiv 0 \quad (3) \quad (\text{all } \mathfrak{p}).$$

Hence

$$mR^{-1}S = EB^3$$

for some unit E and some $B \in k$. If we define r,s,t by

$$r = R, \quad s = sE, \quad rst = d,$$

then we have what is required.

COROLLARY 2. <u>The Selmer group consists of the</u> $(r/s)(k^*)^3$ <u>satisfying (4.28), (4.29) for which there is everywhere locally a point on</u>

$$\mathcal{E}(r,s,t) : rX_1^3 + sX_2^3 + tX_3^3 = 0. \tag{4.31}$$

<u>Proof</u>. The equation (4.23) of $\mathcal{D}(m)$ with $m = r/s$ reduces to (4.31) on putting

$$X_1 = sY_1, \quad X_2 = rY_2, \quad X_3 = rsY_3.$$

<u>Metalemma 4.1</u>. <u>For given</u> d <u>the Selmer group can be determined in a finite number of steps</u>.

<u>Metaproof</u>. There is only a finite number of sets r,s,t satisfying (4.28),(4.29). If $\mathfrak{p}$ is a valuation of k for which $|d|_{\mathfrak{p}} = |3|_{\mathfrak{p}} = 1$, there is certainly a nontrivial solution of (4.31) mod $\mathfrak{p}$ (cf. Appendix D),

which gives a solution in $k_{\mathfrak{p}}$ by Hensel's Lemma. There is only one arch. valuation of k: the corresponding completion of k is $\mathbb{C}$ and so there certainly are local solutions of (4.31). It remains to check whether there are solutions of (4.31) for the finite number of non-arch. places $\mathfrak{p}$ for which $|3|_{\mathfrak{p}} < 1$ or $|d|_{\mathfrak{p}} < 1$; and by Hensel's lemma each of these checks can be carried out in a finite number of steps.

By discussing specific values of d, we require one further general Lemma.

LEMMA 4.6. *Let* $d \neq 2$ *and suppose that there is a point* $\underline{a}$ *on* C *defined over* k *with* $a_3 \neq 0$. *Then there is such a point* $\underline{b}$ *for which*

$$\tilde{m}(\underline{b}) \neq (k^*)^3, \quad d(k^*)^3, \quad d^{-1}(k^*)^3.$$

Proof. Suppose not. By homogeneity we may assume without loss of generality that $a_1, a_2, a_3 \in \mathfrak{o}$ have no common factor. From the equation (4.1) of C and the assumption that d is cubefree, it follows readily that a_1, a_2, a_3 are coprime in pairs. The norm $N_{k/\mathbb{Q}}(a_3)$ is a positive integer. We choose $\underline{a}$ for which it is as small as possible. By (4.8), (4.18) we may suppose without loss of generality that $\tilde{m}(\underline{a}) = (k^*)^3$.

We may take $m = 1$ in Lemma 4.2, so $\mathcal{D}(m) = C$. By homogeneity we may suppose that $\alpha_1, \alpha_2, \alpha_3 \in \mathfrak{o}$ are coprime in pairs, where $\underline{\alpha}$ is the point given by Lemma 4.2 for which $\phi(\underline{\alpha}) = \underline{a}$. Hence and by (4.26) with $m = 1$, $\underline{Y} = \underline{\alpha}$ we obtain, after a little manipulation,

$$ha_1 = -\rho^2\alpha_1^3 + \rho\alpha_2^3, \quad ha_2 = \rho\alpha_1^3 - \rho^2\alpha_2^3, \quad ha_3 = (\rho - \rho^2)\alpha_1\alpha_2\alpha_3 \tag{4.32}$$

for some $h \in \mathfrak{o}$. On eliminating α_1^3, α_2^3 successively from the first two equations, we see that h divides $(\rho - \rho^2)\alpha_1^3$ and $(\rho - \rho^2)\alpha_2^3$. Since α_1, α_2 are coprime, it follows that $N(h) = 1$ or 3. The last equation in (4.32) now gives

$$N(a_3) \geq N(\alpha_1)\, N(\alpha_2)\, N(\alpha_3).$$

By (4.32) and since $a_3 \neq 0$, we have $\alpha_1 \neq 0$, $\alpha_2 \neq 0$, $\alpha_3 \neq 0$. If $N(\alpha_1) = N(\alpha_2) = 1$ we have $d = 2$, the case excluded from the enunciation.

Otherwise we have $N(a_3) > N(\alpha_3) > 0$. But $\underline{\alpha}$ is a point on C, so this contradicts the minimality assumption on $\underline{a}$.

Note The proof is an example of Fermat's "infinite descent".

THEOREM 4.1 *For* $d = 1$ *every point* $\underline{a}$ *on* C *defined over* k, *has* $a_1a_2a_3 = 0$.

Proof. It is left to the reader to find an appropriate modification of the proof of Lemma 4.6.

Note This is, of course, a case of "Fermat's Last Theorem".

THEOREM 4.2 *Let* $d = q$ *or* $d = q^2$ *where* $q > 0$ *is a rational prime, and*

$$q \equiv 2 \quad \text{or} \quad 5 \quad (9). \tag{4.33}$$

Exclude the case $d = 2$. *Then any point* $\underline{a}$ *on* C *defined over* k *has* $a_3 = 0$.

Note It is left to the reader to modify the proof of Lemma 4.6 and of the Theorem to show that the points when $d = 2$ have either $a_3 = 0$ or $a_1^3 = a_2^3 = -a_3^3$.

Proof. By (4.33) q is a prime in k. Further,

$$N_{k/\mathbb{Q}}\, q = q^2 \not\equiv 1 \quad (9)$$

and so ρ is not a cubic residue mod q in $\mathfrak{o}$. It follows that an equation such as

$$\rho X_1^3 + \rho^{-1}X_2^3 + dX_3^3 = 0$$

has no nontrivial solution in k_q. By Lemma 4.5, Corollary 2, the Selmer group consists thus of $(k^*)^3$, $d(k^*)^3$, $d^{-1}(k^*)^3$. The truth of the Theorem now follows from Lemma 4.6.

<u>THEOREM 4.3</u> <u>Let</u> $d = 3p$, <u>where</u> $p > 0$ <u>is a rational prime</u>, $p \equiv 1 \quad (3)$. <u>Suppose that</u> 3 <u>is not a cubic residue</u> <u>mod</u> p. <u>Then any point</u> <u>a</u> <u>on</u> C <u>defined over</u> k <u>has</u> $a_3 = 0$.

<u>Proof</u>. We require some facts about arithmetic in k. First,

$$(\rho - \rho^2)^2 = -3.$$

Next, the p-adic valuation on $\mathbb{Q}$ splits completely in k and so, since k has class number 1,

$$p = \pi \pi'$$

for $\pi, \pi' \in \mathfrak{o}$ where, on taking $\pm \rho^j \pi$ for π we may suppose without loss of generality that

$$\pi \equiv \pi' \equiv 1 \quad (3).$$

With this normalization, π, π' are cubic residues of each other. (This is part of the law of cubic reciprocity, or see Chapter 9, Exercise 11). For the proof we do not need a criterion in terms of π for 3 to be a cubic residue, but it is that $m \equiv 0 \quad (3)$, where $2\pi = \ell + 3m\sqrt{-3}$.

We now turn to the proof of the Theorem and use Lemma 4.5, Corollary 2. In (4.31) precisely one of r,s,t is devisible by 3, say $r = 3u$; so we have an equation

$$3uX_1^3 + sX_2^3 + tX_3^3 = 0, \quad u,s,t \in \mathfrak{o}, \quad ust = p.$$

On taking $\pm X_j$ for X_j we may suppose that

$$u \equiv s \equiv t \equiv 1 \quad (\rho - \rho^2)$$

and on multiplying u,s,t simultaneously by ρ^j that

$$t \equiv 1 \quad (3).$$

On considering solubility in k_3, the completion of k with respect to the unique continuation of the 3-adic valuation, we get

$$s \equiv 1 \quad (3),$$

since $\beta^3 \equiv \pm 1 \ (3)$ for any unit β of k_3. Since $ust = p$, the only possible values for u, s, t are $1, \pi, \pi'$ and p.

We now consider solubility in k_π. If either s or t were divisible by π, solubility implies that 3 is a cubic residue of π (and hence of p), since, as noted above, π' is a cubic residue of π. The hypotheses of the Theorem thus imply that $\pi | u$. Similarly $\pi' | u$. Hence $r = 3u = d$, $r = s = 1$ and Lemma 4.6 applies.

Theorems 4.2, 4.3 are just specimens of a wide variety of types of d for which the determination of the Selmer group suffices to determine the group of k-rational points on C. This is not, however, always the case, as we proceed to show.

LEMMA 4.7. Let $d = 60$. The Selmer group is generated by $2(k^*)^3$, $3(k^*)^3$, $5(k^*)^3$.

Proof. Arguing as in the proof of Theorem 4.3, we are reduced to the consideration of

$$X_1^3 + X_2^3 + 60X_3^3 = 0 \tag{4.35}$$

$$X_1^3 + 3X_2^3 + 20X_3^3 = 0 \tag{4.36}$$

$$X_1^3 + 4X_2^3 + 15X_3^3 = 0 \tag{4.37}$$

$$X_1^3 + 5X_2^3 + 12X_3^3 = 0 \tag{4.38}$$

$$3X_1^3 + 4X_2^3 + 5X_3^3 = 0 \tag{4.39}$$

It is left to the reader to verify that these have points everywhere locally. The Lemma now follows by Lemma 4.5, Corollary 2.

THEOREM 4.4 Let $d = 60$. Any point $\underline{a}$ on C defined over k has $a_3 = 0$.

Proof. By Lemma 9.1 of Chapter 10, there are no points on (4.39) defined over $\mathbb{Q}$. Hence by Lemma 3.1 there is no point defined over k. It is left to the reader to prove similarly that there are no global points on (4.36), (4.37), (4.38). The Theorem now follows from Lemma 4.6.

Similarly we have

THEOREM 4.5 *Let* $d = 66$. *Then*

(i) *The Selmer group is generated by* $2(k^*)^3$, $3(k^*)^3$, $11(k^*)^3$.

(ii) *Any point* $\underline{a}$ *on* C *defined over* k *has* $a_3 = 0$.

Proof. Left to the reader. Compare Lemma 9.2 of Chapter 10.

We conclude by sketching a proof of Mordell's Finite Basis Theorem C for the curve C we have been considering and for the ground field $k = \mathbb{Q}(\rho)$. We define the *height* $H(\underline{a})$ of a point $\underline{a}$ defined over k by

$$H(\underline{a}) = \max_j\{N_{k/\mathbb{Q}}(a_j)\} = \max_j |a_j|^2, \tag{4.40}$$

where $|\ |$ is the absolute value and where the homogeneous co-ordinates a_j are normalized by

$$a_j \in \mathfrak{o} \quad ; \quad \text{g.c.d.}(a_j) = 1. \tag{4.41}$$

LEMMA 4.8. *For any real* H_o *there are only finitely many* $\underline{a}$ *with* $H(\underline{a}) \leq H_o$.

Proof. Clear.

LEMMA 4.9. *There is a real constant* $c = c(d) > 0$, *depending only on* d, *with the following property: let* $\underline{\alpha}, \underline{a}$ *be points on* C *defined over* k *with* $\underline{a} = \phi(\underline{\alpha})$. *Then*

$$H(\underline{a}) \geq c\{H(\underline{\alpha})\}^3. \tag{4.42}$$

Proof. The relation between $\underline{a}$ and $\underline{\alpha}$ is given by (4.32). It was shown in the proof of Lemma 4.6 that $|h|^2 \leq 3$. Hence we can take $c = C^2/3$, where

$$C = \min[\max\{|-\rho^2 z_1^3 + \rho z_2^3|, \ |\rho z_1^3 - \rho^2 z_2^3|, \ 3^{\frac{1}{2}}|z_1 z_2 z_3|\}] \quad (4.43)$$

and where the minimum is taken over all $z_1, z_2, z_3 \in \mathbb{C}$ such that

$$\max\{|z_1|, |z_2|, |z_3|\} = 1, \quad z_1^3 + z_2^3 + dz_3^3 = 0. \quad (4.44)$$

Since this set is compact, the minimum is attained; and so $C > 0$ since the function in [ ] in (4.43) clearly never vanishes on (4.44).

<u>LEMMA 4.10</u>. <u>Let</u> a,b <u>be points on</u> C <u>defined over</u> k. <u>Then</u>

$$H(\underline{a} + \underline{b}) \lesssim 4\{H(\underline{a})\}^2\{H(\underline{b})\}^2. \quad (4.45)$$

<u>Proof</u>. Follows at once from (3.4).

<u>THEOREM 4.6</u> <u>The group of points on</u> C <u>defined over</u> k <u>is finitely generated</u>.

<u>Proof</u>. By Lemma 4.6, Corollary 2 there is only a finite number of cosets $m_i(k^*)^3$ of $(k^*)/(k^*)^3$ for which there are points $\underline{a}_i$ with

$$\tilde{m}(\underline{a}_i) = m_i(k^*)^3.$$

For each i we choose $\underline{a}_i$ arbitrarily once for all. For any point $\underline{b}$ we have $\tilde{m}(\underline{b}) = \tilde{m}(a_i)$ for some i, and so $\tilde{m}(\underline{b} - \underline{a}_i) = (k^*)^3$. As in the proof of Lemma 4.6, there is some point $\underline{\beta} \in \mathfrak{G}$ with $\phi(\underline{\beta}) = \underline{b} - \underline{a}_i$. Now

$$c\{H(\underline{\beta})\}^3 \lesssim H(\underline{b} - \underline{a}_i)$$

$$\lesssim 4\{H(\underline{b})\}^2 \{H(\underline{a}_i)\}^2$$

by Lemmas (4.9) (4.10).

Hence

$$H(\underline{\beta}) \lesssim c_0\{H(\underline{b})\}^{2/3},$$

where

$$c_o = \max_i [4c^{-1}\{H(\underline{a}_i)\}^2]^{1/3}.$$

In particular,

$$H(\underline{\beta}) < H(\underline{b})$$

whenever

$$H(\underline{b}) > c_o^3.$$

It follows that the group of points on C is generated by the $\underline{a}_i$ and the points $\underline{b}$ with $H(\underline{b}) \leq c_o^3$. But the latter are finite in number by Lemma 4.8.

Note The Lemma does not give an algorithm for finding a set of generators since, as already remarked, there is no algorithm for deciding whether a coset $m(k^*)^3$ is represented by a point $\underline{a}$ with $\tilde{m}(\underline{a}) = m(k^*)^3$ or not.

Notes

§1 For theorem A see, for example, O'Meara (1963). There is a proof for $k = \mathbb{Q}$ in Cassels (1978).

Theorem B is due to Hasse (1931), where he also showed that the conclusion already ceases to hold when K/k is normal with non-cyclic galois group of order 4. There is a substantial literature on normal extensions K/k and elements of k which are norms everywhere locally but not globally ("knots"), see Jehne (1979), Cassels-Fröhlich (1967), Exercise 5 and Exercise 5 to this Chapter. The "correct" generalization of Theorem B to normal extensions is to the cohomology group $H^2(\Gamma, K^*)$, where Γ is the galois group. For cyclic Γ, the group H^2 is isomorphic to H^0, which gives Theorem B. (See, for example, Chevalley (1953-4) §20 or §9.6 of Tate's article in Cassels-Fröhlich (1967)).

Theorems of Siegel (to some extent subsuming earlier results) about the number of representations by quadratic forms may be regarded as a quantitative formulation of the Hasse principle. More recently, this has been shown to be a special case of the theory of "Tamagawa numbers". There is some discussion of this, together with further references, in

Cassels (1978). The still largely unproved Birch-Swinnerton-Dyer conjectures about elliptic curves and abelian varieties of higher dimension can be regarded as a further development.

The Hardy-Littlewood circle method is also closely related to the Hasse principle. For example, in Waring's problem it produces the following results. Let $k > 1$, s be positive integers and for positive integer n let

$$r(n) = r(n,k,s)$$

be the number of representations of n as the sum of s k-th powers of non-negative integers:

$$n = \sum_{j=1}^{s} x_j^{k} : x_j \in \mathbb{Z}, \quad x_j \geq 0.$$

If s is larger than a function of k, the circle method shows that

$$r(n) = n^e \, C \, \mathfrak{S}(n) + \rho(n),$$

where

$$e + 1 = s/k,$$

$$C = \{\Gamma(1 + k^{-1})\}^s / \Gamma(s/k),$$

$$\rho(n)/n^e \to 0 \quad (n \to \infty)$$

and $\mathfrak{S}(n)$ is the "singular series". Further,

$$\mathfrak{S}(n) = \prod_p \mathfrak{S}_p(n) \quad (p = \text{prime}),$$

where $\mathfrak{S}_p(n)$ represents in an appropriate sense the p-adic density of representations of n. One can regard C as the contribution of the infinite prime. The situation is so simple only when s is very large. Adepts of the circle method yoga can show e.g. that $r(n) > 0$ for all sufficiently large n when s is considerably smaller, but then the relationship with the Hasse principle fades.

Birch (1962) has used a variant of the circle method to show that a form of given degree satisfies the Hasse principle provided that the number of variables is large enough and that the singular locus is not too large. It will be observed that (2.12) and (2.13) have large singular loci.

§§3,4 There is a survey of the arithmetic of curves of genus 1, together with background information on diophantine equations generally, in Cassels (1966). A later survey is Tate (1974). Yet more recently, there have arisen interesting connections with p-adic L-functions more general than those which will be discussed in Chapter 13: see the survey by Coates in Rankin (1984). The Theory is still in a state of stormy development.

The particular family of curves (4.1) was widely studied in the 19th century and there is an account with extensive tables in Selmer (1951). The author (Cassels (1959)) subsequently proved a conjecture of Selmer which implies, in particular, that if one of the four equations (4.36) - (4.39) has a rational solution, then they all have. This was generalized to all curves of genus 1 (Cassels (1962)) and subsequently, by Tate, to abelian varieties of any dimension.

Exercises.

1. Let $\mathcal{E}$ be a curve of genus 0 given by an equation $F(X,Y,Z) = 0$, where F is a form of odd degree with coefficients in a field k. Show that $\mathcal{E}$ is birationally equivalent over k to a straight line.

[Hint show first that there is a divisor on $\mathcal{E}$ of degree 1 defined over k.]

2. (i) Let $a_1, a_2, a_3,\ b \in \mathbb{Z}$ and put

$$F(X_1,X_2,X_3) = a_1X_1^3 + a_2X_2^3 + a_3X_3^3 + bX_1X_2X_3.$$

Suppose that $a_1a_2a_3$ is square free and that $F(u_1,u_2,u_3) = 0$, where $u_1,u_2,u_3 \in \mathbb{Z}$ have no common factor. Show that $a_1u_1^3$, $a_2u_2^3$, $a_3u_3^3$ are coprime in pairs.

(ii) Suppose, further, that at most one of a_1, a_2, a_3 is ± 1. Let $\underline{x} = (x_1, x_2, x_3)$ be the third point of intersection of the tangent at $\underline{u}$ to $F = 0$, where the homogeneous coordinates of $\underline{x}$ are chosen so that $x_j \in \mathbb{Z}$ and x_1, x_2, x_3 have no common factor. Show that

$$|x_1 x_2 x_3| > |u_1 u_2 u_3|.$$

(iii) Deduce that if there is one rational point on $F = 0$, then there are infinitely many.

3. Show that the only rational point on $X^3 + Y^3 + dZ^3 = 0$ is that with $Z = 0$ in each of the following cases:

(i) $d = q_1 q_2$ where $q_1 > 0$, $q_2 > 0$ are primes and $q_1 \equiv 2 \ (9)$, $q_2 \equiv 5 \ (9)$

(ii) $d = pq \not\equiv \pm 1 \ (9)$, where $p > 0$, $q > 0$ are primes, $p \equiv 1 \ (3)$, $q \equiv 2 \ (3)$ and q is a cubic non-residue mod p.

4. Let $F_j(X,Y,Z)$ $(1 \leq j \leq J)$ be quadratic forms defined over the global field k. Suppose that for every localization $k_{\mathfrak{p}}$ (including the archimedean ones) they have a common nontrivial zero defined over $k_{\mathfrak{p}}$. Show that they have a common zero defined over k.

[Schinzel (1983). Hint The intersection of the $F_j(X,Y,Z) = 0$ is either (i) a conic (ii) a line or (iii) a set of $n \leq 4$ points, conjugate over k. Compare Chapter 10, Exercise 13.]

5. The following exercise constructs a "knot" (see Notes to Chapter). Let $K = \mathbb{Q}(\sqrt{d}, \sqrt{e})$, where d, e, de are not squares, so the Galois group Γ is $1, \sigma, \tau, \sigma\tau$ with $\sigma^2 = \tau^2 = 1$, $\sigma\tau = \tau\sigma$. As is usual, we write the elements of Γ as indexes, so $\sigma : A \to A^\sigma$ and we extend this to write $A^{1+\sigma} = A.A^\sigma$ etc.

(i) Let $\mathbb{A}$ be a K-divisor and suppose that

$$\text{Norm } \mathbb{A} = \mathbb{A}^{1+\sigma+\tau+\sigma\tau} = [1].$$

Show that

$$\mathfrak{A} = \mathfrak{B}^{1-\sigma}\mathfrak{C}^{1-\tau}$$

for some divisors $\mathfrak{B},\mathfrak{C}$.

[Hint this is a semilocal question].

(ii) Suppose, further, that the class-number of K is 2. Show that $\mathfrak{A}$ is principal.

(iii) Let $K = \mathbb{Q}(\sqrt{-3},\sqrt{13})$ and suppose, if possible, that there is an $A \in K$ with $N_{K/\mathbb{Q}}A = 3$. Show that the principal divisor

$$[A] = \mathfrak{p}_3\,\mathfrak{A},$$

where $\mathfrak{p}_3$ is a prime divisor of 3 and Norm $\mathfrak{A} = [1]$. Show that the conditions of (ii) apply, and deduce a contradiction.

[Hint cf. Chapter 10, Exercise 14].

(iv) Show that 3 is a norm for $K/\mathbb{Q}$ everywhere locally.

[Note This is the original example of Hasse (1931), but a different argument].

6. This exercise provides a cubic surface for which the Hasse principle fails.

(i) If M is a primitive 7-th root of unity, show that $C = 2 - M - M^{-1}$ satisfies

$$C^3 - 7C^2 + 14C - 7 = 0.$$

Deduce (a) that the ring of integers of $\mathbb{Q}(C)$ is $\mathbb{Z}[C]$ (b) that 7 ramifies completely, but no other prime ramifies, and (c) that the norm of an integral divisor prime to 7 is $\equiv \pm 1 \pmod 7$.

Let S be the cubic surface

$$t(t+x)(2t+x) = \mathrm{Norm}(x + yC + zC^2) \qquad (*)$$

in variables (x,y,z,t).

(ii) Show that there are points on S everywhere locally.

(iii) If there is a rational point on S, show that without loss of generality x,y,z,t are integers without common divisor.

(iv) Then the absolute values of t, $t + x$, $2t + x$ are all norms of integral divisors of $\mathbb{Q}(C)$.

(v) Hence if the common value of (*) is prime to 7, we have t, $t + x$, $2t + x$ all $\equiv \pm 1 \ (7)$; a contradiction.

(vi) If the common value of (*) is divisible by 7 we have $x \equiv 0 \Rightarrow t \equiv 0 \Rightarrow y \equiv z \equiv 0$; a contradiction.

[Hints Chapter 9, Exercise 10 ter; Chapter 10, Exercise 9 bis; Appendix D, Theorem D.1.

Note Due to Swinnerton-Dyer (1962).].

Note For Exercise 3 cf. Selmer (1951).

CHAPTER TWELVE: ADVANCED ANALYSIS

1 INTRODUCTION

There has developed in recent years a p-adic analysis comparable in scope to the classical analysis of real and complex functions, but with many salient differences. A particular object of study has been the so-called p-adic L-functions, which appear to hold the key to several deep unsolved problems. In this chapter we shall develop the theory sufficiently far to explain the slightly sophisticated relationship between p-adic L-functions and the classical L-functions of analytic number theory and to prove one or two nontrivial theorems. Incidentally, this will put the proof of the von Staudt-Clausen relations of Chapter 1, §3 into a natural setting. The final section, which is somewhat divorced from the rest, explains a theorem of Mahler about a development of continuous functions on $\mathbb{Z}_p$ as infinite series. Necessarily, this chapter is rather a <u>pot pourri</u>.

The natural domain to work is the field $\Omega_p \supset \mathbb{Q}_p$ introduced in Chapter 8, §3. All we shall need here, however, is that Ω_p is both complete and algebraically closed.

2 ELEMENTARY FUNCTIONS

If a function of a complex variable is defined by a power series with coefficients in $\mathbb{Q}$ then it is natural to use the same power series to define a function of a variable in Ω_p (or, rather, in the domain of convergence). One might expect the new function to have properties analogous to the old, and in some cases this is an immediate consequence of the "permanence of algebraic form" (an old Victorian mathematical tool which fell into disrepute because it was mishandled, but which remains a most potent weapon in the mathematician's arsenal).

Consider the formal power series

$$F(X) = \sum (-)^{j+1} X^j/j. \tag{2.1}$$

If Y is a second indeterminate, we show now that

$$F(X + Y + XY) = F(X) + F(Y), \tag{2.2}$$

as an equality of formal power series in X and Y. If x is a complex number, $|x| < 1$, the series $F(x)$ converges to the function $\log(1 + x)$. If also $y \in \mathbb{C}$, $|y| < 1$, $|x + y + xy| < 1$, then, from the theory of functions of a complex variable, (2.2) holds when x,y are substituted for X,Y respectively. But the coefficients of a complex power series which converges in a neighbourhood of the origin are determined uniquely by the values taken there. Hence (2.2) holds as an identity in formal power series.

We now go over to Ω_p. For $x \in \Omega_p$ the series on the right hand side of (2.1) converges precisely when $|x| < 1$, as is readily verified (we write $|\ |$ for $|\ |_p$). If, further, $y \in \Omega_p$, $|y| < 1$ then $|x + y + xy| < 1$ by the ultrametric inequality, and (2.2) holds with x,y for X,Y: the necessary rearrangements of the formal series being justified by Lemma 4.1 of Chapter 4. We have thus defined the p-adic _logarithm_ function

$$\log_p z \tag{2.3}$$

in

$$|z - 1| < 1 \tag{2.4}$$

by

$$\log_p(1 + x) = F(x); \tag{2.5}$$

and we have verified that

$$\log_p(wz) = \log_p w + \log_p z \tag{2.6}$$

for $|z - 1| < 1$, $|w - 1| < 1$.

Let $\zeta^{p^N} = 1$ for some positive integer N. Then ζ lies in

(2.4), and so $\log_p \zeta$ is defined. By (2.6) we have

$$p^N \log_p \zeta = \log_p 1 = 0,$$

and so

$$\log_p \zeta = 0 \quad (\zeta^{p^N} = 1). \tag{2.7}$$

LEMMA 2.1. *There is an* Ω_p*-valued function* $\log_p x$ *defined for all nonzero* $x \in \Omega_p$ *with the following properties:*

(i) *if* $|x - 1| < 1$, *then* $\log_p x$ *is the function just defined*

(ii)

$$\log_p xy = \log_p x + \log_p y$$

for all nonzero $x,y \in \Omega_p$

(iii)

$$\log_p p = 0.$$

Further, this definition is unique.

Note Some additional condition is necessary, and (iii) is that conventionally adopted.

Proof. Every $x \neq 0$ can be written

$$x = uvw,$$

where

(i) $u^m = p^n$

for some $m > 0$ and n in $\mathbb{Z}$.

(ii) v is a root of unity

(iii)

$|w - 1| < 1.$

We <u>define</u>

$$\log_p x = \log_p w,$$

where $\log_p w$ has already been defined.

It is left to the reader to complete the proof.

In a similar way, one defines the p-adic <u>exponential function</u> $\exp_p(z)$ by the power series

$$\exp_p(z) = \sum_{0}^{\infty} z^n/n! \ , \tag{2.8}$$

whose domain of convergence is

$$|z| < p^{-1/(p-1)} \tag{2.9}$$

by Lemma 5.1 of Chapter 4. Further,

$$\exp_p(z + w) = \exp_p z \exp_p w \tag{2.10}$$

if both z,w are in the domain of convergence, again by the permanence of algebraic form. Finally,

$$\log_p(\exp_p z) = z \tag{2.11}$$

if $|z|$ is small enough for both the relevant power series to converge; and similarly for

$$\exp_p(\log_p z) = z. \tag{2.12}$$

We could thus have used the $\exp_p$ and $\log_p$ functions to obtain the power series in Lemma 5.2 of Chapter 4.

There can be no analogue of the complex function identity

$\exp(2\pi i) = 1$. For if $\exp_p(z_o) = 1$, $z_o \neq 0$, then by (2.10) we should have

$$\exp_p(mz_o) = 1 \quad (\text{all } m \in \mathbb{Z}),$$

in contradiction to Strassmann's Theorem 4.1 of Chapter 4.

It is at first sight somewhat surprising that $\exp_p$ has a smaller domain of convergence than that of $\log_p$. We now introduce a modification (the Artin-Hasse exponential function) which converges in a larger domain. We shall not need it later, but the ideas behind the definition have some interest. (See also Exercises).

From the power series definition we have

$$-x = \sum(\mu(n)/n)\log(1 - x^n), \tag{2.13}$$

where μ is the Moebius function, so

$$\sum_{n|N} \mu(n) = 1 \text{ if } N = 1$$

$$= 0 \text{ otherwise.}$$

[See e.g. Hardy and Wright (1938), Chapter 16, §3]. Hence

$$\exp_p(x) = \prod_{n=1}^{\infty} (1 - x^n)^{-\mu(n)/n} \tag{2.14}$$

[The argument is, of course, also valid in $\mathbb{C}$].

The Artin-Hasse exponential $E_p(x)$ is defined by

$$E_p(x) = \prod_{\substack{n=1 \\ p\nmid n}}^{\infty} (1 - x^n)^{-\mu(n)/n} \tag{2.15}$$

(i.e. taking only $p \nmid n$ in (2.14)). On expanding the

$$(1 - x^n)^{\mu(n)/n}$$

with $p \nmid n$ by the binomial theorem, we get a power series in x with coefficients in $\mathbb{Z}_p \cap \mathbb{Q}$ (Chapter 4, Exercise 8), and so (2.15) gives an expansion

$$E_p(x) = \sum_{j=o} e_j x^j, \tag{2.16}$$

where

$$e_j \in \mathbb{Z}_p \cap \mathbb{Q}.$$

Hence (or directly from (2.15)) $E_p(x)$ is defined for $|x| < 1$.

Finally, on comparing (2.14) and (2.15) we have

$$E_p(x) = \exp_p(x + x^p/_p + x^{p^2}/_{p^2} + \ldots + x^{p^j}/_{p^j} + \ldots) \tag{2.17}$$

valid for $|x| < p^{-1/(p-1)}$

3 ANALYTIC CONTINUATION

As has already been remarked, the technique of analytic continuation in terms of power series which is used in complex function theory does not transfer in a naive way to p-adic theory. There are, however, other techniques for extending the domain of definition of a function in a unique way. The older technique, due to Krasner, is fairly elaborate and operates on Swiss cheeses. The more modern technique ("rigid analytic spaces"), which is due to Tate, requires sophisticated commutative algebra. We content ourselves with a simple case, which brings out some of the ideas involved, and which we shall need to invoke later.

Let W be the ring of power series convergent on the integers $\mathbb{O}$ of Ω_p and taking values in Ω_p: that is, W is the ring of

$$f(X) = \sum_{n=o}^{\infty} f_n X^n \tag{3.1}$$

with

$$f_n \in \Omega_p, \quad f_n \to 0 \quad (n \to \infty). \tag{3.2}$$

Then W becomes an ultrametric space under the norm

$$||f|| = \max_n |f_n|, \tag{3.3}$$

where

$$||f + g|| \leq \max\{ ||f|| , ||g||\} , \quad (3.4)$$

$$||fg|| = ||f|| \, ||g|| . \quad (3.5)$$

LEMMA 3.1. W *is complete with respect to* $|| \; ||$.

Proof. This was already proved in Chapter 6, §5.

We now relate the norms of $f \in W$ to the values taken by them.

LEMMA 3.2.

$$||f|| = \sup_{x \in \mathfrak{O}} |f(x)|. \quad (3.6)$$

for $f \in W$.

Proof. Clearly $|f(x)| \leq ||f||$, so it is enough to show that equality can occur. By homogeneity we can suppose that

$$||f|| = 1.$$

Let a bar denote the map onto the residue class field. Then

$$\bar{f}(\bar{x}) = \sum \bar{f}_n \bar{x}^n \quad (3.7)$$

is a non-zero polynomial in $\bar{x}$. Since the residue class field is infinite, we can choose $\bar{x}$ so that $\bar{f}(\bar{x}) \neq 0$. Then $|f(x)| = 1$ for any lift of $\bar{x}$ to $\mathfrak{O}$.

COROLLARY. *Let* $f^{(n)}$ $(n = 0,1,2, \ldots)$ *be a sequence of elements of* W. *A necessary and sufficient condition that it be convergent (with respect to the* $|| \; ||$ *topology) is that the sequence of functions* $f^{(n)}(x)$ *converge uniformly in* $\mathfrak{O}$.

Breweis. klar.

LEMMA 3.3. *Let* $f \in W$ *and suppose that* $f(x)$ *vanishes in some nonempty open set* $S \subset \mathcal{O}$. *Then* $f = 0$.

Proof. Suppose, first, that $0 \in S$, so that $f(x) = 0$ for all $|x| < \delta$ for some $\delta > 0$. Let $\zeta \in \mathcal{O}$, $\zeta \neq 0$ and $|\zeta| < \delta$, and put

$$F(x) = f(\zeta x).$$

Then $F \in W$ and

$$\sup_{x \in \mathcal{O}} |F(x)| = 0.$$

Hence $\|F\| = 0$, that is $f_n = 0$ for all n, as required.

If $0 \notin S$, let $x_o \in S$. The transformation $x \to x - x_o$ takes W into itself with preservation of norm $\| \ \|$, and we are reduced to the previous case.

COROLLARY. *Let* $f, g \in W$ *and suppose that their values coincide on some nonempty open* $S \subset \mathcal{O}$. *Then* $f = g$.

Beweis. klar.

LEMMA 3.4. *A necessary and sufficient condition that a function* $f(x)$ *defined in* $\mathcal{O}$ *belong to* W *is that it be the uniform limit of a sequence* $\{h^{(n)}(x)\}$ *of rational functions all having their singularities outside* $\mathcal{O}$.

Proof. Suppose, first, that $f(x) \in W$, say given by (3.1), (3.2). Then $f(x)$ is the uniform limit of the polynomials

$$f^{(n)}(x) = \sum_{j \leq n} f_j x^j,$$

whose only singularity is the point at infinity.

Now suppose that $f(x)$ is the uniform limit of rational functions $h^{(n)}(x)$ with no singularities in $\mathcal{O}$. The $h^{(n)}(x)$ are in W, as is seen at once by taking a decomposition into partial fractions

$$h^{(n)}(x) = \sum_{r,s} c_{r,s}(1 - \theta_r x)^s$$

with $\theta_r \in \Omega_p$, $|\theta_r| < 1$. Now $f(x) \in W$ by Lemma 3.1, as required.

The $f \in W$ are the <u>Krasner analytic functions</u> for the domain $\mathcal{O}$. We shall need a slightly different case. Let $\mathcal{D}$ be the domain

$$\mathcal{D} = \{x \in \Omega_p : |x - 1| \geq 1\}. \tag{3.8}$$

We say that a function f defined on $\mathcal{D}$ is <u>Krasner analytic</u> there if it is the uniform limit on $\mathcal{D}$ of rational functions which have all their singularities in the complement $|x - 1| < 1$ of $\mathcal{D}$.

<u>LEMMA 3.5</u>. <u>Suppose that two Krasner analytic functions on</u> $\mathcal{D}$ <u>are equal on a non-empty open subset of</u> $\mathcal{D}$. <u>Then they are equal</u>.

<u>Proof</u>. We map $\mathcal{D}$ into $\mathcal{O}$ by putting

$$t = 1/(x - 1). \tag{3.9}$$

The image consists of the whole of $\mathcal{O}$ except the origin. A uniformly convergent sequence of rational functions of x gives a sequence of rational functions defined for $t \in \mathcal{O}$, which is uniformly convergent on the whole of $\mathcal{O}$ (including the origin) by an easy standard argument. The transformation (3.9) therefore gives a correspondence between the Krasner analytic functions on $\mathcal{D}$ and those on $\mathcal{O}$. Lemma 3.5 now follows at once from Lemma 3.3, Corollary.

<u>Note</u> The <u>Swiss cheese</u> (3.8) has only one hole. In the general case considered by Krasner there may be several holes.

4 MEASURE ON $\mathbb{Z}_p$

We say that a set in a metric space is <u>clopen</u> if it is both open and closed. It is easy to see (Chapter 4, Exercise 20) that the clopen subsets of $\mathbb{Z}_p$ are precisely the union of <u>finite</u> families of disjoint sets

$$T(a,N) = \{b \in \mathbb{Z}_p : b \equiv a \quad (p^N)\}, \tag{4.1}$$

where $a \in \mathbb{Z}$ and, without loss of generality,

$$0 \leq a < p^N. \tag{4.2}$$

A <u>measure</u> μ on $\mathbb{Z}_p$ is a bounded Ω_p-valued function on the clopen sets which satisfies

$$\mu(S \cup T) = \mu(S) + \mu(T) \tag{4.3}$$

for disjoint S,T. It is clearly enough to define μ on the sets (4.1). Since

$$T(a,N) = \bigcup_{j=o}^{p-1} T(a + jp^N, N + 1), \tag{4.4}$$

there is the consistency condition

$$\mu(T(a,N)) = \sum_j \mu(T(a + jp^N, N + 1). \tag{4.5}$$

It would be nice if we could choose μ to be translation-invariant, i.e., so that $\mu(T(a,N)) = \nu(N)$ (say) is independent of a. Then (4.5) gives $\nu(N + 1) = p^{-1}\ \nu(N)$; $\nu(N) = p^{-N}\nu(0)$ and so μ would be unbounded, contrary to definition, except in the unexciting case when μ is identically zero.

We can integrate continuous functions $f(x)$, $x \in \mathbb{Z}_p$ with respect to a measure μ. If f is locally constant (which in the ultrametric world is consistent with being non-constant), it takes a finite set of values f_j on clopen sets S_j and we define

$$\int_{\mathbb{Z}_p} f\mu = \sum f_j\ \mu(S_j). \tag{4.6}$$

It is easy to see that any continuous Ω_p-valued functions $f(x)$ on $\mathbb{Z}_p$ is the uniform limit of continuous locally constant functions $g^{(n)}(x)$, say (cf. Chapter 4, Exercise 21). We define

$$\int_{\mathbb{Z}_p} f\mu = \lim_{n\to\infty} \int_{\mathbb{Z}_p} g^{(n)}\mu. \tag{4.7}$$

A particular case of (4.7) is

$$\int f\mu = \lim_{N\to\infty} \sum_{a=o}^{p^N-1} f(a)\ \mu(T(a,N)). \tag{4.8}$$

It will be easy to check that this definition of integral has those properties, suggested by the notation, which will be invoked below.

We have already shown that there is no translation-invariant measure. As a next best thing, we take $z \neq 0$ in Ω_p and define

$$\mu_z(T(a,N)) = z^a/(1 - z^{p^N}). \tag{4.9}$$

It is readily seen that μ_z satisfies the compatibility condition (4.5) [cf. Exercise 7]. We make the further condition that

$$|z^{p^N} - 1| \geq 1 \quad \text{(all N)}. \tag{4.10}$$

This ensures that μ_z is bounded by 1. In fact

$$|\mu_z(T(a,N))| \leq 1. \tag{4.11}$$

follows at once from (4.9), (4.10) if $|z| = 1$, while for $|z| \neq 1$ it follows from (4.2), (4.9).

<u>LEMMA 4.1</u>. With z, μ_z as above, we have

$$\left|\int_{\mathbb{Z}_p} f\mu_z\right| \leq \max_x |f(x)| . \tag{4.12}$$

<u>Proof</u>. Follows from (4.11) and the definition of integral.

With an application to the next section in mind, we compute a specific integral.

<u>LEMMA 4.2</u>. <u>Let</u> $t \in \Omega_p$, $|t| < p^{-1/(p-1)}$. <u>Then</u>

$$\int_{\mathbb{Z}_p} \exp_p(tx)\mu_z(x) = \frac{1}{1 - z\exp_p(t)} . \tag{4.13}$$

<u>Note</u> The notation on the left hand side indicates that x is the variable of integration.

<u>Proof</u>. We note first that the series (2.8) defining $\exp_p(tx)$ converges for all $x \in \mathbb{Z}_p$.

By (4.8) we have

$$\int_{\mathbb{Z}_p} \exp_p(tx)\mu_z(x) = \lim_{N\to\infty}(1 - z^{p^N})^{-1} \sum_{a=o}^{p^N-1} \exp_p(ta)z^a$$
$$= \frac{1}{1 - z\exp_p(t)} \lim_N \frac{1 - \exp_p(p^N t)z^{p^N}}{1 - z^{p^N}} . \tag{4.14}$$

But $p^N \to 0$, $\exp_p(p^N t) \to 1$.

COROLLARY. *For rational integers* $k \geq 0$ *the value of* $\int_{\mathbb{Z}_p} x^k \mu_z(x)$ *is* $k!$ *times the coefficient of* t^k *in the formal expansion of* $(1 - z\exp_p t)^{-1}$ *in powers of* t.

Proof. Both sides have expansions in powers of t valid in some neighbourhood of the origin, so the coefficients may be equated.

By the integral of a continuous function f on a clopen subset S of $\mathbb{Z}_p$ we shall mean the integral over $\mathbb{Z}_p$ of a function equal to f on S and to zero elsewhere. We shall need the formula

$$\int_{p\mathbb{Z}_p} f(x)\,\mu_z(x) = \int_{\mathbb{Z}_p} f(px)\,\mu_Z(x), \tag{4.15}$$

where

$$Z = z^p. \tag{4.16}$$

This follows at once from the definitions (4.8), (4.9).

5 THE ZETA FUNCTION

We first recall some properties of the classical Riemann function $\zeta(s)$ of a complex variable s. It is defined when the real part of s is > 1 by

$$\zeta(s) = \sum_1^\infty n^{-s} = \prod_{q \text{ prime}} (1 - q^{-s})^{-1}, \tag{5.1}$$

and can be continued analytically over the entire complex s plane, the only singularity being a pole at $s = 1$. It is conjectured that the values of $\zeta(s)$ at positive integral s are transcendental, but the values at negative integral s are rational, and it is these which are used to construct the p-adic analogue.

We shall require the properties of the "twisted" zeta function

$$\zeta(s,z) = \sum_{n=1}^{\infty} z^n n^{-s}, \tag{5.2}$$

where $z \in \mathbb{C}$ and

$$|z|_\infty \leqslant 1. \tag{5.3}$$

The series again converges for $\mathrm{Re}(s) > 1$ and defines a function analytic everywhere except possibly for a pole at $s = 1$.

LEMMA 5.1. *The value of* $\zeta(s,z)$ *at rational integers* $-k \leqslant 0$ *is given by equating the coefficients of powers of* t *in the identity*

$$\sum_{k=0}^{\infty} \zeta(-k,z)t^k/k! = z\exp(t)/\{1 - z\exp(t)\}. \tag{5.4}$$

The proof is given in Appendix E. On putting $z = 1$, we have the

COROLLARY.

$$\zeta(-k) = -B_{k+1}/(k+1) \tag{5.5}$$

for $k > 0$, *where* B_{k+1} *is the Bernoulli number.*

Proof. Follows from the definition of the Bernoulli numbers ((3.1) of Chapter 1) since

$$e^t/(1 - e^t) = -1 - 1/(e^t - 1).$$

In considering the p-adic analogue it is better to take out the factor in (5.1) corresponding to $q = p$, and so to work with

$$\zeta^*(s) = (1 - p^{-s})\zeta(s). \tag{5.6}$$

[cf. the Artin-Hasse exponential of §1].

THEOREM 5.1 *For* $1 \leq \ell < p$ *there is a continuous function* $\zeta_p^{(\ell)}(w)$ *of* $w \in \mathbb{Z}_p$, *except that* $\zeta_p^{(1)}(1)$ *is not defined. They interpolate* $\zeta^*(-k)$ *for integers* k *in the sense that*

$$\zeta_p^{(\ell)}(-k) = \zeta^*(-k) \tag{5.7}$$

whenever

$$-k \equiv \ell \quad (p - 1). \tag{5.8}$$

For $\ell \neq 1$ *we have*

$$|\zeta_p^{(\ell)}(w)| \leq 1 \quad (\ell \neq 1) \tag{5.9}$$

and

$$|\zeta_p^{(\ell)}(w_1) - \zeta_p^{(\ell)}(w_2)| < |w_1 - w_2| \quad (\ell \neq 1) \tag{5.10}$$

for all $w_1, w_2 \in \mathbb{Z}_p$. *Further, the* $\zeta_p^{(\ell)}$ *are uniquely defined by (5.7), (5.8).*

COROLLARY. *Let* m,n *be positive rational integers and suppose that*

$$m \equiv n \not\equiv 0 \quad (p - 1); \quad m \equiv n \quad (p^N), \tag{5.11}$$

for some N. *Then*

$$(1 - p^{m-1})\, B_m/m \equiv (1 - p^{n-1})\, B_n/n, \quad (p^{N+1}), \tag{5.12}$$

where both sides are in $\mathbb{Z}_p \cap \mathbb{Q}$.

Proof. Follows from (5.5),(5.10) on taking $w_1 = 1 - m$, $w_2 = 1 - n$ and $\ell - 1 \equiv m \equiv n \quad (p - 1)$.

The congruences (5.12) are examples of Kummer's congruences. When they were discovered in the nineteenth century they were regarded as curiosities.

It will be apparent that Witt's argument, which we reproduced in Chapter 1, §3, has close affinities with the present argument. Indeed it could be said to work as if there were a translation-invariant measure on $\mathbb{Z}_p$.

Proof of Theorem 5.1. If continuous functions $\zeta_p^{(\ell)}$ exist on $\mathbb{Z}_p$ satisfying (5.7), then they are certainly unique, since the k satisfying (5.8) are dense in $\mathbb{Z}_p$. Hence all that is needed is a proof of existence.

The key to the proof is the close similarity between the formulae (4.13) and (5.4). We require a rational prime $r \neq p$ which may be chosen arbitrarily. Let z be a primitive r-th root of unity. The field $\mathbb{Q}(z)$ can be embedded both into $\mathbb{C}$ and into Ω_p; and we suppose this done one for all. On comparing the coefficients of t^k in (4.13) and (5.4), we have

$$\int_{\mathbb{Z}_p} x^k \mu_z = \zeta(-k,z) \quad (\text{all } k > 0), \tag{5.13}$$

both sides beong elements of $\mathbb{Q}(z)$.

We now sum over all primitive r-th roots z of unity. On the left hand side we simply introduce the new measure

$$\mu^{(r)} = \sum_{\substack{z^r=1 \\ z\neq 1}} \mu_z. \tag{5.14}$$

On the right hand side we note that

$$\sum_{\substack{z^r=1 \\ z\neq 1}} \zeta(s,z) = (r^{-s+1} - 1)\, \zeta(s)$$

for all $s \in \mathbb{C}$, since this is trivial when $\operatorname{Re}(s) > 1$ by (5.2) and extends to all s by analytic continuation. Hence

$$\int_{\mathbb{Z}_p} x^k \mu^{(r)} = (r^{k+1} - 1)\, \zeta(-k). \tag{5.15}$$

We want to interpret k as a p-adic variable and (5.15) is still not quite satisfactory. Although x^k behaves well as a function of k if $|x| = 1$ (at least if k is restricted to a fixed class mod $(p - 1)$), it does not do so if $|x| < 1$. We shall therefore replace $\mathbb{Z}_p$ by the units $U_p \subset \mathbb{Z}_p$:

$$\int_{U_p} x^k \mu_z = \int_{\mathbb{Z}_p} x^k \mu_z - \int_{p\mathbb{Z}_p} x^k \mu_z ,$$

and, by (4.15),

$$\int_{p\mathbb{Z}_p} x^k \mu_z = \int_{\mathbb{Z}_p} (px)^k \mu_Z \qquad (Z = z^p)$$

$$= p^k \int x^k \mu_Z .$$

Now Z runs through all the primitive r-th roots of unity when z does, and so finally

$$\int_{U_p} x^k \mu^{(r)} = (1 - p^k) \int_{\mathbb{Z}_p} x^k \mu^{(r)}$$

$$= (1 - p^k)(r^{1+k} - 1)\, \zeta(-k)$$

$$= (r^{1+k} - 1)\, \zeta^* (-k)$$

by (5.6): that is

$$\zeta^*(-k) = \frac{1}{(r^{1+k} - 1)} \int_{U_p} x^k \mu^{(r)} . \tag{5.16}$$

We are now in a position to define the functions $\zeta_p^{(\ell)}$ of the enunciation. If $k(1)$, $k(2)$ are positive integers and

$$k(1) \equiv k(2) \qquad ((p - 1)p^N), \tag{5.17}$$

then

$$x^{k(1)} \equiv x^{k(2)} \qquad (p^{N+1}) \tag{5.18}$$

for all $x \in U_p$ (and so, in particular, for $x = r$). For $\ell = 1,2,\ldots, p-1$ and $w \in \mathbb{Z}_p$, we define $\zeta_p^{(\ell)}(w)$ to be the limit of the right hand side of (5.16) for

$$-k \equiv \ell \quad (p-1), \qquad -k \to w. \tag{5.19}$$

The limit exists and is continuous by (5.18). The only case in which this definition breaks down is $\ell = 1$, $w = 1$ since then $r^{1+k} \to 1$.

As already explained, the $\zeta_p^{(\ell)}(w)$ are independent of the choice of the auxiliary prime r. We can choose r to be a primitive root mod p. For $\ell \neq 1$, it follows that $r^{1+k} - 1$ is a unit in the limiting process (5.19). This gives (5.9): and then (5.10) follows from (5.17), (5.18).

6 L-FUNCTIONS

In this section we construct a p-adic analogue of the Dirichlet L-series $L(s,\chi)$ along similar lines to those of the previous section for the zeta function. There is an explicit formula for $L(1,\chi)$ which occurs in the analytic formulae for the class numbers of cyclotomic fields. A somewhat similar explicit formula was found by Leopoldt in the p-adic case. We shall give a proof here, as it gives a simple but nontrivial application of the p-adic analytic continuation of §3.

We first recall the classical theory. A (Dirichlet) character modulo the positive integer d is a function $\chi(m)$ defined on the integers m prime to d and depending only on m modulo d and which gives a homomorphism from the multiplicative group of residue classes mod d prime to d into the roots of unity (i.e. $\chi(\ell m) = \chi(\ell)\chi(m)$). If d' is a proper divisor of d and χ' is a character modulo d', it induces a character modulo d in an obvious way. We say that the character χ is primitive if it is not induced: and then d is the conductor. It is conventional to extend the domain of χ to the whole of $\mathbb{Z}$ by putting $\chi(m) = 0$ if d and m have a common factor.

The Dirichlet L-function of the complex variable s is defined by

$$L(s,\chi) = \sum_{n=1}^{\infty} \chi(n)n^{-s} = \prod_{q \text{ prime}} (1 - \chi(q)q^{-s})^{-1} \tag{6.1}$$

when $\text{Re}(s) > 1$ and for all complex s by analytic continuation. It has no singularities except when $d = 1$, $\chi = 1$ identically and we have the Riemann zeta-function: a case which we exclude.

The twisted L-function is given for $\text{Re}(s) > 1$ by

$$L(s,\chi,z) = \sum_{n=1}^{\infty} z^n \chi(n)n^{-s}, \tag{6.2}$$

where

$$|z|_{\infty} \leq 1, \tag{6.3}$$

and, again, it continues analytically to all complex s.

<u>LEMMA 6.1.</u>

$$\sum_{k=o}^{\infty} L(-k,\chi,z)t^k/k! = \sum_{a=1}^{d} \frac{\chi(a)z^a \exp(at)}{z^d \exp(dt) - 1} \tag{6.4}$$

<u>in some neighbourhood of the origin</u> $t = 0$.

The proof is sketched in Appendix E.

Let θ be a primitive d-th root of unity. The <u>Gauss sum</u> is

$$G(\chi,\theta) = \sum_{a=1}^{d} \chi(a)\theta^a. \tag{6.5}$$

It is elementary and well-known that

$$\sum_a \chi(a)\theta^{ab} = \bar{\chi}(b)\ G(\chi,\theta) \tag{6.6}$$

for $b \in \mathbb{Z}$, where $\bar{\chi}$ is the complex conjugate. This is valid also when b,d have a common factor (and so $\bar{\chi}(b) = 0$).

<u>LEMMA 6.2.</u>

$$G(\bar{\chi},\theta)L(1,\chi) = -\sum_{a \bmod d} \bar{\chi}(a) \log(1 - \theta^a). \tag{6.7}$$

<u>Proof</u>. We give a proof, as we shall want to mimic some of its features below.

With the notation (5.2), we have for $\text{Re}(s) > 1$ that

$$\sum_{a \bmod d} \bar{\chi}(a)\zeta(s,\theta^a) = \sum_{n=1}^{\infty} (\sum_a \bar{\chi}(a)\theta^{na})n^{-s}$$

$$= G(\bar{\chi},\theta)L(s,\chi)$$

by (6.6). If $\theta^a = 1$ we have $\bar{\chi}(a) = 0$. But for $\theta^a \neq 1$ we have

$$\zeta(1,\theta^a) = \sum_n \theta^{an}/n$$

$$= -\log(1 - \theta^a);$$

and we are done.

Finally, before initiating the p-adic discussion, we introduce the notation

$$L^*(s,\chi) = \{1 - \chi(p)p^{-s}\} L(s,\chi). \tag{6.8}$$

We commence our discussion of the p-adic theory by recalling that the Teichmüller representative $\omega(a)$ is a map from the units U_p of $\mathbb{Q}_p$ to the $(p - 1)$ - th roots of unity. Define $\langle x\rangle$ by

$$x = \langle x\rangle\, \omega(x), \tag{6.9}$$

where

$$\langle x\rangle \equiv 1 \quad (p). \tag{6.10}$$

(Chapter 8, §2). The case $p = 2$ is exceptional in that $\omega(x) = \pm 1$ and $\langle x\rangle \equiv 1$ (4), but we shall not make the modifications below to include $p = 2$.

The Teichmüller representative $\omega(a)$ is a character of conductor p. If χ is a character of conductor d and $k \in \mathbb{Z}$, we denote by χ_k the primitive character for which

$$\chi_k(a) = \chi(a)\, \{\omega(a)\}^{-k} \tag{6.11}$$

when

$$\text{g.c.d.}(pd,a) = 1.$$

In comparing p-adic and complex theory we fix once for all embeddings of the algebraic closure of $\mathbb{Q}$ into $\mathbb{C}$ and Ω_p. Our objective is

THEOREM 6.1 *Let* χ *be a character of conductor* $d > 1$. *There is a continuous* Ω_p*-valued function* $L_p(w,\chi)$ *of* $w \in \mathbb{Z}_p$ *such that*

$$L_p(-k,\chi) = L^*(-k,\chi_{k+1}) \tag{6.12}$$

for all positive rational integers k.

Note 1 Since the $-k$ are dense in $\mathbb{Z}_p$, the function $L_p(w,\chi)$, if it exists, is unique.

Note 2 The Theorem continues to hold for the trivial character which is identically equal to 1, except that there is a pole at $w = 1$. This is the case treated in the previous section.

Proof. The argument is a mild generalization of that in the last section. Let $e \geq 1$ be a rational integer, $p \nmid e$. We work in the space

$$X = \lim_{\substack{\leftarrow \\ N}} \mathbb{Z}/ep^N\mathbb{Z} \tag{6.13}$$

(so $X = \mathbb{Z}_p$ if $e = 1$). A basis for clopen sets is

$$S(a,N) = \{b \in X : b \equiv a \ (ep^N)\}, \tag{6.14}$$

where

$$0 \leq a < ep^N. \tag{6.15}$$

For $z \in \Omega_p$ with

$$\left| z^{ep^N} - 1 \right| \geq 1 \quad (\text{all } N), \tag{6.16}$$

we define a measure ν_z by

$$\nu_z(S(a,N)) = z^a/(1 - z^{ep^N}). \tag{6.17}$$

It is readily verified that this satisfies the consistency condition corresponding to (4.5) and that

$$\int_X f\nu_z = \lim_{N\to\infty} \{(1 - z^{ep^N})^{-1} \sum_{a=o}^{ep^N-1} z^a f(a)\} \tag{6.18}$$

for any continuous function f on X.

The space X contains e copies of $\mathbb{Z}_p$ as clopen sets corresponding to the values of a mod e in (6.14). We shall also use the natural map

$$X \to \mathbb{Z}_p \tag{6.19}$$

induced by the identity map on $\mathbb{Z}$ in

$$\lim_{\overleftarrow{N}} \mathbb{Z}/ep^N\mathbb{Z} \to \lim_{\overleftarrow{N}} \mathbb{Z}/p^N\mathbb{Z} = \mathbb{Z}_p \text{ .} \tag{6.20}$$

(The "forgetful functor" which forgets about the modulus e). Any continuous function f on $\mathbb{Z}_p$ induces a function on X by "pull back", which we shall also denote by f. Later, we shall need

LEMMA 6.3. *Let $f(x)$ be a continuous function on X which is independent of x mod e, and so derived from a function f on $\mathbb{Z}_p$ by pull back. Then*

$$\int_X f\nu_z = \int_{\mathbb{Z}_p} f\mu_z. \tag{6.21}$$

Proof. Readily follows by comparison of (4.8), (4.9) with (6.18). Alternatively, it is enough to verify the Lemma for the characteristic function of a basic clopen set $T(a,N)$ of $\mathbb{Z}_p$ and its pull back

$$\bigcup_{j=o}^{e-1} S(a + jp^N, N)$$

in X.

We now return to the proof of Theorem 6.1. Let χ be a primitive character to modulus

$$d = ep^m \quad (p \nmid e) \quad (m \geq 0). \tag{6.22}$$

By (6.15) the domain of definition of χ naturally extends to the whole of X. As in Lemma 4.2, for all sufficiently small $t \in \Omega_p$ we have

$$\int_X \exp_p(tx)\chi(x)\nu_z(x) = \sum_{a=o}^{d-1} \frac{\chi(a)z^a \exp_p(at)}{1 - z^d \exp_p(dt)} . \tag{6.23}$$

Let r be any prime such that $r \neq p$, $r \nmid e$, and let z be a primitive r-th root of unity. On comparing (6.23) with Lemma 6.1 and equating coefficients of powers of t, we have

$$\int_X x^k \chi(x)\nu_z(x) = - L(- k,\chi,z) \quad (k > 0). \tag{6.24}$$

For

$$\nu^{(r)} = \sum_{\substack{z^r=1 \\ z\neq 1}} \nu_z \tag{6.25}$$

it follows, as for (5.15), that

$$\int_X x^k \chi(x)\nu^{(r)}(x) = (\chi(r)r^{1+k} - 1) L(- k,\chi). \tag{6.26}$$

The analogue in X of the set U_p of units of $\mathbb{Z}_p$ is the set X^* of x with $p \nmid x$ (which makes sense in terms of (6.14)). Exactly as for (5.16), we deduce from (6.26) that

$$\int_{X^*} x^k \chi(r)\nu^{(r)}(x) = (\chi(r)r^{1+k} - 1) L^*(- k,\chi) \tag{6.27}$$

We now introduce the

DEFINITION 6.1

$$L_p(w,\chi) = \frac{1}{<r>^{1-w} \chi(r) - 1} \int_{X^*} <x>^{-w} \chi_1(x) \nu^{(r)}(x) \tag{6.28}$$

for all $w \in \mathbb{Z}_p$, where χ_1 is as in (6.11).

Note This is independent of r, see Note 1 to Theorem 6.1.

Theorem 6.1 follows at once from (6.27) on taking χ_{k+1} for χ and using (6.9), (6.11).

COROLLARY 1. *Suppose that* $p \neq 2$ *and that the conductor is not a power of* p. *Then* $L_p(w,\chi)$ *is given by a power series*

$$L_p(w,\chi) = \gamma_o + \gamma_1 w + \ldots \tag{6.29}$$

with

$$|\gamma_j| < 1 \quad (j > 0), \qquad \gamma_j \to 0. \tag{6.30}$$

Proof. We can choose the auxiliary prime r so that $\chi(r)$ is a root of unity of order prime to p. Then

$$|\langle r\rangle^{1-w}\chi(r) - 1| \geq 1$$

for all w. The expansion (6.29) with the estimates (6.30) now follows at once from (6.28) on expanding $\langle x\rangle^{-w}$ as a power series in w and integrating term by term.

COROLLARY 2. *Under the condition of Corollary 1,*

$$|L_p(w_1,\chi) - L_p(w_2,\chi)| < |w_1 - w_2|$$

for all $w_1, w_2 \in \mathbb{Z}_p$.

Proof. Follows from the estimates (6.30).

We now embark on the proof of Leopoldt's analogue of Lemma 6.2.

THEOREM 6.2. *Let* $p \neq 2$, *let* χ *be a primitive character of conductor* d, *and let* $\theta \in \Omega_p$ *be a primitive* d-th *root of* 1. *Then*

$$G(\bar{\chi},\theta)L_p(1,\chi) = -(1 - p^{-1}\chi(p)) \sum_{a \bmod d} \bar{\chi}(a)\log_p(1-\theta^a). \tag{6.31}$$

Note 1 $\bar{\chi}$ is defined purely algebraically: $\bar{\chi}(b) = \{\chi(b)\}^{-1}$ if $\chi(b) \neq 0$. Recall that $G(\chi,\theta)$ is the Gauss sum given by (6.5).

Note 2 The factor $(1 - p^{-1}\chi(p))$ may be regarded as a counterpart of the factor in (6.8).

Note 3 Here $\log_p$ is the p-adic logarithm extended to all Ω_p by Lemma 2.1. It may be observed, however, that Theorem 6.2 does not use the normalization (iii) of that Lemma.

We require some lemmas. The first operates with the integral over $\mathbb{Z}_p$ introduced in §4.

LEMMA 6.4. Suppose $p \neq 2$. Then

$$\int_{U_p} x^{-1}\,\mu_z = -\frac{1}{p}\log_p\left\{\frac{(1-z)^p}{1-z^p}\right\} \tag{6.32}$$

for all z for which the left hand side is defined (i.e., satisfying (4.10)).

Proof. The left hand side is, by definition,

$$\lim_{N\to\infty} \sum_{\substack{0\leq j<p^N \\ p\nmid j}} \frac{j^{-1}z^j}{1-z^{p^N}} \tag{6.33}$$

(cf. (4.8), (4.9). The zeros of $1 - z^{p^N}$ all lie in $|z - 1| < 1$, and so the left hand side of (6.32) is Krasner analytic for the Swiss cheese

$$\mathcal{D} = \{z \in \Omega_p : |z - 1| \geq 1\} \tag{6.34}$$

discussed in Lemma 3.5.

We have

$$\frac{(1-z)^p}{1-z^p} = 1 + U,$$

where

$$U = \frac{-\binom{p}{1}z + \binom{p}{2}z^2 \ldots + \binom{p}{p-1}z^{p-1}}{1-z^p},$$

so

$$|U| \leq p^{-1} \quad (\text{all } z \in \mathcal{D}).$$

Hence

$$\log_p \{\frac{(1-z)^p}{1-z^p}\} = \lim_{N\to\infty} \{U - \tfrac{1}{2}U^2 + \tfrac{1}{3}U^3 \ldots \pm \frac{1}{N} U^N\}$$

is also Krasner analytic in $\mathcal{D}$.

By Lemma 3.5 it is enough to prove (6.32) for the z in some open subset of $\mathcal{D}$. For $|z| < 1$ the limit (6.33) is equal to

$$\lim_{N\to\infty} \sum_{\substack{0 \leq j < p^N \\ p \nmid j}} j^{-1} z^j$$

$$= \lim \sum_{1 \leq j \leq p^N} j^{-1} z^j - \lim \sum_{1 \leq j \leq p^{N-1}} (pj)^{-1} z^{pj}$$

$$= -\log_p(1 - z) + p^{-1} \log_p(1 - z^p).$$

Hence (6.32) is true for $|z| < 1$, and so by analytic continuation for all of $\mathcal{D}$.

We now return to the more general integral over X given by (6.13), (6.18).

LEMMA 6.5. *Let θ be a root of unity whose order divides ep^N for some N. Then*

$$\int_X f\nu_{\theta z} = \int_X f(x)\theta^x \nu_z(x), \tag{6.35}$$

for all continuous functions f on X.

Note Here θ^x is well-defined for $x \in X$ by (6.13).

Proof. Follows immediately from (6.18), or alternatively from a consideration of the characteristic functions of the basic clopen sets.

COROLLARY. *In the notation of Theorem 6.2,*

$$G(\bar{\chi},\theta)\int_X f(x)\chi(x)\nu_z = \sum_{a \bmod d} \bar{\chi}(a)\int_X f\nu_{z(a)}, \tag{6.36}$$

where

$$z(a) = z\theta^a. \tag{6.37}$$

Proof. By (6.35) the right hand side is

$$\int_X \{\sum_{a \bmod d} \bar{\chi}(a)\theta^{ax}\} f(x)\ \nu_z(x),$$

where $\{\ \}$ is equal to $\chi(x)\ G(\bar{\chi},a)$. (cf. proof of Lemma 6.2).

To prove Theorem 6.2 we put

$$f(x) = \begin{cases} x^{-1} & (|x| = 1) \\ 0 & (\text{otherwise}) \end{cases} \tag{6.38}$$

in (6.36) and sum over the primitive r-th roots of unity z. By Definition 6.1 the left hand side gives

$$G(\bar{\chi},\theta)(\chi(r) - 1)\ L_p(1,\chi). \tag{6.39}$$

On the other hand, by Lemma 6.3

$$\int_X f\nu_{z(a)} = \int_{U_p} x^{-1}\ \mu_{z(a)}$$

$$= -\frac{1}{p}\log_p \{\frac{(1 - z\theta^a)^p}{1 - z^p\ \theta^{ap}}\} \tag{6.40}$$

by Lemma 6.4 and (6.37). Now

$$\prod_{\substack{z^r=1 \\ z\neq 1}} (1 - z\theta^a) = (1 - \theta^{ar})/(1 - \theta^a).$$

Hence and by (6.40),

$$p\sum_z \int_X f\nu_{z(a)} = -p\log_p(1 - \theta^{ar}) + p\log_p(1 - \theta^a)$$

$$-\log_p(1 - \theta^{apr}) + \log_p(1 - \theta^{ap}). \tag{6.41}$$

Now

$$\sum_{a \bmod d} \bar{\chi}(a) \log_p(1 - \theta^{ar}) = \chi(r) \sum_{a \bmod d} \bar{\chi}(a) \log_p(1 - \theta^{a}).$$

On multiplying (6.41) by $\bar{\chi}(a)$ and summing we thus get

$$\{ - p\chi(r) + p - \chi(pr) + \chi(p)\} \sum_{a \bmod d} \bar{\chi}(a) \log_p(1 - \theta^{a}).$$

The term in $\{ \}$ is

$$- (\chi(r) - 1)(p - \chi(p)).$$

Hence and by comparison with (6.39) we obtain the required equation (6.31) by summing (6.36) over z and then dividing by $(\chi(r) - 1)$.

7 MAHLER's EXPANSION

For fixed non-negative rational integer k, the function

$$\binom{x}{k} = \frac{x(x - 1) \ldots (x - k + 1)}{k!} \tag{7.1}$$

takes values in $\mathbb{Z}_p$ for $x \in \mathbb{Z}_p$ (Chapter 4, Exercise 8), and it is clearly continuous. It follows that

$$f(x) = \sum_{k=o}^{\infty} a_k\binom{x}{k} \tag{7.2}$$

is a continuous $\mathbb{Z}_p$-valued function of $x \in \mathbb{Z}_p$ provided that

$$a_k \in \mathbb{Z}_p, \quad a_k \to 0 \quad (k \to \infty). \tag{7.3}$$

The coefficients a_k are uniquely determined by the

$$b_n \text{ (say)} = f(n) \quad (n = 0,1,2, \ldots) \tag{7.4}$$

Indeed by putting $n = 0,1,2, \ldots$ in (7.2.) we recover $a_o,a_1,a_2, \ldots$ in order. This is, of course, not unexpected as the non-negative integers are dense in $\mathbb{Z}_p$, and so a continuous function is determined by the values it takes on them.

There is a converse:

THEOREM 7.1 (Mahler) *A necessary and sufficient condition that a $\mathbb{Z}_p$-valued function $f(x)$, $x \in \mathbb{Z}_p$ be continuous, is that it can be written in the form (7.2), where the a_k satisfy (7.3).*

COROLLARY. *A necessary and sufficient condition that a $\mathbb{Z}_p$-valued function b_n of the non-negative integer n be p-adically continuous is that the $a_k \in \mathbb{Z}_p$ given recursively by*

$$b_n = \sum_{k=o}^{n} a_k \binom{n}{k} \tag{7.5}$$

should satisfy $a_k \to 0$.

The proof requires

LEMMA 7.1. *Let $k < p^N$ for some N. Then $\binom{x}{k}$ mod p has period p^N as a function of x.*

Proof. Follows at once from considering the coefficient of T^k in the expansions of

$$\begin{aligned}(1 + T)^{x+p^N} &= (1 + T)^x \ (1 + T)^{p^N} \\ &\equiv (1 + T)^x \ (1 + T^{p^N}) \quad (p).\end{aligned}$$

We now prove Theorem 7.1. It is enough to show that a continuous function $f(x)$ has an expansion (7.2) with (7.3). Define b_n by (7.4). Since $f(x)$ is continuous, there is some N such that

$$f(x) \equiv f(y) \quad (p) \quad \text{if} \quad x \equiv y \quad (p^N). \tag{7.6}$$

Put

$$F_o(x) = \sum_{o \leq k < p^N} a_k \binom{x}{k}, \tag{7.7}$$

where the a_k are given by (7.5), (7.4). Then

$$f(n) = F_o(n) \quad 0 \leq n < p^N ,$$

and so

$$f(x) \equiv F_o(x) \quad (p) \quad (\text{all } x),$$

by Lemma 7.1 and (7.6). Hence

$$f_1(x) \text{ (say) } = p^{-1}(f(x) - F_o(x))$$

is a continuous $\mathbb{Z}_p$-valued function of x. We repeat the process on $f_1(x)$, and so on. In this way we have

$$f(x) = \sum_{\ell=o}^{\infty} p^{\ell} F_{\ell}(x),$$

where the F_ℓ are all similar to (7.7). This clearly does what is required.

Notes

§1 For an account of the analysis of a p-adic variable at undergraduate level, see Schikhof (1985).

§3 Krasner developed his theory of functions of a $\mathfrak{p}$-adic variable and of analytic continuation in a long series of notes in Comptes Rendus. He gives a connected account in Krasner (1974). A fundamentally different approach was adopted by Tate (1971), see also Bosch, Günter and Remmert (1984).

§§4-6 The p-adic ζ and L functions were introduced by Kubota and Leopoldt (1964). Our account follows that of Koblitz (1980). As was shown by Iwasawa, the p-adic L-functions are a powerful tool for studying cyclotomic fields, see Iwasawa (1972) or the accounts in Koblitz (1977, 1980), Lang (1978), Washington (1980). For a (largely conjectural) generalization to elliptic curves, see the article by Coates in Rankin (1984). An integral closer in spirit to Chapter 1, §5 was introduced by Volkenborn (1972), see also Schikhof (1985).

§7 Mahler (1973) is largely devoted to an examination of this expansion and the relation between the sequence of coefficients and the properties of the function. See also Schickof (1985).

Exercises

1. Why does (2.7) not contradict Strassman's Theorem 4.1 of Chapter 4?

1. bis. If $|x - 1| < 1$ and $\log_p(x) = 0$ show that $x^{p^N} = 1$ for some positive integer N.

[Hint (i) If $|y - 1| < p^{-1/(p-1)}$ and $\log_p y = 0$ show that $y = 1$.

(ii) $x^{p^N} \to 1$ as $N \to \infty$].

2. In the notation of §3 let $f \in W$ and $\zeta \in \mathbb{O}$, $\zeta \neq 0$. For $n \geq 0$ show that

$$f_n = \lim_{\substack{N\to\infty \\ p\nmid N}} N^{-1} \sum_{\omega^N=1} (\omega\zeta)^{-n} f(\omega\zeta). \tag{*}$$

Use this to give alternative proofs of Lemmas 3.2, 3.3.

[Note This is an analogue of a theorem of Cauchy for complex variables. The expression on the right hand side of (*) is an example of a Shnirelman integral. cf. Schikhof (1985).]

3. Show, by analytic continuation or otherwise, that

$$\{E_p(x)\}^p = \exp_p(px)\ E_p(x^p),$$

for $|x| < 1$.

4. Fill in the details of the following argument which is at variance with naive expectations from analytic continuation ($p > 2$):

(i) Let π be any solution of

$$\pi^{p-1} = -p.$$

Defining the functions $\exp_p(\cdot)$ and $E_p(\cdot)$ by (2.8) and (2.15), show that there is some $\delta > 0$ such that all the terms on the right of

$$G(x) = E_p(\pi x) \prod_{j\geq 2} \exp_p(-(\pi x)^{p^j}/p^j)$$

are defined in $|x| < 1 + \delta$ and the product is convergent there.

(ii) Show that $G(x)$ is given by a power series

$$g_o + g_1 x + \ldots + g_n x^n + \ldots, \quad (g_n \in \mathbb{Q}_p(\pi))$$

convergent in $|x| < 1 + \delta$.

(iii) Put

$$H(x) = \exp_p\{\pi(x - x^p)\}.$$

Show that $H(x)$ is defined in $|x| < 1$ and that

$$H(x) = G(x) \quad (|x| < 1).$$

[Hint Use (2.17)].

(iv) Show that

$$G(1) \equiv 1 + \pi \quad (\pi^2),$$

while $H(1) = 1$.

(v) For $|x| \lesssim 1$ show that

$$\{G(x)\}^p = \exp_p(p\pi x)\exp_p(-p\pi x^p)$$
$$= \exp_p\{p\pi(x - x^p)\}.$$

Deduce that $\lambda^p = 1$, where $G(1) = \lambda$.

(vi) Show, similarly that if $a^{p-1} = 1$, then

$$G(a) \equiv 1 + a\pi \quad (\pi^2)$$

and

$$\{G(a)\}^p = 1.$$

(vii) Deduce that

$$G(\omega(b)) = \lambda^b$$

for all $b \in \mathbb{Z}$, where $\omega(b)$ is the Teichmüller representative.

[Note cf. Dwork (1960)].

5. If $x \in \Omega_p$, $|x - 1| < 1$, show that

$$\log_p x = \lim_{n\to\infty}\{p^{-n}(x^{p^n} - 1)\}.$$

6. Let

$$F(X) = 1 + \sum_{n\geq 1} a_n X^n \in 1 + X\mathbb{Q}_p[[X]].$$

Show that a necessary and sufficient condition that

$$F(X) \in 1 + X\mathbb{Z}_p[[X]]$$

is that

$$F(X^p)/\{F(X)\}^p \in 1 + p\mathbb{Z}_p[[X]].$$

Hence show that

$$F(X) = \sum_{n\geq 0} Y^n/n!, \quad Y = X + X^p/p + X^{p^2}/p^2 + \ldots$$

is in $\mathbb{Z}_p[X]$.

[Note The double square bracket denotes a formal power series, cf. Chapter 2, §8. Compare (2.17). The criterion, which extends in the obvious way to several variables, is due to Dwork (1960)].

7. If $|z| < 1$ and μ_z is the measure given by (4.9) show that

$$\mu_z(S) = \sum z^n \quad (n \in \mathbb{N} \cap S),$$

for any clopen set $S \subset \mathbb{Z}_p$. Deduce that $\mu = \mu_z$ satisfies (4.3) and (4.5).

8. For a given sequence $b_o, b_1, b_2, \ldots$, show that the a_k defined by (7.5) are

$$a_k = \sum_{n=o}^{k} (-)^{k-n} b_n \binom{k}{n}.$$

CHAPTER THIRTEEN: A THEOREM OF BOREL AND DWORK

1 INTRODUCTION

One of the themes of this book has been that one can sometimes derive information about an object defined over a global field k from information about the behaviour at all the localizations. In this Chapter we give a rather different specimen. It is a generalization by Dwork of a theorem of E. Borel.

Let k be an algebraic number field. For a place $\mathfrak{p}$ we denote $|\ |_{\mathfrak{p}}$ the renormalized valuation. We recall that for $a \in k$, $a \neq 0$ we have $|a|_{\mathfrak{p}} = 1$ for almost all $\mathfrak{p}$, and that

$$\prod_{\mathfrak{p}} |a|_{\mathfrak{p}} = 1, \tag{1.1}$$

where the product extends over all places, archimedean and non-archimedean (Chapter 10, §2).

The object of this Chapter is

THEOREM 1.1 *Let*

$$f(X) = f_o + f_1 X + \dots + f_n X^n + \dots \tag{1.2}$$

be a formal power series, where $f_n \in k$. *Suppose that there is a finite set* S *of the places of* k *such that*:

(i) $|f_n|_{\mathfrak{p}} \leqslant 1$ *for all* $\mathfrak{p} \notin S$ *and all* n.

(ii) *For each* $\mathfrak{p} \in S$ *there is a polynomial*

$$g_{\mathfrak{p}}(X) = 1 + g_{1\mathfrak{p}} X + \dots + g_{m\mathfrak{p}} X^m \in k_{\mathfrak{p}}[X] \tag{1.3}$$

such that the series $g_{\mathfrak{p}}(X) f(X)$, *considered as a series with coefficients in* $k_{\mathfrak{p}}$, *has radius of convergence* $R_{\mathfrak{p}}$, *where*

$$\prod_{\mathfrak{p} \in S} R_{\mathfrak{p}} > 1. \tag{1.4}$$

Then $f(X)$ is the expansion of an element of $k(X)$.

Note 1 A priori the degree m of $g_{\mathfrak{p}}(X)$ may depend on $\mathfrak{p}$, but we can treat it as independent of $\mathfrak{p}$ by allowing terms with zero coefficients.

Note 2 One can say that $f(X)$ represents a meromorphic function of a $\mathfrak{p}$-adic variable in the domain of convergence of $g_{\mathfrak{p}}(X)f(X)$.

2 SOME LEMMAS

Let A be the determinant of the matrix

$$\begin{pmatrix} a_{00} & a_{01} & \cdots & a_{0s} \\ a_{10} & a_{11} & \cdots & a_{1s} \\ & & \cdots & \\ a_{s0} & a_{s1} & \cdots & a_{ss} \end{pmatrix}, \tag{2.1}$$

let A_{ij} be the elements of the adjoint matrix (A_{ij} is the cofactor of a_{ji}), and let a be the determinant of the $(s - 1) \times (s - 1)$ matrix obtained by deleting the first and last row and column. Then

$$A_{00} A_{ss} - A_{0s} A_{s0} = Aa. \tag{2.2}$$

Note This is a special case of "Jacobi's Theorem on the minors of the adjugate".

Proof. Follows by taking determinants in the matrix identity

$$\begin{pmatrix} A_{00} & A_{01} & \cdots & A_{0,s-1} & A_{0s} \\ 0 & 1 & \cdots & 0 & 0 \\ & & \cdots & & \\ & & \cdots & & \\ 0 & 0 & \cdots & 1 & 0 \\ A_{s0} & A_{s1} & \cdots & A_{s,s-1} & A_{ss} \end{pmatrix} \begin{pmatrix} a_{00} & a_{01} & \cdots & a_{0,s-1} & a_{0s} \\ a_{10} & a_{11} & \cdots & a_{1,s-1} & a_{1s} \\ & & \cdots & & \\ & & \cdots & & \\ a_{s-1,0} & a_{s-1,1} & \cdots & a_{s-1,s-1} & a_{s-1,s} \\ a_{s0} & a_{s1} & \cdots & a_{s,s-1} & a_{ss} \end{pmatrix}$$

$$= \begin{pmatrix} A & 0 & \cdots & 0 & 0 \\ a_{10} & a_{11} & & a_{1,s-1} & a_{1s} \\ & & \cdots & & \\ & & \cdots & & \\ a_{s-1,0} & a_{s-1,1} & \cdots & a_{s-1,s-1} & a_{s-1,s} \\ 0 & 0 & \cdots & 0 & A \end{pmatrix} \tag{2.3}$$

Here the first matrix has 1 on the diagonal, 0 elsewhere, except in the first and last rows. The second matrix is just (2.1).

We shall be concerned with the ("Hankel") determinant

$$F(n,s) = \det(a_{ij}) \tag{2.4}$$

for which

$$a_{ij} = f_{n+i+j} \quad (0 \leq i \leq s, \quad 0 \leq j \leq s) \tag{2.5}$$

and the f_n are the coefficients in (1.2):

COROLLARY.

$$F(n, s-1)F(n+2,s-1) - \{F(n+1, s-1)\}^2$$

$$= F(n,s)\, F(n+2,s-2). \tag{2.6}$$

LEMMA 2.2. Suppose that there is an s such that $F(n,s) = 0$ for all sufficiently large n. Then $f(X)$ is the expansion of a rational function.

Proof. Without loss of generality, s is minimal. There is thus some n_o such that

$$F(n,s) = 0 \quad (\text{all } n \geq n_o) \tag{2.7}$$

$$F(n_o + 1, s - 1) \neq 0. \tag{2.7 bis}$$

By (2.7) the right hand side of (2.6) is 0 for all $n \geq n_o$. It now follows from (2.7 bis) by induction on n that

$$F(n, s - 1) \neq 0 \quad (\text{all } n \geq n_o + 1) \tag{2.8}$$

By (2.7), (2.8), for any $n \geq n_o + 1$ there are unique $c_r = c_r(n)$ such that

$$f_{\ell+s} + c_1 f_{\ell+s-1} + \dots + c_s f_\ell = 0 \quad (n \leq \ell \leq n + s). \tag{2.9}$$

On comparing (2.9) for n and $n + 1$ and using (2.8), we readily deduce that $c_r(n + 1) = c_r(n)$: that is, the c_r are independent of n. Hence

$$(1 + c_1 X + \dots + c_s X^s) f(X)$$

is a polynomial in X, as required.

LEMMA 2.3. (i) Let A be the determinant of the matrix (2.1), where $a_{ij} \in k$, let $| \;|_{\mathfrak{p}}$ be a valuation and let

$$\alpha_i = \max_j |a_{ij}|_{\mathfrak{p}} \quad (0 \leq i \leq s, \;\; 0 \leq j \leq s).$$

Then

$$|A|_{\mathfrak{p}} \leq M(\mathfrak{p}, s) \prod_i \alpha_i,$$

where $M(\mathfrak{p},s)$ depends only on $\mathfrak{p}$ and s.

(ii) If $\mathfrak{p}$ is non-arch., one can take $M(\mathfrak{p},s) = 1$ for all s.

Proof.

$$A = \sum \pm a_{o,j(o)} a_{1,j(1)} \dots a_{s,j(s)};$$

where $j(0), \dots, j(s)$ runs through the permutations of $0, \dots, s$. If $\mathfrak{p}$ is non-arch., (ii) of the Lemma is immediate. If $| \ |_{\mathfrak{p}}$ satisfies the triangle inequality, we can take $M = (s + 1)!$. The truth of the Lemma now follows in general, since every valuation is equivalent to one satisfying the triangle inequality.

3. Proof. By (1.4) we can choose $T_{\mathfrak{p}}$ $(\mathfrak{p} \in S)$ such that

$$T_{\mathfrak{p}}R_{\mathfrak{p}} > 1 \quad (\mathfrak{p} \in S); \quad \prod_{\mathfrak{p}\in S} T_{\mathfrak{p}} < 1. \tag{3.1}$$

Then the coefficients of

$$h_{\mathfrak{p}}(X) = g_{\mathfrak{p}}(X)f(X) = h_o + h_1X + \dots \tag{3.2}$$

satisfy

$$|h_n|_{\mathfrak{p}} \leq T_{\mathfrak{p}}^n \tag{3.3}$$

for all sufficiently large n.

The formal expansion of $\{g_{\mathfrak{p}}(X)\}^{-1}$ as a power series clearly converges in some neighbourhood of the origin. Hence $f(X)$, considered as a power series with coefficients in $k_{\mathfrak{p}}$, converges in some neighbourhood of the origin. There is thus some $t_{\mathfrak{p}} > 0$ such that

$$|f_n|_{\mathfrak{p}} \leq t_{\mathfrak{p}}^n \tag{3.4}$$

for all sufficiently large n.

We now estimate $\mathfrak{p}$-adically for $\mathfrak{p} \in S$ the determinants $F(n,s)$ given by (2.4), (2.5). If $s > m$, which we suppose, we can use the recurrence relations furnished by (3.2) to replace the f_{n+i+j} for

$i \geq m$ by h_{n+i+j}, so

$$F(n,s) = \det(b_{ij}),$$

where

$$b_{ij} = \begin{cases} f_{n+i+j} & \text{for } 0 \leq i < m \\ h_{n+i+j} & \text{for } m \leq i \leq s. \end{cases}$$

On using the estimates (3.3), (3.4) in Lemma 2.3 we readily obtain

$$|F(n,s)|_{\mathfrak{p}} \leq L_{\mathfrak{p}} t_{\mathfrak{p}}^{nm} T_{\mathfrak{p}}^{n(s-m)} \tag{3.5}$$

for all sufficiently large n, where $L_{\mathfrak{p}} = L_{\mathfrak{p}}(s)$ is independent of n.

The $\mathfrak{p} \notin S$ are non-arch. by hypothesis, and so, by (i) of the enunciation of the Theorem and by Lemma 2.3. (ii), we have

$$|F(n,s)|_{\mathfrak{p}} \leq 1 \quad (\text{all } \mathfrak{p} \notin S, \text{ all } n,s).$$

Hence and by (3.5),

$$\prod_{\text{all } \mathfrak{p}} |F(n,s)|_{\mathfrak{p}} \leq A\, B^{nm}\, C^{n(s-m)}, \tag{3.6}$$

for all sufficiently large n, where

$$A = A(s) = \prod_{\mathfrak{p} \in S} L_{\mathfrak{p}},$$

$$B = \prod_{\mathfrak{p} \in S} t_{\mathfrak{p}}$$

and

$$C = \prod_{\mathfrak{p} \in S} T_{\mathfrak{p}} < 1, \tag{3.7}$$

by (3.1).

By (3.7) we can choose s so that

$$D \text{ (say) } = B^m C^{s-m} < 1$$

and then, by (3.6),

$$\prod_{\text{all } \mathfrak{p}} |F(n,s)|_{\mathfrak{p}} \leq AD^n < 1$$

for all sufficiently large n. By (1.1) this implies that $F(n,s) = 0$. The truth of the Theorem now follows from Lemma 2.2.

Notes

Dwork (1960) used Theorem 1.1. to prove an important result about varieties over finite fields. It was generalized by Bertrandias (1963-4). Both theorems are treated in Amice (1975).

For Lemma 2.2, which is due to E. Borel, and for related criteria, see that invaluable compendium Pólya and Szegő (1976), Section VII.

Exercises

1. Let the power series $f(X) = \sum f_n X^n$, $f_n \in k$ have radius of convergence $r_{\mathfrak{p}}$ in $k_{\mathfrak{p}}$. Suppose that $\prod r_{\mathfrak{p}} > 1$. Show that $f(X)$ is a polynomial.

APPENDIX A: RESULTANTS AND DISCRIMINANTS

RESULTANTS AND DISCRIMINANTS

Let A be a commutative ring (with 1) and put

$$\left.\begin{aligned} f(X) &= f_o + \ldots + f_m X^m \in A[X], \\ g(X) &= g_o + \ldots + g_n X^n \in A[X], \end{aligned}\right\} \quad \text{(A.1)}$$

where X is an indeterminate. Multiply $f(X)$ successively by $1, X, \ldots, X^{n-1}$ and $g(X)$ by $1, X, \ldots, X^{m-1}$. We have now $m + n$ linear forms in $1, X, \ldots, X^{m+n-1}$. Their determinant is the resultant $R = R(f,g)$. It is defined up to sign [the usual convention being that the term $f_m^n g_o^n$ has coefficient +1]. On eliminating $X, X^2, \ldots, X^{m+n-1}$ from the $m + n$ linear forms, we obtain

$$\left.\begin{aligned} u(X) &= u_o + \ldots + u_{n-1} X^{n-1} \in A[X] \\ v(X) &= v_o + \ldots + v_{m-1} X^{m-1} \in A[X] \end{aligned}\right\} \quad \text{(A.2)}$$

such that

$$u(X)f(X) + v(X)g(X) = R. \quad \text{(A.3)}$$

If $R \neq 0$, then clearly

$$\text{either} \quad u(X) \neq 0 \quad \text{or} \quad v(X) \neq 0\ . \quad \text{(A.4)}$$

If $R = 0$, the values of $u_o, \ldots, u_{n-1}, v_o, \ldots, v_{m-1}$ given as cofactors in the $(m + n) \times (m + n)$ matrix of the linear forms may all be 0. We show that we can still choose $u(X)$, $v(X)$ satisfying (A.2), (A.3) and (A.4). If $f(X) = 0$ take $u(X) = 1$, $v(X) = 0$. Otherwise there is some $r < m + n$ such that there is at least one minor M of order r which is not 0 but all minors of order $r + 1$ vanish. On using determinantal elimination of the variables between the r linear forms involved in M together with any further linear form, we obtain a non-trivial relation, which gives us our $u(X)$, $v(X)$.

LEMMA 1. *Let $A = k$ be a field.*

(i) *If f,g have a (non-constant) common factor in $k[X]$, then $R = 0$.*

(ii) *Suppose that either $f_m \neq 0$ or $g_n \neq 0$. If also $R = 0$, then f,g have a common factor.*

Proof. (i) Clear since by (A.3) any common factor of f,g must divide R.

(ii) Suppose $f_m \neq 0$ and $R = 0$. If $v(X) = 0$ then (A.3) implies $u(X) = 0$ contrary to (A.4). Hence $v(X) \neq 0$ and (A.3) implies that $f(X)$ divides $v(X)g(X)$. But $f(X)$ has degree exactly m, so cannot divide $v(X)$. Hence f,g have a common factor.

LEMMA 2. *Let $A = k$ be a field and suppose $f_m \neq 0$. Let $\theta = \theta_1, \ldots, \theta_m$ be the roots of $f(X)$ in the algebraic closure $\bar{k}$ of k. Then*

$$R(f,g) = \pm f_m^n \prod_\theta g(\theta). \qquad (A.5)$$

Proof. We have

$$f(X) = f_m \prod_\theta (X - \theta).$$

By Lemma 1, R vanishes whenever $g(\theta) = 0$ for some θ. On the other hand, if we take f_m, $\theta_1, \ldots, \theta_m$, and the coefficients $g_o, \ldots, g_n$ of $g(X)$ to be indeterminates, it is easy to see that the degree of $R(f,g)$ in each of $\theta_1, \ldots, \theta_m$ is at most n.

COROLLARY. *Suppose also that $g_n \neq 0$ and let $\phi = \phi_1, \ldots, \phi_n \in \bar{k}$ be the roots of g. Then*

$$R = \pm f_m^n g_n^m \prod_{\theta,\phi} (\theta - \phi). \tag{A.6}$$

We now revert to a general ring A but consider the special case when

$$g(X) = f'(X)$$

is the (formal) derivative of $f(X)$, so $n = m - 1$. In the linear forms considered in the definition of a resultant the monomial $X^{m+n-1} = X^{2m-2}$ occurs only with the coefficients f_m or mf_m. We define the _discriminant_ $D(f)$ of f to be the determinant of the $n + m = 2m - 1$ linear forms in $1, X, \ldots, X^{2m-3}$ and $f_m X^{2m-2}$, with a sign convention to be explained below. Hence $D(f) \in A$ and

$$f_m D(f) = \pm R(f,f'). \tag{A.7}$$

LEMMA 2 bis. _There are_ $u(X), v(X) \in A[X]$ _of degrees at most_ $m - 2, m - 1$ _respectively such that_

$$u(X)f(X) + v(X)f'(X) = D(f). \tag{A.7 bis}$$

Proof. As for (A.3).

LEMMA 3. _Let_ $A = k$ _be a field and suppose_ $f_m \neq 0$. _A necessary and sufficient condition for_ $f(X)$ _to have a multiple root_ (_in_ $\bar{k}$) _is that_ $D(f) = 0$.

Proof. For a necessary and sufficient condition for $f(X)$ to have a multiple root is that f and f' have a common factor.

Note If k has prime characteristic p then it is possible that $f'(X) = 0$ (identically): this happens precisely when f is a polynomial in X^p (_inseparable_ polynomial). An inseparable polynomial can be irreducible. It follows from Lemma 3 that (if $f_m \neq 0$) the condition $D(f) = 0$ is necessary and sufficient for $f(X)$ to have either a multiple factor or an inseparable factor in $k[X]$.

LEMMA 4. Let $A = k$, $f_m \neq 0$ and let $\theta_1, \ldots, \theta_m \in \bar{k}$ be the roots of $f(X)$. Then

$$D(f) = \pm f_m^{2m-2} \prod_{i<j} (\theta_i - \theta_j)^2. \tag{A.8}$$

Note It is usual to define $D(f)$ so that the $+$ sign holds in (A.8).

Proof. Follows from (A.5) with $g = f'$ and from

$$f'(\theta_i) = f_m \prod_{j \neq i} (\theta_i - \theta_j).$$

LEMMA 5. Let $f, g \in A[X]$. Then

$$D(fg) = D(f)\, D(g)\, \{R(f,g)\}^2. \tag{A.9}$$

Proof. What is asserted is an identity between polynomials with coefficients in $\mathbb{Z}$ in the coefficients $g_0, \ldots, g_n, f_0, \ldots, f_m$. It is therefore enough to prove the Lemma when $A = \mathbb{Q}$ and $f_m \neq 0$, $g_n \neq 0$. Let $\theta_1, \ldots, \theta_m$; $\phi_1, \ldots, \phi_n$ be the roots of $f(X)$, $g(X)$ respectively in $\bar{\mathbb{Q}}$. Then together they are the roots of fg. The assertion of the Lemma now follows from (A.8) applied to fg, f and g and from (A.6).

We conclude by giving the formulae for some discriminants (taken from van der Waerden (1949)).

If $f = f_2X^2 + f_1X + f_0$, then

$$D(f) = f_1^2 - 4f_0f_2.$$

If $g = g_3X^3 + g_2X^2 + g_1X + g_0$, then

$$D(g) = g_1^2g_2^2 - 4g_0g_2^3 - 4g_1^3g_3 - 27g_0^2g_3^2 + 18g_0g_1g_2g_3.$$

If $h = X^4 + h_2X^2 + h_1X + h_0$, then

$$D(h) = 16h_2^4h_0 - 4h_2^3h_1^2 - 128h_2^2h_0^2$$

$$+ 144h_2h_1^2h_0 - 27h_1^4 + 256h_0^3.$$

The discriminant of a general quartic may be obtained by making a substitution $X \to X + c$ in h, but is quite unwieldy, as are the discriminants of polynomials of higher degree. The form of $D(h)$ may be checked on noting that the cubic whose roots are

$$y_1 = (x_1 + x_2)(x_3 + x_4), \quad y_2 = (x_1 + x_3)(x_2 + x_4),$$

$$y_3 = (x_1 + x_4)(x_2 + x_3)$$

(where the x_j are the roots of h) is

$$Y^3 - 2h_2Y^2 + (h_2^2 + 4h_0)Y + h_1^2.$$

For discriminants of forms of higher degree, see Sasaki et al. (1981).

APPENDIX B: NORMS, TRACES AND CHARACTERISTIC POLYNOMIALS

Let A be a commutative ring with 1. By a vector space V of dimension n over A we shall mean a free A-module on n generators $\underline{v}_1,\ldots,\underline{v}_n$ (say). The set of A-linear maps $V \to V$ (<u>endomorphisms</u>) is denoted by $\mathrm{End}_A(V)$.

Let $\alpha \in \mathrm{End}_A(V)$. We call its determinant the <u>norm</u> of α, written $N_{V/A}(\alpha)$. The <u>trace</u> of α is written $S_{V/A}(\alpha)$. Clearly

$$N_{V/A}(\alpha\beta) = N_{V/A}(\alpha)\, N_{V/A}(\beta) \qquad (B.1)$$

and

$$S_{V/A}(\alpha + \beta) = S_{V/A}(\alpha) + S_{V/A}(\beta). \qquad (B.1\ bis)$$

We identify $a \in A$ with multiplication by a as an element of $\mathrm{End}_A(V)$, so

$$N_{V/A}(a) = a^n; \quad S_{V/A}(a) = na \quad (a \in A). \qquad (B.2)$$

Let T be an indeterminate over A and put $V[T] = V \otimes_A A[T]$, so $V[T]$ is an $A[T]$ - vector space, and we can identify $\mathrm{End}_A(V)$ as a subring of $\mathrm{End}_{A[T]}(V[T])$ in an obvious way. For $\alpha \in \mathrm{End}_A(V)$ we call

$$F_\alpha(T) = N_{V[T]/A[T]}(T - \alpha) = T^n - S_{V/A}(\alpha)T^{n-1} + \ldots + (-)^n N_{V/A}(\alpha) \qquad (B.3)$$

the <u>characteristic polynomial</u> of α.

LEMMA 1. (Cayley-Hamilton Theorem). $F_\alpha(\alpha) = 0$.

Proof. Let

$$\alpha \underline{v}_i = \sum_j a_{ij} \underline{v}_j \quad (a_{ij} \in A),$$

so

$$\sum_j (\delta_{ij}\alpha - a_{ij})\underline{v}_j = 0, \qquad \text{(B.3 bis)}$$

where δ_{ij} is the "Kronecker delta" (= 1 if $i = j$; = 0 otherwise). Working in the commutative ring $A[\alpha]$, eliminate $\underline{v}_2,\ldots,\underline{v}_n$ from (B.3 bis) by "multiplying" by the "cofactors" of the "coefficients" of $\underline{v}_1$ and adding. We get $F_\alpha(\alpha)\underline{v}_1 = 0$. Similarly $F_\alpha(\alpha)\underline{v}_j = 0$ for $j = 2,\ldots,n$. Hence $F_\alpha(\alpha) = 0$, as asserted.

Now let B be a commutative algebra (with 1) over A: that is, B is both a commutative ring with 1 and a finite-dimensional vector space over A. We regard A as a subring of B in the obvious way. Let U be a finite-dimensional vector space over B. Then U has an induced structure as a finite-dimensional vector space over A, and

$$\mathrm{End}_B(U) \subset \mathrm{End}_A(U).$$

LEMMA 2. Let $\alpha \in B$. Then (identifying α with an element of $\mathrm{End}_B(U)$) we have

$$N_{U/A}(\alpha) = \{N_{B/A}(\alpha)\}^n, \qquad \text{(B.4)}$$

where n is the dimension of U over B.

Note This is a special case of Lemma 4.

Proof. Let $\underline{u}_1,\ldots,\underline{u}_n$ be generators for U over B. As an A-vector space, U is the direct sum of modules $B\underline{u}_j$, each of which is A-isomorphic to B.

We now specialize to fields. Let K be an extension of finite degree of the field k. Then K is a finite-dimensional vector space over k. Further, as above, we can identify $\alpha \in K$ with the element $\xi \mapsto \alpha\xi$ of $\mathrm{End}_k(K)$. By $N_{K/k}(\alpha)$ and $S_{K/k}(\alpha)$ we shall mean the norm and trace of this endomorphism.

LEMMA 3. *Let* K, k *be as above, and let* $\alpha \in K$. *Denote by* $F_\alpha(T)$, $f_\alpha(T)$ *the characteristic polynomial and the minimum polynomial of* α *over* k. *Then*

$$F_\alpha(T) = \{f_\alpha(T)\}^t, \tag{B.5}$$

where r *is the degree of* K *over* $k(\alpha)$.

Proof. By Lemma 1, $f_\alpha(T)$ divides $F_\alpha(T)$. If $r = 1$, it follows that the two polynomials are the same.

In the general case, let $L = k(\alpha)$. Then by Lemma 2 and the above remark, we have

$$\begin{aligned} F_\alpha(T) &= N_{K[T]/k[T]}(T - \alpha) \\ &= \{N_{L[T]/k[T]}(T - \alpha)\}^t \\ &= \{f_\alpha(T)\}^r. \end{aligned}$$

COROLLARY. *Let* $\alpha_1, \ldots, \alpha_m$ *be the roots of* $f_\alpha(T)$ *in some splitting field. Then*

$$N_{K/k}(\alpha) = (\alpha_1, \ldots, \alpha_m)^r; \quad S_{K/k}(\alpha) = r(\alpha_1 + \ldots + \alpha_m). \tag{B.6}$$

"Transitivity of the norm". We do not actually have to invoke Lemma 4 below, but it illuminates the situation, particularly in Chapter 7. Perhaps surprisingly, there appears to be no really transparent proof. We revert to the general situation, in which U is a B-vector space and B is an A-algebra.

LEMMA 4. *Let* $\alpha \in \mathrm{End}_B(U) \subset \mathrm{End}_A(U)$. *Then*

$$N_{U/A}(\alpha) = N_{B/A}(N_{U/B}(\alpha)). \tag{B.7}$$

<u>Proof</u>. We note, first, that the right hand side makes sense, since $N_{U/B}(\alpha) \in B$. Clearly, both sides of (B.7) are multiplicative in α in the sense that their values when $\alpha\beta$ is substituted for α ($\alpha, \beta \in \mathrm{End}_B(U)$) are the products of the values for α and β. We shall verify (B.7) first for α of special kinds, and then use this multiplicativity to prove it for general α.

We use induction on the dimension n of U over B. Clearly (B.7) is true for $n = 1$. Let $\underline{u}_1, \ldots, \underline{u}_n$ be a set of B-generators of U.

<u>First Case</u>.

$$\beta\underline{u}_1 = b_{11}\underline{u}_1 + b_{12}\underline{u}_2 + \ldots + b_{1n}\underline{u}_n,$$

$$\beta\underline{u}_i = b_{ii}\underline{u}_i \quad (i \neq 1)$$

where $b_{ij} \in B$. Then both sides of (B.7) for $\alpha = \beta$ are clearly $\prod_{i=1}^{n} N_{B/A}(b_{ii})$.

<u>Second Case</u>.

$$\gamma\underline{u}_1 = c_{11}\underline{u}_1,$$

$$\gamma\underline{u}_i = \sum_{j \geq 1} c_{ij}\underline{u}_j \quad (i \neq 1).$$

Then

$$N_{U/B}(\gamma) = c_{11}N_{W/B}(\theta),$$

where W is the B-vector space with basis $\underline{u}_2, \ldots, \underline{u}_n$ and θ is $\underline{u}_i \to \sum_{j \neq 1} c_{ij}\underline{u}_j$. By the induction hypothesis,

$$N_{W/A}(\theta) = N_{B/A}(N_{W/B}(\theta)).$$

On the other hand, clearly

$$N_{U/A}(\gamma) = \{N_{B/A}(c_{11})\}\ \{N_{W/A}(\theta)\}.$$

Hence (B.7) holds for $\alpha = \gamma$.

<u>General Case</u>. Let

$$\alpha \underline{u}_i = \sum_j a_{ij} \underline{u}_j \quad (a_{ij} \in B).$$

Define $\beta \in \mathrm{End}_B(U)$ by

$$\beta \underline{u}_1 = \underline{u}_1 - \sum_j a_{1j} \underline{u}_j,$$

$$\beta \underline{u}_j = a_{11} \underline{u}_j \quad (j > 1).$$

Put $\gamma = \beta\alpha$, so

$$\gamma \underline{u}_1 = a_{11} \underline{u}_1,$$

$$\gamma \underline{u}_i = \sum_j c_{ij} \underline{u}_j \quad (i \neq 1),$$

where

$$c_{i1} = a_{i1}; \quad c_{ij} = a_{11} a_{ij} - a_{i1} a_{1j} \quad (j \neq 1).$$

Then β, γ are the special types discussed above, and so, using the multiplicativity of both sides of (B.7) for $\gamma = \alpha\beta$, we obtain

$$\{N_{B/A}(a_{11})\}^{n-1} N_{U/A}(\alpha) = \{N_{B/A}(a_{11})\}^{n-1} N_{B/A}(N_{U/B}(\alpha)).$$

This gives (B.7) provided that $N_{B/A}(a_{11})$ is not a zero-divisor.

If $N_{B/A}(a_{11})$ is a zero-divisor, we adjoin an indeterminate T and consider $T + \alpha$, $A[T]$, $B[T]$ instead of α, A, B. Then the analogue of (B.7) holds by the above remark, and on putting $T = 0$ we get (B.7) in general.

<u>COROLLARY 1</u>. <u>Let</u> $\alpha \in \mathrm{End}_B(U)$ <u>have characteristic polynomials</u> $F_\alpha(T)$, $\Phi_\alpha(T)$ <u>considered as an element of</u> $\mathrm{End}_A(U)$, $\mathrm{End}_B(U)$ <u>respectively</u>. <u>Then</u>

$$F_\alpha(T) = N_{B[T]/A[T]}(\Phi_\alpha(T)). \qquad \text{(B.8)}$$

Proof. Apply lemma 4 to $T - \alpha$.

COROLLARY 2.

$$S_{U/A}(\alpha) = S_{B/A}(S_{U/B}(\alpha)). \tag{B.9}$$

Proof. Follows by equating coefficients in the previous Corollary.

APPENDIX C: MINKOWSKI'S CONVEX BODY THEOREM

In this appendix we prove Minkowski's theorem, which is a useful tool in many contexts. We follow it with some results on finitely generated $\mathbb{Z}$-modules.

We start with what can be regarded as a continuous analogue of Dirichlet's Schubfachprinzip (pigeon hole principle).

THEOREM 1. (Blichfeldt). *Let* $S \subset \mathbb{R}^n$ *have volume* $V(S) > 1$. *Then there are* $\underline{a}, \underline{b} \in S$ *such that* $\underline{a} \neq \underline{b}$ *and* $\underline{a} - \underline{b} \in \mathbb{Z}^n$.

Note Volume is meant here in the sense of Lebesgue measure. We shall, however, be applying the theorem only in contexts where S is very well behaved.

Proof. Denote by $\mathcal{W}$ the unit cube

$$\mathcal{W} = \{\underline{x} = (x_1,\ldots,x_n);\quad 0 \leq x_j < 1 \quad (1 \leq j \leq n)\} \tag{C.1}$$

and let $\chi(\underline{x})$ be the characteristic function of S (= 1 if $\underline{x} \in S$ and = 0 otherwise). Then

$$V(S) = \int_{\mathbb{R}^n} \chi(\underline{x})\, d\underline{x} = \int_{\mathcal{W}} \{ \sum_{g \in \mathbb{Z}^n} \chi(x + \underline{g})\}\, d\underline{x}. \tag{C.2}$$

Since $V(S) > 1$, there is some $\underline{x}^* \in \mathcal{W}$ such that $\sum_{\underline{g}} \chi(\underline{x}^* + \underline{g}) > 1$. Hence there are $\underline{g}, \underline{g}^* \in \mathbb{Z}^n$ with $\underline{g} \neq \underline{g}^*$ such that $\underline{a} = \underline{x}^* + \underline{g}$, $\underline{b} = \underline{x}^* + \underline{g}^*$ are in S.

We shall say that a set $C \in \mathbb{R}^n$ is *symmetric* if $-\underline{a} \in C$ whenever $\underline{a} \in C$. The set C is *convex* if the line segment

$$\ell\underline{a} + (1 - \ell)\underline{b} \quad (0 \leq \ell \leq 1) \tag{C.3}$$

is in C whenever $\underline{a},\underline{b} \in C$.

THEOREM 2. (Minkowski's convex body theorem) *Let* $C \subset \mathbb{R}^n$ *be symmetric and convex, and suppose that*

$$V(C) > 2^n. \tag{C.4}$$

Then there is a $\underline{c} \neq \underline{0}$ *in* $\mathbb{Z}^n \cap C$.

Note $V(C) = \infty$ is not excluded.

Proof. Define $\frac{1}{2}C$ to be the set of $\frac{1}{2}\underline{x}$, $\underline{x} \in C$. By Blichfeldt's theorem 1, there are $\underline{a},\underline{b} \in \frac{1}{2}C$ with

$$\underline{0} \neq \underline{c} = \underline{a} - \underline{b} \in \mathbb{Z}^n .$$

Now $2\underline{a}$, $2\underline{b} \in C$ and so $-2\underline{b} \in C$ by symmetry. Hence $\underline{c} = \frac{1}{2}(2\underline{a} + 2(-\underline{b})) \in C$ by convexity.

COROLLARY 1. *Suppose, further, that* C *is compact* (*i.e., closed and bounded*). *Then the conclusion of the theorem holds if* (C.4) *is replaced by*

$$V(C) \geq 2^n. \tag{C.5}$$

Proof. For every $\varepsilon > 0$ the theorem applies to $(1 + \varepsilon)C$ and so there is a $\underline{c}^{(\varepsilon)} \neq 0$ with $\underline{c}^{(\varepsilon)} \in \mathbb{Z}^n$ such that $\underline{c}^{(\varepsilon)} \in (1 + \varepsilon)C$. Hence $(1 + \varepsilon)^{-1}\underline{c}^{(\varepsilon)} \in C$. By compactness, the $(1 + \varepsilon)^{-1}\underline{c}^{(\varepsilon)}$ have a limit point $\underline{c}$ as $\varepsilon \to 0$ and clearly $\underline{c}$ does what is required. [Alternatively, if $\varepsilon \leq 1$ we have $\underline{c}^{(\varepsilon)} \in 2C$ and so $\underline{c}^{(\varepsilon)}$ belongs to a finite set independent of ε].

COROLLARY 2. Let

$$L_j(\underline{g}) = \ell_{j1}y_1 + \ldots + \ell_{jn}y_n \quad (1 \leq j \leq n) \tag{C.6}$$

be linear forms with real coefficients, and let

$$\Delta = \pm \det(\ell_{ji}) \geq 0. \tag{C.7}$$

Let C be symmetric and convex and suppose that

$$V(C) > 2^n \Delta. \tag{C.8}$$

Then there is an $\underline{e} \neq \underline{0}$ in $\mathbb{Z}^n$ such that

$$(L_1(\underline{e}), \ldots, L_n(\underline{e})) \in C. \tag{C.9}$$

Further, (C.8) can be replaced by

$$V(C) \geq 2^n \Delta \tag{C.10}$$

provided that C is compact and $\Delta \neq 0$.

Proof. Let $\mathcal{D}$ be the set of $\underline{y} \in \mathbb{R}^n$ such that

$$(L_1(\underline{y}), \ldots, L_n(y)) \in C. \tag{C.11}$$

Clearly $\mathcal{D}$ is symmetric and convex and

$$V(\mathcal{D}) = \Delta^{-1} V(C) \tag{C.12}$$

(= ∞ if $\Delta = 0$ and $V(C) > 0$). Further, $\mathcal{D}$ is compact if C is compact and $\Delta \neq 0$. Thus the theorem and Corollary 1 apply to $\mathcal{D}$.

Before applying Minkowski's theorem, we need two lemmas.

LEMMA 1. Let G be a free $\mathbb{Z}$-module with basis $\underline{g}_1, \ldots, \underline{g}_n$ and let H be a submodule of index m. Then H is free and has a basis

$$\left.\begin{aligned} \underline{h}_1 &= b_{11}\, \underline{g}_1 \\ \underline{h}_2 &= b_{21}\, \underline{g}_1 + b_{22}\, \underline{g}_2 \\ \underline{h}_n &= b_{n1}\, \underline{g}_1 + \ldots + b_{nn}\, \underline{g}_n. \end{aligned}\right\} \tag{C.13}$$

Here $b_{ij} \in \mathbb{Z}$, $b_{jj} > 0$ and $\Pi b_{jj} = m$.

Note A free $\mathbb{Z}$-module is, of course, an alias for a free abelian group.

Proof. For $1 \leq J \leq n$ denote by $H(J)$ the set of elements of H which are of the form

$$\underline{h} = a_1\underline{g}_1 + \ldots + a_J\underline{g}_J \quad (a_1,\ldots,a_J \in \mathbb{Z}) . \tag{C.14}$$

There are elements of H of the form (C.14) with $a_J > 0$, since $mg_J \in H$. Let $\underline{h}_J$ be any element of $H(J)$ for which a_J is positive and minimal. Then for any $\underline{h} \in H(J)$ there is a unique $c \in \mathbb{Z}$ such that

$$\underline{h} - c\underline{h}_J \in H(J - 1) \tag{C.15}$$

[$\underline{h} = c\underline{h}_1$ if $J = 1$].

Now, for any $\underline{h} \in H = H(n)$, there are uniquely determined integers $c_n, c_{n-1}, \ldots, c_1$ such that

$$\underline{h} - c_n\underline{h}_n - \ldots - c_{J+1}\, h_{J+1} \in H(J) \tag{C.16}$$

for $J = n - 1,\ldots, 0$. Hence $\underline{h}_1,\ldots,\underline{h}_n$ is a basis for H. By (C.14) the $\underline{h}_j$ are of the shape (C.13). Finally, by (C.13) the index of H in G is Πb_{jj}, so this must equal m.

As a consequence, we have the following generalization of Minkowski's Convex Body Theorem 2.

COROLLARY. Let H be a subgroup of $\mathbb{Z}^n$ of index m and let $C \subset \mathbb{R}^n$ have volume

$$V(C) > m2^n.$$

Then there is a $\mathbf{c} \neq 0$ in $H \cap C$.

Proof. In Lemma 1 we can take for $\underline{g}_j$ the unit vector $(0,0,-0,1,0,\ldots 0)$ $(1 \leq j < n)$. Then

$$\sum y_j \underline{h}_j = (L_1(\underline{y}), L_2(\underline{y}), \ldots, L_n(\underline{y})),$$

where

$$L_i(\underline{y}) = \sum_j b_{ji} y_i$$

($b_{ji} = 0$ if $i < j$). Hence Corollary 2 to Theorem 2 applies.

LEMMA 2. *Let G be a free $\mathbb{Z}$-module with basis $\underline{g}_1, \ldots, \underline{g}_n$ and let F be a torsion-free $\mathbb{Z}$-module. Suppose that G is a submodule of F of index m. Then F is free. It has a basis $\underline{f}_1, \ldots, \underline{f}_n$ such that*

$$\left.\begin{aligned} \underline{g}_1 &= c_{11}\underline{f}_1 \\ \underline{g}_2 &= c_{21}\underline{f}_1 + c_{22}\underline{f}_2 \\ &\cdots \\ \underline{g}_n &= c_{n1}\underline{f}_1 + \cdots \qquad + c_{nn}\underline{f}_n \end{aligned}\right\}, \tag{C.17}$$

where $c_{ij} \in \mathbb{Z}$, $c_{jj} > 0$ and $\Pi c_{jj} = m$.

Note "Torsion-free" means that no element $\neq \underline{0}$ is of finite order.

Proof. We have H (say) $= mF \subset G$ and H is of finite index in G since $mG \subset H$. By Lemma 1 $H = mF$ has a basis $\underline{h}_j = m\underline{f}_j$ $(1 \leq j \leq n)$ of the shape (C.13) for some $b_{ij} \in \mathbb{Z}$ with $b_{jj} > 0$. On solving for $\underline{g}_1, \underline{g}_2, \ldots, \underline{g}_n$ in order, we get (C.17), where the $\underline{c}_{ij} \in \mathbb{Z}$ since the $\underline{g}_j \in F$. Finally, Πc_{jj} is the index m, as before.

THEOREM 3. *A discrete $\mathbb{Z}$-module $G \subset \mathbb{R}^n$ is free on at most n generators. Further, any elements of G which are $\mathbb{Z}$-independent are $\mathbb{R}$-independent.*

Proof. We may suppose without loss of generality that G spans $\mathbb{R}^n$ as an $\mathbb{R}$-module, and will show that G is free on n generators.

For $\underline{x} = (x_1, \ldots, x_n) \in \mathbb{R}^n$, we write

$$||\underline{x}|| = \max_j |x_j| , \tag{C.18}$$

where $|\ |$ is the absolute value. By hypothesis, there is a $\delta > 0$ such that

$$\underline{x} \in G, \ ||\underline{x}|| < \delta \Rightarrow \underline{x} = \underline{0}. \tag{C.19}$$

Since G spans $\mathbb{R}^n$, there are $\mathbb{R}$-independent $\underline{g}_1^{(o)}, \ldots, \underline{g}_n^{(o)} \in G$. If they generate G as a $\mathbb{Z}$-module, we are done. Otherwise, there is some $\underline{g}^* \in G$ which is not a $\mathbb{Z}$-combination of the $\underline{g}_j^{(o)}$. By Theorem 1, Corollary 2 (with $n+1$ for n and $\Delta = 0$), there are $e_1, \ldots, e_n, e^* \in \mathbb{Z}$, not all zero, such that

$$||e_1\underline{g}_1^{(o)} + \ldots + e_n\underline{g}_n^{(o)} + e^*\underline{g}^*|| < \delta.$$

By (C.19) we have

$$e_1\underline{g}_1^{(o)} + \ldots + e_n\underline{g}_n^{(o)} + e^*\underline{g}^* = \underline{0}.$$

Clearly $e^* \neq 0$. Hence the module $M^{(1)}$ generated by the $\underline{g}_j^{(o)}$ together with $\underline{g}^*$ contains as a submodule of finite index m (say) > 1 the module $M^{(o)}$ generated by the $\underline{g}_j^{(o)}$ alone. By Lemma 2 there is a basis $\underline{g}_j^{(1)}$ $(1 \leq j \leq n)$ of $M^{(1)}$, and

$$\left|\det (\underline{g}_1^{(1)}, \ldots, \underline{g}_n^{(1)})\right| = m^{-1}\left|\det (\underline{g}_1^{(o)}, \ldots, \underline{g}_n^{(o)})\right|.$$

If $M^{(1)} = G$, we are done. If not, we repeat the process. And so on. Either we find a $\mathbb{Z}$-basis of G or we ultimately find $\mathbb{R}$-independent $\underline{g}_1^*, \ldots, \underline{g}_n^* \in G$ such that

$$\left|\det (\underline{g}_1^*, \ldots, \underline{g}_n^*)\right| < \delta^n.$$

By Theorem 2 Corollary 2 with $\Delta = \delta^n$, there are $e_1, \ldots, e_n \in \mathbb{Z}$, not all 0, such that

$$||e_1\underline{g}_1^* + \ldots + e_n\underline{g}_n^*|| < \delta.$$

This is a contradiction to (C.19).

APPENDIX D: SOLUTION OF EQUATIONS IN FINITE FIELDS

In this appendix ρ is the finite field with $q = p^f$ elements, where p is the characteristic. We recall that the multiplicative group of ρ is cyclic, that

$$a^{q-1} = \begin{cases} 0 & \text{if} \quad a = 0 \\ 1 & \text{otherwise} \end{cases} \tag{D.1}$$

and that

$$\sum_{a\in\rho} a^m = 0 \qquad (0 \leq m < q - 1). \tag{D.2}$$

At several places in the book, we have required the existence of a solution $\underline{a} = (a_1,\dots,a_n)$ of an equation

$$F(\underline{a}) = 0,$$

where

$$F(X_1,\dots,X_n) \in \rho[X\ ,\dots,X_n].$$

What has been required is a <u>nonsingular</u> solution; that is one for which at least one of the partial derivatives $\partial F/\partial X_j$ does not vanish at $\underline{a}$. Then Hensel's Lemma gives a zero of the polynomial in the local field from which F was obtained by going to the residue class field.

There is now a deep and extensive theory which deals with algebraic varieties over finite fields, and in particular with the number of points on them. In this appendix we shall first discuss some elementary techniques and then state without proof some deeper results.

THEOREM D.1 (Warning (1935)). *Let* $F(\underline{X}) \in \rho[X]$ *have (total) degree* d *in* $n > d$ *variables. Then the number of* $\underline{a}$ *for which* $F(\underline{a}) = 0$ *is divisible by* p.

Proof. It is enough to show that the number of $\underline{a}$ for which $F(\underline{a})$ does *not* vanish is divisible by p. By (D.1) this number modulo p is

$$\sum_{a_1,\ldots,a_n} F(\underline{a})^{q-1} \tag{D.3}$$

On expanding $F(\underline{X})^{q-1}$, we have to evaluate sums of the type

$$\sum_{a_1,\ldots,a_n} a_1^{m(1)} a_2^{m(2)} \ldots a_n^{m(n)} ,$$

where

$$m(1) + \ldots + m(n) \leq (q-1)d < (q-1)n.$$

Hence there is some j for which $m(j) < q - 1$. Now

$$\sum_{a_j} a_j^{m(j)} = 0$$

by (D.2). It follows that (D.3) is 0, as required.

COROLLARY. (Chevalley) *Suppose, further, that* $F(X)$ *is homogeneous. Then* $F(\underline{a}) = 0$ *for some* $\underline{a} \neq \underline{0}$.

Proof. $F(\underline{0}) = 0$ and so by the theorem there are at least $p - 1$ vectors $\underline{a} \neq \underline{0}$ for which $F(\underline{a}) = 0$.

A particular case is when $F(\underline{x})$ is a ternary quadratic. Note that the corollary does not assert that the $\underline{a}$ obtained is nonsingular. When the quadratic itself is nonsingular, any $\underline{a} \neq \underline{0}$ is nonsingular. For $p \neq 2$ this is clear on considering the diagonalized form

$$F(\underline{X}) = b_1 X_1^2 + b_2 X_2^2 + b_3 X_3^2 \tag{D.4}$$

when $b_1 b_2 b_3 \neq 0$. If, however, $b_3 = 0$, there is no reason why all the $\underline{a}$ provided by the lemma should not have $a_1 = a_2 = 0$. For example, this is the case when $p = 3$ and $F(X_1,X_2,X_3) = X_1^2 + X_2^2$. Such $\underline{a}$ are, of course, entirely useless as a starting point for the application of Hensel's Lemma.

Another useful technique is that of finite Fourier series. Let $e_1,\ldots,e_f$ be a basis for $\rho/\mathbb{F}_p$, where $\mathbb{F}_p$ is the field of p elements. The additive characters of ρ (with values in $\mathbb{C}$) are

$$\psi(a) = \psi(\sum \alpha_j e_j) = \exp(2\pi i \sum_j \ell_j \alpha_j / p) \tag{D.5}$$

where by abuse of language we have identified $\alpha_j \in \mathbb{F}_p$ with an element of $\mathbb{Z}$ mod p. The ℓ_j are elements of $\mathbb{Z}$ mod p which specify ψ. The trivial additive character ψ_o is that which is identically 1. There are $p^f = q$ additive characters in all.

We denote by χ a character of the multiplicative group ρ^* of non-zero elements of ρ (which, as already mentioned, is cyclic of order $q - 1$). The trivial multiplicative group character χ_o is that which is identically 1. We define

$$\chi(0) = 0 \tag{D.6}$$

(even for χ_o).

We write

$$\tau(\chi,\psi) = \sum_{a\in\rho} \chi(a)\psi(a). \tag{D.7}$$

These are the (generalized) <u>Gauss sums</u>: essentially they give the coefficients of the Fourier expansion of each set of characters in terms of the other.

<u>LEMMA D.1</u>. <u>Suppose that</u> χ,ψ <u>are not the trivial characters. Then</u>

$$|\tau(\chi,\psi)| = q^{\frac{1}{2}}, \tag{D.8}$$

<u>where</u> $|\ |$ <u>is the absolute value</u>.

<u>Proof</u>.

$$|\tau(\chi,\psi)|^2 = \tau(\chi,\psi)\tau(\bar{\chi},\bar{\psi})$$

$$= \sum_{a,b} \chi(a)\bar{\chi}(b)\psi(a)\bar{\psi}(b), \tag{D.9}$$

where we need sum only over $a \neq 0$, $b \neq 0$ by (D.6). Put $c = b^{-1}a$. Then

$$\chi(a)\bar{\chi}(b) = \chi(ab^{-1}) = \chi(c)$$

and

$$\psi(a)\bar{\psi}(b) = \psi(a - b) = \psi\{b(c - 1)\}.$$

Hence (D.9) is

$$\sum_{c \neq o} \chi(c) \sum_{b \neq o} \psi\{b(c - 1)\}.$$

The inner sum is $q - 1$ when $c = 1$ and -1 otherwise. Hence, finally, we obtain

$$q\chi(1) - \sum_{c \neq o} \chi(c) = q,$$

as required.

<u>COROLLARY</u>. <u>Let</u> $\ell|(q - 1)$ <u>and let</u> $b \in \rho$, $b \neq 0$. <u>Then</u>

$$\left|\sum_{\ell} \psi(be^{\ell})\right| \leq (\ell - 1)q^{\frac{1}{2}},$$

<u>for a non-trivial additive character</u> ψ.

<u>Proof</u>. For $c \in \rho$, $c \neq 0$ we have

$$\sum_{\chi^{\ell}=\chi_o} \bar{\chi}(b)\chi(c) = \begin{cases} \ell \text{ if } c = be^{\ell} \text{ for some } e \in \rho \\ 0 \text{ otherwise} \end{cases} \tag{D.10}$$

Further, if $c = be^{\ell}$ has one solution e, then it has precisely ℓ solutions. Hence

$$\sum_{\chi^{\ell}=\chi_o} \bar{\chi}(b)\tau(\chi,\psi) = \sum_{e\neq o} \psi(be^{\ell}). \tag{D.11}$$

Clearly $\tau(\chi_o,\psi) = -1$, and so we have the required estimate on adding 1 to both sides and using the Lemma.

As an example, consider the number of solutions $x,y,z \in \rho$ of

$$ax^3 + by^3 + cz^3 = 0, \tag{D.12}$$

where $a,b,c \neq 0$. Suppose, first, that $3 \nmid (q - 1)$. Then $x \to x^3$ gives a bijection of ρ with itself, so the number of solutions is the same as that of

$$au + bv + cw = 0,$$

which is easy. Now, suppose that $3|(q - 1)$ and let N be the number of solutions. Then

$$qN = \sum_{x,y,z} \sum_{\psi} \psi(ax^3 + by^3 + cz^3).$$

The terms with $\psi = \psi_o$ give a contribution of q^3 to the right hand side. Hence

$$qN - q^3 = \sum_{\psi\neq\psi_o} \sum_{x,y,z} \psi(ax^3 + by^3 + cz^3)$$

$$= \sum_{\psi\neq\psi_o} \{\sum_{x} \psi(ax^3)\} \{\sum_{y} \psi(by^3)\} \{\sum_{z} \psi(cz^3)\}.$$

Estimating the sums by the Corollary we have

$$|qN - q^3| \leq (q - 1) (2q^{\frac{1}{2}})^3,$$

and so

$$|N - q^2| < 8q^{3/2}. \qquad (D.13)$$

We should have done better, in fact, to estimate the number N_1 of solutions of (D.12) with $z = 1$. A similar argument gives

$$|N_1 - q| < 4q^{1/2}, \qquad (D.14)$$

and so $N_1 > 0$ as soon as $q \geq 16$. A more careful argument confirms for (D.12) the "Riemann hypothesis for function fields" estimate, to which we now return.

We now turn to results which we shall not prove. Let

$$F(X_1,X_2,X_3) \in \rho[X_1,X_2,X_3] \qquad (D.15)$$

be homogeneous and suppose that it is irreducible over the algebraic closure of ρ. Then (D.15) can be regarded as the equation of a curve C in the projective plane. As in the characteristic 0 case, a curve has a genus, which is a non-negative integer. We have

THEOREM A. The number N of points defined over ρ on a curve C of genus g satisfies

$$|N - q| \lesssim 2gq^{1/2}.$$

Note 1 The proof of this theorem is associated with an analogue of the classical Riemann Hypothesis, and so Theorem A is often referred to as the Riemann Hypothesis for function fields (but this "hypothesis" has been proved!).

Note 2 Here "point" means place of the function field of the curve. The theorem applies more generally to curves in projective space of any dimension.

As a first application of Theorem A consider

$$F(X_1,X_2,X_3) = a_1X_1^3 + a_2X_2^3 + a_3X_3^3.$$

If $p \neq 3$, the curve $F = 0$ has genus 1, and so the number N_0 of points satisfies

$$|N_o - q| \leq 2q^{1/2}.$$

In particular $N_o > 0$ whenever $q \geq 5$. Hence under this condition there is a nontrivial solution of $F = 0$.

As a second application, consider

$$F(X_1,X_2,X_3) = a_1X_1^4 + a_2X_2^4 + a_3X_3^4.$$

If $p \neq 2$, the curve $F = 0$ has genus 3, and so the number N of points defined over ρ satisfies

$$|N - q| \leq 6q^{1/2}.$$

In particular, $N \geq 1$ provided that $q \geq 37$; and so this is a sufficient condition for a nontrivial solution of $F = 0$.

Theorem A also applies to curves such as

$$Y^2Z^2 = X^4 + aX^3Z + bX^2Z^2 + cXZ^3 + dZ^4,$$

which cannot be dealt with by the methods introduced earlier. They have genus 1 in general.

Finally, we enunciate

THEOREM B. *There are constants* $A(n,d,r)$ *depending only on* n,d,r *with the following property:*

Let V *be a variety defined over* ρ *of dimension* r *and degree* d *in* n*-dimensional projective space. Then the number* N *of points on* V *satisfies*

$$|N - q^r| \leq (d-1)(d-2)\, q^{r-\frac{1}{2}} + A(n,d,r)\, q^{r-1}.$$

Note Here points of V means a solution of the defining equations.

Notes

For more on the method of finite Fourier series and its relation to the general theory, see Weil (1949).

The "Riemann Hypothesis for function fields" (of curves) was proved by Hasse in genus 1 and by Weil in general. For a long time it seemed that no "elementary" proof existed, but one was found by Stepanov; it was however complicated and intransparent. Perhaps the most attractive "elementary" treatment is that of Bombieri (1972-3).

Theorem B is due to Lang and Weil (1954). The appropriate generalization of the "Riemann Hypothesis" to varieties of higher dimension was conjectured by Weil and proved by Deligne.

For an account of elementary methods, see Schmidt (1976).

Exercises

1. In the notation of Theorem D.1, let $G(\underline{X}) \in \rho[\underline{X}]$ be a further polynomial whose (total) degree is $e < (q-1)(n-d)$. Show that

$$\sum_{F(\underline{a})=0} G(\underline{a}) = 0.$$

2. Let $F_j(\underline{X})$ be homogeneous of degree d_j $(1 \le j \le J)$, where $\underline{X} = (X_1,\ldots,X_n)$. Show that the F_j have a common non-trivial zero in ρ provided that $\Sigma\, d_j < n$.

[Hint Consider

$$\sum_{\underline{a}} \prod_j \{1 - (F_j(\underline{a}))^{q-1}\}.$$
]

3. By the method of finite Fourier series, show that the number N of solutions $x,y \in \rho$ of

$$y^2 = ax^4 + b \qquad (a,b \in \rho)$$

satisfies

$$|N - q| \le Cq^{1/2},$$

where C is independent of a,b and q.

4. Let $p \equiv 1\ (3)$. Use the result of Chapter 9, Ex. 11 together with those of this Appendix to obtain a formula for the number of solutions of (D.12). More precisely, justify the following statements:

Let

$$4p = \ell^2 + 27m^2$$

where the sign of ℓ is given by $\ell \equiv 1\ (3)$ and that of m is arbitrary (but fixed). Put

$$\pi = \tfrac{1}{2}(\ell + 3m\sqrt{-3}).$$

Let χ be the character of order 3 taking values

$$1,\ \tfrac{1}{2}(-1 \pm \sqrt{-3})$$

and defined for $u \not\equiv 0\ (p)$ by

$$\chi(u) \equiv u^{(p-1)/3} \quad (\text{mod } \pi).$$

Let $a,b,c \in \mathbb{Z}$ prime to p and let N be the number of solutions of

$$ax^3 + by^3 + cz^3 \equiv 0 \quad (p),$$

considered as a homogeneous equation. Then

$$N = p + 1 + \bar{\chi}(d)\pi + \chi(d)\bar{\pi},$$

where $d = abc$. More explicitly,

$$N = \begin{cases} p + 1 + \ell & \text{if } d^{(p-1)/3} \equiv 1 \quad (p) \\ p + 1 + \tfrac{1}{2}(-\ell + 9m) & \text{if } \ell + 3m + 6md^{(p-1)/3} \equiv 0\ (p) \\ p + 1 + \tfrac{1}{2}(-\ell - 9m) & \text{otherwise.} \end{cases}$$

5. As in the previous exercise, but making use of Chapter 9, Ex 12, find explicit form value when $p \equiv 1\ (4)$ for the number of solutions of

$$y^2 \equiv ax^4 + b \quad (p)$$

and

$$y^4 \equiv ax^4 + b \quad (p)$$

for given a,b (which are curves over $\mathbb{F}_p$ of genus 1,3 respectively).

APPENDIX E: ZETA AND L-FUNCTIONS AT NEGATIVE INTEGERS

In this appendix we obtain the values of certain zeta and L-functions at negative integral values of the argument. They are obtained by analytic continuation of functions of a complex variable s which are defined for $\mathrm{Re}(s) > 1$ by certain series.

We work in the complex t-plane cut along the negative real axis. The argument θ of t is 0 on the positive real axis, so

$$\log t = \log r + i\theta,$$

where $r > 0$ is real and

$$-\pi < \theta < \pi.$$

By definition,

$$t^s = \exp(s \log t)$$

for all complex s.

Let C be a contour which comes from infinity along the bottom of the cut, passes round the origin, and then goes off to infinity along the top of the cut:

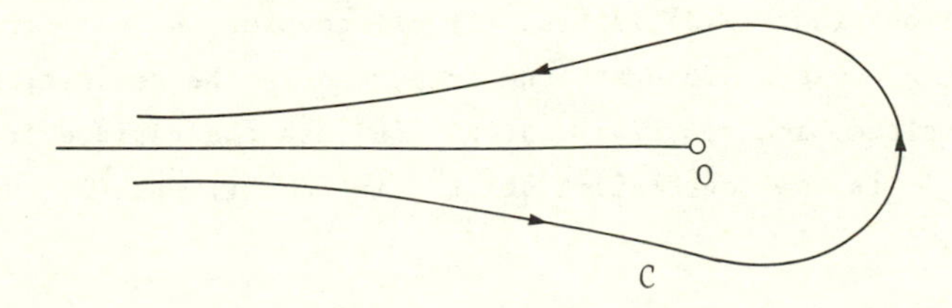

For any complex s the integral

$$I(s) = \int_C t^{s-1} e^t \, dt$$

is well-defined. For $\mathrm{Re}(s) > 1$ we have

$$I(s) = 2i \sin(s\pi) \int_0^\infty r^{s-1} e^{-r} \, dr$$

$$= 2i \sin(s\pi) \, \Gamma(s), \qquad \text{(E.1)}$$

as is readily checked by deforming the contour so that it runs right along the slit. Hence (E.1) holds for all complex s.

Now let z be any complex number with $|z| \lesssim 1$. Then for $\mathrm{Re}(s) > 1$ we have

$$J(s) = \int_C \frac{t^{s-1} z e^t}{1 - ze^t} dt = \int_C t^{s-1} \sum_{n=1}^{\infty} z^n e^{nt} \, dt$$

$$= \sum_n z^n \int_C t^{s-1} e^{nt} \, dt$$

$$= \left(\sum_{n=1}^{\infty} z^n n^{-s} \right) I(s). \qquad \text{(E.2)}$$

Here the sum defines the "twisted" zeta-function $\zeta(s,z)$. Hence by (E.1)

$$\zeta(s,z) = \frac{\Gamma(s) \, J(s)}{2i \sin(s\pi)}$$

$$= \Gamma(1 - s) \, J(s)/2\pi i, \qquad \text{(E.3)}$$

on using the gamma function identity

$$\Gamma(s)\Gamma(1 - s) = \pi/\sin(\pi s). \qquad \text{(E.4)}$$

It follows that (E.3) is true for all complex s. If $s = -k$, where $k \gtrsim 0$ is an integer, the integrand in the definition of $J(-k)$ is single-valued, and so $(2\pi i)^{-1} J(-k)$ is just the residue at the origin: that is, it is the coefficient of t^k in $ze^t/(1 - ze^t)$. Hence and by (E.3)

$$ze^t/(1 - ze^t) = \sum_{k=0}^{\infty} \zeta(-k,z) \, t^k/k! \qquad (z \neq 1). \qquad \text{(E.5)}$$

This is the required formula. For $z = 1$ we must modify (E.5) to take account of the pole of $e^t/(1 - e^t)$ at $t = 0$ but the coefficients of t^k in (E.5) on both sides remain the same for $k \geq 1$.

We also require a similar formula for the twisted L-function

$$L(s,\chi,z) = \sum_{n=1}^{\infty} \chi(n)\, z^n\, n^{-s}, \tag{E.6}$$

where χ has period d. On working as above with

$$K(s) = \int_C t^{s-1} \{ \sum_{a=1}^{n} \chi(a)\, z^a\, e^{at}\} \{1 - z^d\, e^{dt}\}\, dt$$

one readily obtains

$$\frac{\sum_{a=1}^{d} \chi(a)\, z^a\, e^{at}}{1 - z^d\, e^{dt}} = \sum_{k=o}^{\infty} L(-k,\chi,z)\, t^k/k! \tag{E.7}$$

APPENDIX F: CALCULATION OF EXPONENTIALS

In several places, e.g., Chapter 4 §6, we have to compute expressions $b^c \pmod{n}$, where b, c, n are positive integers and c is fairly large. At first sight this requires c multiplications mod n but a well-known algorithm shows that $O(\log c)$ multiplications suffice.

We have

$$c = c_0 + 2c_1 + \dots + 2^s c_s$$

for some s, where each c_j is 0 or 1. Now compute $b_0, b_1, \dots, b_s$ by

$$b_0 = b; \quad b_{j+1} \equiv b_j^2 \pmod{n}.$$

Clearly

$$b^c \equiv \prod_{c_j=1} b_j \pmod{n}.$$

A convenient way of arranging the calculation is as follows:

1. Set $X := 1$, $Y := b$, $Z := c$.

2. If $Z = 0$, stop.

3. Let $Z = 2T + E$, $E = 0$ or 1.

4. If $E = 0$, go to 6.

5. Set $X :\equiv XY \pmod{n}$.

6. Set $Y :\equiv Y^2 \pmod{n}$.

7. Set $Z := T$

8. Go to 2.

The final value of X is congruent to b^c mod n.

REFERENCES

Amice, Y. (1975). Les nombres p-adiques. Collection SUP. Presses universitaires de France.

Arhipov, G.I. & Karacuba, A.A. (1981). Local representation of zero by a form (in Russian). Izv. akad. nauk SSSR ser. mat. 45, 948-961. (American translation: Math. USSR izvestia 19 (1982), 231-240).

Arhipov, G.I. & Karacuba, A.A. (1982). A problem of comparison theory (in Russian). Uspehi mat. nauk 37.5, 161-162. (English translation: Russian math. surveys 37 (1982), 157-158.)

Artin, E. (1951). Algebraic numbers and algebraic functions. Duplicated lecture notes, Princeton University. Reprinted by Gordon and Breach, New York, 1967.

Artin, E. (1959). Theory of algebraic numbers. Striker, Göttingen.

Artin, E. & Tate, J. (1968). Class field theory. Duplicated lecture notes, Institute for Advanced Study, Princeton (n.d.). Reprinted by W.A. Benjamin, New York, 1968.

Ax, J. & Kochen, S. (1965). Diophantine problems over local fields, I. Amer. J. math. 87, 605-630.

Bachmann, P. (1872). Die Lehre von der Kreistheilung. Teubner, Leipzig.

Bachmann, P. (1921). Grundlehren der neueren Zahlentheorie (2-te Auflage). de Gruyter, Berlin and Leipzig.

Baker, A. (1975). Transcendental number theory. Cambridge University Press. Second edn., 1979.

Baker, A. & Davenport, H. (1969). The equations $3x^2 - 2 = y^2$ and $8x^2 - 7 = z^2$. Quart. J. math. Oxford 20, 129-137. (= Davenport, Collected papers IV, 1748-1756).

Baumgart, O. (1885). Über das quadratische Reprocitätsgesetz: eine vergleichende Darstellung. Teubner, Leipzig.

Bertrandias, F. (1963-64). Diamètre transfini dans un corps valué, application au prolongement analytique. Sém. Delange-Pisot, Nr. 3.

Beukers, F. (1980). The multiplicity of binary recurrences. Comp. math. 40, 251-267.

Birch, B.J. (1962). Forms in many variables, Proc. roy. soc. A265, 245-263.

Bombieri, E. (1972/3). Counting points on curves over finite fields (d'après S.A. Stepanov). Sém. Bourbaki, 25e annee, exposé 430. Springer lecture notes in math. 383 (1974), 234-241.

Borevich, Z. & Shafarevich, I. (1966). Number theory (translated by Newcomb Greenleaf). Academic Press, New York, London.

Bosch, S, Günter, U. & Remmert, R. (1984). Non-archimedean analysis. Grundlehren d. math. Wiss. 261. Springer, Berlin &c.

Bourbaki, N. (1972). Éléments de mathematique, Fasc. 37. Groupes et algébres de Lie, Chaps. II, III. Hermann, Paris.
Bremner, A., Lewis, D.J. & Morton, P. (1984). Some varieties with points only in a field extension. Arch. math. 43, 344-350.
Brownawell, W.D. (1984). On p-adic zeros of forms. J. number theory 18, 342-349.
Burnside, W. (1911). Theory of groups of finite order. Cambridge. Second edition.
Cassels, J.W.S. (1959). Arithmetic on curves of genus 1. I. on a conjecture of Selmer. J. reine angew. Math. 202, 52-99.
Cassels, J.W.S. (1962). Arithmetic on curves of genus 1. IV. Proof of the Hauptvermutung. J. reine angew. Math. 211, 95-112.
Cassels, J.W.S. (1966). Diophantine equations with special reference to elliptic curves. J. London math. soc. 41, 193-291 and 42(1967), 183.
Cassels, J.W.S. (1976). An embedding theorem for fields. Bull. Austral. math. soc. 14, 193-198 & 479-480.
Cassels, J.W.S. (1978). Rational quadratic forms. Academic Press, London.
Cassels, J.W.S. (1985). The arithmetic of certain quartic curves. Proc. roy. soc. Edinburgh 100A, 201-218.
Cassels, J.W.S. & Fröhlich, A. eds. (1967). Algebraic number theory ("The Brighton book"). Academic Press, London, 1967.
Chevalley, C. (1953-4). Class field theory. Nagoya University.
Clausen, Th. (1840). (untitled notice). Astronomische Nachrichten 17, 351.
Delaunay (= Delone), B. (1928). Vollständige Lösung der unbestimmten Gleichung $X^3q + Y^3 = 1$. Math. Zeit. 28, 1-9.
Denef, J. (1984). The rationality of the Poincaré series associated with the p-adic points of a variety. Invent. math. 77, 1-23.
Deuring, M. (1935). Algebren. Ergeb. d. Math. 4.1. Springer, Berlin.
Dwork, B. (1960). On the rationality of the zeta function of an algebraic variety. Amer. J. math. 82, 631-648.
Dwork, B. (1982). Lectures on p-adic differential equations. Springer, New York &c.
Eisenstein, G. (1975). Mathematische Werke (2 vols.). Chelsea, New York.
Ellison, F., Ellison, W.J., Pesek, J., Stall, C.E. & Stall, D.S. (1972). The diophantine equation $y^2 + k = x^3$. J. number theory 4, 107-117.
Evertse, J.H. (1983). On the representation of integers by binary cubic forms of positive discriminant. Invent. math. 73 (1983), 117-138.
Fröhlich, A. (1960). Discriminants of algebraic number fields. Math. Zeit. 74, 18-28.
Fröhlich, A., ed. (1977). Algebraic number fields (L-functions and galois properties). Academic Press, London.
Fujiwara, M. (1973). Hasse principle in algebraic equations. Acta arith. 22, 267-276.
Gross, B. & Koblitz, N. (1979). Gauss sums and the p-adic gamma function. Ann. math. 109, 569-581.
Hardy, G.H. & Wright, E.M. (1938). An introduction to the theory of numbers. Oxford University Press. 5th ed., 1979.

Hasse, H. (1926). Bericht über neuere Untersuchungen und Probleme aus der Theorie der algebraischen Zahlkörper. ("Klassenkörperbericht"). Jber. deutsch. Math.-Verein. 35, 1-55: 36 (1927), 233-311: Ergänzungsband 6 (1930), 1-204. Reprints: Teubner, Leipzig, 1930 and Physica-Verlag, Würzburg, Wien, 1965.

Hasse, H. (1931). Beweis eines Satzes und Widerlegung einer Vermutung über das allgemeine Normenrestsymbol. Nachr. Gesell. Wiss. Göttingen, math. phys. Kl. 1931, 64-69 (= Math. Abh. I, 155-160).

Hasse, H. (1949). Zahlentheorie ("Das blaue Hasse"). Akademie-Verlag, Berlin. 2nd ed., 1963.

Hasse, H. (1975). Mathematische Abhandlungen (3 vols.). de Gruyter, Berlin.

Hensel, K. (1902). Über die Entwicklung der algebraischen Zahlen in Potenzreihen. Math. Ann. 55, 301-336.

Hensel, K. (1904). Neue Grundlagen der Arithmetik. J. reine angew. Math. 127, 51-84.

Hensel, K. (1905). Über eine neue Begründung der Theorie der algebraischen Zahlen. J. reine angew. Math. 128, 1-32.

Hensel, K. (1908). Theorie der algebraischen Zahlen. Teubner, Leipzig und Berlin.

Hensel, K. (1913). Zahlentheorie. Göschen, Berlin und Leipzig.

Hensel, K. & Landsberg, G. (1902). Theorie der algebraischen Funktionen einer Variabeln und ihre Anwendung auf algebraische Kurven und abelsche Integrale. Teubner, Leipzig.

Hilbert, D. (1897). Die Theorie der algebraischen Zahlkörper ("Zahlbericht"). Jber. deutsch. Math.-Verein, 4, 175-546 (= Ges. Abh. I, 63-363).

Iwasawa, K. (1972). Lectures on p-adic L-functions. Annals of math. studies, No. 74., Princeton University Press.

Jehne, W. (1979). On knots in algebraic number theory. J. reine angew. Math. 311/312, 215-254.

Koblitz, N. (1977). p-adic numbers, p-adic analysis and zeta-functions. Springer, New York &c. 2nd ed., 1984.

Koblitz, N. (1980). p-adic analysis: a short course on recent work. LMS lecture notes, 46. Cambridge University Press.

Krasner, M. (1974). Rapport sur le prolongement analytique... . Bull. soc. math. France, Mémoire 39-40, 131-254.

Kubota, T. & Leopoldt, H.W. (1964). Eine p-adische Theorie der Zetawerte: I. Einführung der p-adischen Dirichletschen L-Funktionen. J. reine angew. Math. 214/215, 328-339.

Kürschák, J. (1913). Über Limesbildung und allgemeine Körpertheorie. J. reine angew. Math. 142, 211-263.

Landau, E. (1927). Vorlesungen über Zahlentheorie (3 vols.). Hirzel, Leipzig.

Lang, S. (1978). Cyclotomic fields (2 vols.). Springer, New York &c. Vol I, 1978; Vol II, 1980.

Lang, S. & Weil. A. (1954). Number of points of varieties in finite fields. Bull. Amer. math. soc. 76, 819-827. (= Weil, Collected papers II, 165-173).

Lech, C. (1954). A note on recurring series. Ark. mat. 2, 417-421.

Lenstra, A.K., Lenstra, H.W., Jr. & Lovasz, L. (1982). Factoring polynomials with rational coefficients. Math. Ann. 261, 515-534.

Lenstra, H.W. Jr. & Tijdeman, R. (1982). Computational methods in number theory (2 vols.). Mathematical centre tracts 154 & 155. Mathematisch Centrum, Amsterdam.

Leveque, W.J. (1956). Topics in number theory (2 vols). Addison Wesley.

Lewis, D.J. & Montgomery, H.L. (1983). On zeros of p-adic forms. Mich. math. J. 30, 83-87.

Ljunggren, W. (1942). Zur Theorie der Gleichung $x^2 + 1 = Dy^4$. Avh. norske videnskaps-akad. Oslo I. mat.-naturv. kl. No 5 (27 pp.)

Ljunggren, W. (1946). Solution complète de quelques équations du sixième degré à deux indeterminées. Arch. math. naturvid. 48, 177-211.

Ljunggren, W. (1971). On the representation of integers by certain binary cubic and biquadratic forms. Acta arith. 17, 379-387.

Loomis, L.H. (1953). An introduction to abstract harmonic analysis. van Nostrand, Toronto.

Loxton, J.H. & van der Poorten, A.J. (1977). On the growth of recurrent sequences. Math. proc. Cambridge philos. soc. 81, 369-376.

Lutz, E. (1955). Sur les approximations diophantiennes linéaires P-adiques. Actualités sci. ind. 1224. Hermann, Paris.

Mahler, K. (1935). Eine arithmetische Eigenschaft der Taylor-Koeffizienten rationaler Funktionen. Proceedings k. akad. wet. Amsterdam, 38, 50-60.

Mahler, K. (1973). p-adic numbers and their functions. Cambridge tracts, 64. Cambridge University Press. Second edn: Cambridge tracts, 76, 1984.

Matthews, C.R. (1979). Gauss sums and elliptic functions I. The Kummer sum. Invent. math. 52, 163-185.

Matthews, C.R. (1979a). Gauss sums and elliptic functions II. The quartic sum. Invent. math. 54, 23-52.

Mignotte, M. (1973/4). Suites récurrentes linéaires. Sém. Delange-Pisot-Poitou 15e année No 14 (9 pp.)

Mordell, L.J. (1961). The congruence $(p-1/2)! \equiv \pm 1 \pmod p$. American math. monthly 68, 145-146.

Mordell, L.J. (1969). Diophantine equations. Academic Press, London &c.

Nagell, T. (1925). Über einige kubische Gleichungen mit zwei Unbestimmten. Math. Zeit. 24, 422-447.

Nagell, T. (1928). Darstellung ganzer Zahlen durch binäre kubische Formen mit negativer Diskriminante. Math. Zeit 28, 10-29.

Nagell, T. (1960). The diophantine equation $x^2 + 7 = 2^n$. Arkiv för matematik 4, 185-187.

Narkiewicz, (1974). Elementary and analytic theory of algebraic numbers. Monografie matematyczne No. 57. PWN, Warsaw.

Neukirch, J. (1969). Klassenkörpertheorie. Bibliographisches Institut, Mannheim.

O'Meara, O.T. (1963). Introduction to quadratic forms. Grundlehren d. math. Wiss. 117. Springer, Berlin &c.

Ostrowski, A. (1918). Über einige Lösungen der Funktionalgleichung $\phi(x)\phi(y) = \phi(xy)$. Acta math. 41, 271-284.

Pólya, G. & Szegő, G. (1976). Exercises and theorems from analysis (2 vols.). Springer, New York &c. The German original, also published by Springer, was called: Aufgaben und Lehrsätze.

Rankin, R.A., ed. (1984). Modular forms. Ellis Harwood, Chichester.

Sasaki, T., Kanada, Y. & Watanabe, S. (1981). Calculation of discriminants of high degree equations. Tokyo J. math. 4, 493-500.

Schikhof, W.H. (1984). Ultrametric calculus. Cambridge University Press.

Schilling, O.F.G. (1950). The theory of valuations. Mathematical surveys, 4. New York, American Mathematical Society.

Schinzel, A. (1983). Hasse's principle for systems of ternary quadratic forms and for one biquadratic form. Studia math. 77, 103-109.

Schmidt, W.M. (1976). Equations over finite fields, an elementary approach. Springer lecture notes in math., 536.

Schur, I. (1905). Über eine Klasse von endlichen Gruppen linearer Substitutionen. Sitzungsberichte d. preuss. Akad. d. Wiss., phys.-math. Kl., 77-91 (= Gesammelte Abhandlungen I, 128-142.)

Schur, I. (1921). Über die Gaussschen Summen. Nachr. d. k. Ges. Göttingen, math.-phys. Kl. 1921, 147-153. (= Ges. Abh. II, 327-333).

Selmer, E.S. (1951). The diophantine equation $ax^3 + by^3 + cz^3 = 0$. Acta math. 85, 203-362 & 92 (1954), 191-197.

Serre, J.-P. (1962). Corps Locaux. Actualités sci. ind. 1296. Hermann, Paris. 2nd ed., 1968.

Shafarevich, I.R. (1951). Novoe dokazatel'stvo teoremy Kronekera-Vebera. Trudy mat. inst. im. Steklova 38, 382-387.

Skolem, Th. (1933). Einige Sätze über gewisse Reihenentwicklungen und exponentiale Beziehungen mit Anwendungen auf diophantische Gleichungen. Skr. norske vid-akad. Oslo Nr 6 (61 pp.)

Skolem, Th. (1934). En metode til behandling av ubestemte ligninger. Chr. Michelsens institutt beretninger 4, Nr 6 (35 pp.).

Skolem, Th. (1938). Diophantische Gleichungen. Ergebniss d. Math. 4. Springer, Berlin.

Skolem, Th. (1955). The use of p-adic methods in the theory of diophantine equations. Bull. soc. math. Belg. 7,83-95.

Smith, H.J.S. (1859). Report on the theory of numbers. Report of the british association, 1859, 228-267; 1860,120-169: 1861, 292-340; 1862, 503-526; 1863, 768-786; 1865, 322-375. (= Coll. math. papers I, 38-364).

Staudt, K.G. Chr. von (1840). Beweis eines Lehrsatzes den Bernoullischen Zahlen betreffend. J. reine angew. Math. 21, 372-4.

Strassmann, R. (1928). Über den Wertevorrat von Potenzreihen im Gebiet der $\mathfrak{p}$-adischen Zahlen. J. reine angew. Math. 159, 13-28 & 65-66.

Swinnerton-Dyer, H.P.F. (1962). Two special cubic surfaces. Mathematika 9, 54-56.

Tate, J. (1971). Rigid analytic spaces. Invent. math. 12, 257-289.

Tate, J. (1974). The arithmetic of elliptic curves. Invent. math. 23, 179-206.

Terjanian, G. (1966). Un contre-example à une conjecture d'Artin. Comptes rendus, Paris 262A, 612.

Tornheim, L. (1952). Normed fields over the real and complex fields. Mich. math. J. 1, 61-68.

Tzanakis, N. (1984). The diophantine equation $x^3 - 3xy^2 - y^3 = 1$ and related equations. J. number theory 18, 192-205.

Volkenborn, A. (1972). Ein P-adisches Integral und seine Anwendungen I. Manus. math. 7, 341-373 : II., ibid. 12 (1974), 17-46.

Waerden, B.L. van der (1949). Modern algebra (2 vols). Translated by Fred Blum. Ungar, New York. The German original entitled Moderne Algebra was published by Springer, 1930. Latest editions are entitled simply: Algebra.

Wang, S. (1950). On Grunwald's theorem. Annals of math. 51, 471-484.

Warning, E. (1935). Bemerkung zur vorstehenden Arbeit von Herrn Chevalley. Abh. math. Sem. Hamburg 11, 76-83.

Washington, L.C. (1980). Introduction to cyclotomic fields. Springer, New York &c.

Weil, A. (1949). Numbers of solutions of equations in finite fields. Bull amer. math. soc. 55, 497-508. (= Collected papers I, 399-410).

Weil, A. (1967). Basic number theory. Springer, Berlin &c.

Weiss, E. (1963). Algebraic number theory. McGraw Hill, New York.

Witt, E. (1937). Zyklische Körper und Algebren der Charakteristik p. J. reine angew. Math. 176, 126-140.

INDEX